Using Geochemical Data:
Evaluation, Presentation, Interpretation

Other titles in the series

Using Geochemical Data:
Evaluation, Presentation, Interpretation

Hugh R. Rollinson

PEARSON

Prentice
Hall

Harlow, England • London • New York • Boston • San Francisco • Toronto • Sydney • Singapore • Hong Kong
Tokyo • Seoul • Taipei • New Delhi • Cape Town • Madrid • Mexico City • Amsterdam • Munich • Paris • Milan

Pearson Education Limited
Edinburgh Gate
Harlow
Essex CM20 2JE
England

and Associated Companies throughout the world

Visit us on the World Wide Web at:
www.pearsoned.co.uk

First published 1993

British Library Cataloguing-in-Publication Data
A catalogue record for this book is available from British Library

ISBN 978-0-582-06701-1

Library of Congress Cataloging-in-Publication Data
A catalog entry for this title is available from the library of Congress.

16 15 14 13
10 09 08

Set by 8RR in 10/12 pt Ehrhardt.
Printed in Malaysia, PP

For
A, E, O and P

Contents

Preface

Pascal once wrote, 'The last thing that we discover in writing a book is to know what to put at the beginning'. In my case this is particularly so, for what excuse have I as a single author for venturing into the fields of so many experts in an attempt to explain the methodologies of the main sub-disciplines of geochemistry? This same problem was acknowledged by Goldschmidt in the preface to his classic 1954 text on geochemistry. Today of course the problem is compounded, for geochemistry has a multitude of relatively new and comparatively narrow fields. We find experts in X-ray fluorescence and plasma emission techniques, those who specialize in the lead isotopes of zircon and the isotope systematics of argon. This specialization is necessary, for the technologies of such techniques are complex. Nevertheless, for some workers there needs to be an overview of the main range of techniques, for there are many who wish to understand the potential of geochemical techniques and interpret the results intelligently. That such an overview does not exist is my chief defence. This text is not original in the ideas that it conveys; rather it is original in the sense that it brings together a wide range of ideas and methods from the geochemical literature.

The principal emphasis in this book is on 'whole-rock' chemistry; the equally large area of mineral chemistry has only been touched upon tangentially. Furthermore, it has not been possible to cover some of the more novel and esoteric techniques currently being applied to geochemical investigations.

This text was conceived as a work to be put into the hands of a graduate student embarking upon a geochemical project. As it has evolved, however, it has become apparent that it serves many more purposes. It may, for example, be used as a text in final-year and graduate-student geochemistry courses. It will be useful to the professional geochemist who has worked chiefly in one sub-discipline of the subject and needs to look more broadly at a problem. It will also be of use to the non-geochemist, whether working in academia, industry or a geological survey, who has access to geochemical data and needs to interpret them.

This book has, therefore, two main goals. The first is to put into the hands of a non-expert, who needs to make use of geochemical data, a summary of the methods and techniques currently used in geochemistry, and yet a text which will enable the user to obtain something of geological significance from the data. The second goal is to put within one cover the disparate techniques and methodologies currently in use by geochemists. Thus this text may be read at two levels. Firstly, it may be read by a geochemist who wishes to evaluate and interpret the data. Secondly, it may be read by a geologist or geochemist who wants to understand some of the current geochemical jargon and make sense of the geochemical literature.

The reader will detect a number of biases in this book which are an inevitable consequence of the author's geological interests. The first bias is towards examples chosen from the Archaean, which is the principal area of geology in which I have worked and which is evident also from the place of writing. The second bias is towards igneous and metamorphic petrology, which again are my fields of interest, but also the area in which many of the methods described were first applied.

I am grateful to many colleagues for their assistance during the preparation of this book. Particular thanks go to Jan Kramers, Gordon Lampitt, Alex Woronow, David Lowry, Ken Eriksson, Kevin Walsh and my late colleague Thorley Sweetman for reading various sections of the text. Final thanks must go to Patricia, my wife, for her tolerance of the back of my head for so many months while seated at this keyboard and to Amy, Oliver and Edward for their patient encouragment of 'How many chapters to go, Dad?'

Hugh R. Rollinson
University of Zimbabwe
May 1992

Acknowledgements

We are indebted to the following for permission to reproduce copyright material:

American Geophysical Union for Fig. 5.4 (Butler & Woronow, 1986) copyright by the American Geophysical Union; Comptes Reudues Acad Sci for Fig. 5.11 (Cabanis & Lecolle, 1989); the author, Dr K G Cox for Figs 3.20 & 3.29 (Cox *et al.*, 1979); Elsevier Science Publishers and the respective authors for Figs 3.4 (Rickwood, 1989), 3.10 (de la Roche *et al.*, 1980), 3.30 (Baker & Eggler, 1983), 4.4c & d (Green & Pearson, 1986), 5.1, 5.2a & 5.3 (Pearce & Cann, 1973), 5.8 (Meschede, 1986), 5.14 (Bailey, 1981), 5.16 & 5.17 (Winchester & Floyd, 1976), 5.18 (Floyd & Winchester, 1975), 5.20 (Pearce *et al.*, 1977), 5.24 (Leterrier *et al.*, 1982), 5.31 & 5.32 (Roser & Korsch, 1988), 6.20 (Whitehouse, 1989a), 7.4 (Farver, 1989); Geological Society and respective authors for Figs 5.6 (Pearce & Gale, 1977), 5.27a & b (Harris *et al.*, 1986), 6.6 (Cliff, 1985) reproduced by permission of the Geological Society; Geologists' Association for Figs 3.7 & 3.28 (Thompson, 1984); the author, Professor R W Le Maitre for Figs 3.1 & 3.2 (Le Maitre *et al.*, 1989); Macmillan Magazines Ltd for Figs 6.2 (Hamilton *et al.*, 1979b), 6.21 (Zindler *et al.*, 1982) reprinted by permission from NATURE, copyright © 1979, 1982 Macmillan Magazines Ltd; Oxford University Press and the respective authors for Figs 5.19a & b (Pearce, 1976), 5.25a & b, 5.26 (Pearce *et al.*, 1984) by permission of Oxford University Press; Pergamon Press PLC for Fig. 4.6 (Drake & Weill, 1975) copyright 1975 Pergamon Press PLC; Society for Sedimentary Geology for Figs 3.12 & 3.13 (Herron, 1988); South African Journal of Science for Fig. 7.13 (Miller & Fairbanks, 1985); Springer-Verlag (Heidelberg) and the respective authors for Figs 4.4a & b (Green & Pearson, 1985b), 5.5a (Pearce & Norry, 1979), The Open University and the editor, Professor C Hawkesworth for Figs 5.5b & 4.25 (Pearce, 1983); The Royal Society, London and authors for Fig. 3.22 (Langmuir & Hanson, 1980); The University of Chicago Press for Figs 5.28 & 5.29 (Bhatia, 1983), 5.30 (Roser & Korsch, 1986) © 1983 and 1986 by the University of Chicago Press, all rights reserved; John Wiley & Sons Ltd and the author for Figs 5.2b, 5.7, 5.12a & b, 5.13 (Pearce, 1982) copyright 1982 John Wiley & Sons Ltd; John Wiley & Sons Ltd and the author for Fig. 7.25 (Ohmoto & Rye, 1979) copyright 1979 John Wiley & Sons Ltd; the author, Dr M Wilson for Fig. 3.3 (Wilson, 1989).

Whilst every effort has been made to trace the owners of copyright material, in a few cases this has proved impossible and we take this opportunity to offer our apologies to any copyright holders whose rights we may have unwittingly infringed.

Glossary

G.1 Abbreviations of mineral names used in the text

Ab	Albite
Ac	Acmite
An	Anorthite
Ap	Apatite
Bi	Biotite
C	Corundum
Cc	Calcite
Cpx	Clinopyroxene
Di	Diopside
En	Enstatite
Fa	Fayalite
Fo	Forsterite
Fs	Ferrosilite
Gt	Garnet
Hbl	Hornblende
He	Hematite
Hed	Hedenbergite
Hy	Hypersthene
Il	Ilmenite
Jd	Jadeite
Ka	Kaolinite
Kp	Kaliophilite
Ks	Kalsilite
Kspar, Ksp	Potassium feldspar
Lc	Leucite
Ms	Muscovite
Mt	Magnetite
Ne	Nepheline
Ol, Oliv, Olv	Olivine
Opx	Orthopyroxene
Or	Orthoclase
Pl, Plag	Plagioclase
Px	Pyroxene
Q, q, Qz	Quartz
Sil	Quartz
Sp	Spinel
Wo	Wollastonite

G.2   Other abbreviations and symbols used in the text

Note: Box 2.1 contains the definitions of statistical terms used in the text and Box 4.1 gives the definition of symbols used in the equations in Chapter 4.

AAS	Atomic Absorption Spectrophotometry
ACM	Active Continental Margin
AFC	Assimilation and Fractional Crystallization
AFM	A triangular variation diagram showing Alkalis (Na_2O + K_2O), FeO and MgO
Alk.	Alkali basalt
ARC	Ocean Island-Arc
atm	Pressure measured in atmospheres; 1 atm = 101 325 Pa
BAB	Back-Arc Basin basalt
BABI	Basaltic Achondrite Best Initial (ratio for primordial $^{87}Sr/^{86}Sr$) – the estimated Sr isotopic composition of the solar system at the time of planetary formation
BE	The composition of the Bulk Earth
Bon	Boninite
BSE	Bulk Silicate Earth — the composition of the bulk earth without the core
CA (CAB)	Calc-Alkaline Basalt
CHUR	CHondritic Uniform Reservoir — the chondritic model for the composition of the bulk earth
CIA	Chemical Index of Alteration — a measure of the degree of chemical weathering
CIPW	Cross, Iddings, Pirrson and Washington — the originators of the currently used norm calculation
CMAS	A projection into $CaO–MgO–Al_2O_3–SiO_2$ space
COLG	Collisional Granite
DM	Depleted Mantle
DS	Data-Set
EM (I and II)	Enriched Mantle
ES	The composition of average European Shale
fO_2	The activity (or fugacity) of oxygen
Ga	Billion (10^9) years
HFS	High Field Strength trace element
HIMU	High μ mantle source region (see μ)
IAT	Island-Arc Tholeiite
ICP	Inductively Coupled Plasma emission spectrometry — used in trace and major element analysis
ICP–MS	Inductively Coupled Plasma emission Mass Spectrometry
IDMS	Isotope Dilution Mass Spectrometry
INAA	Instrumental Neutron Activation Analysis
kb (kbar)	pressure expressed in kilobars; 1 kb = 0.1 GPa
Kd	The Nernst distribution coefficient (partition coefficient) for a trace element distributed between a mineral and a melt

LIL (LILE)	Large Ion Lithophile Element
LFS	Low Field Strength element
Ma	Million (10^6) years
MORB	Mid-Ocean Ridge Basalt
MSWD	Mean Squares of Weighted Deviates — used as a measure of the goodness of fit of an isochron
NAA	Neutron Activation Analysis
NASC	North American Shale Composite — an average shale composition
NHRL	Northern Hemisphere Reference Line — a line against which enrichment in Pb isotopes may be measured in the mantle source of oceanic basalts
OIA	Ocean-Island Alkali Basalt
OIB	Ocean-Island Basalt
OIT	Ocean-Island Tholeiite
ORG	Ocean-Ridge Granite
PAAS	Post-Archaean Australian average Sedimentary rock
PGE	Platinum Group (trace) Element
PM	Passive continental Margin
ppb	Parts per billion (1 in 10^9)
ppm	Parts per million (1 in 10^6)
PREMA	PREvalent MAntle reservoir — a dominant mantle source for oceanic basalts
REE	Rare Earth Element
RNAA	Radiochemical Neutron Activation Analysis
RTF	A magma chamber which is periodically Replenished, periodically Tapped and continuously Fractionated
Sho	Shoshonitic basalt
SMOW	Standard Mean Ocean Water — the standard used in oxygen and hydrogen stable isotopic measurements
SSMS	Spark Source Mass Spectrometry
syn-COLG	syn-COLlisional Granite
TAS	Total Alkalis–Silica diagram — a means of classifying volcanic rocks on the basis of their ($Na_2O + K_2O$) and SiO_2 content
Thol.	Tholeiitic basalt
Trans.	Basalt of transitional chemical composition between tholeiitic and alkaline
VAB	Volcanic-Arc Basalt
VAG	Volcanic-Arc Granite
WPB	Within-Plate Basalt
WPG	Within-Plate Granite
XRF	X-Ray Fluorescence spectrometry
α	The fractionation factor for the distribution of stable isotopes between two species
δ	The stable isotope ratio expressed relative to a standard
ε	A measure of Nd isotopic composition relative to a mantle reservoir

κ	The isotopic ratio ^{232}Th/^{238}U
λ	The decay constant for radioactive decay
μ	The isotopic ratio ^{238}U/^{204}Pb
Δ	The difference in stable isotope ratio (δ) for two coexisting minerals

Chapter 1

Geochemical data

1.1 Introduction

This book is about geochemical data and how they can be used to obtain information about geological processes. Conventionally geochemical data are subdivided into four main categories: the major elements, trace elements, radiogenic isotopes and stable isotopes (see Table 1.1). These four types of geochemical data each form the subject of a chapter of this book. Each chapter shows how the particular form of geochemical data can be used and how it provides clues to the origin of the suite of rocks in question. Different methods of data presentation are discussed and their relative merits evaluated.

Table 1.1 Whole-rock geochemistry of komatiite flows from the Belingwe greenstone belt, Zimbabwe (Data from Nisbet *et al.*, 1987)

	ZV14	ZV85	ZV10		ZV14	ZV85	ZV10
Major element oxides (wt %)				*Selected trace elements (ppm)*			
SiO_2	48.91	45.26	45.26	Ni	470	1110	1460
TiO_2	0.45	0.33	0.29	Cr	2080	2770	2330
Al_2O_3	9.24	6.74	6.07	V	187	140	118
Fe_2O_3	2.62	2.13	1.68	Y	10	6	6
FeO	8.90	8.66	8.70	Zr	21	16	14
MnO	0.18	0.17	0.17	Rb	3.38	1.24	1.38
MgO	15.32	22.98	26.21	Sr	53.3	32.6	31.2
CaO	9.01	6.94	6.41	Ba	32	12	10
Na_2O	1.15	0.88	0.78	Nd	2.62	1.84	2.31
K_2O	0.08	0.05	0.04	Sm	0.96	0.68	0.85
P_2O_5	0.03	0.02	0.02				
S	0.04	0.05	0.05	*Radiogenic isotope ratios*			
H_2O+	3.27	3.41	2.20	εNd	+2.4	+2.4	+2.5
H_2O-	0.72	0.57	0.28	$^{87}Sr/^{86}Sr$	0.7056	0.70511	0.70501
CO_2	0.46	0.84	1.04	*Stable isotope ratios ($^{0}/_{00}$)*			
Total	100.38	99.03	99.20	$\delta^{18}O$	+7.3	+7.0	+6.8

Note: Major elements and Ni, Cr, V, Y, Ba determined by XRF; FeO determined by wet chemistry; H_2O and CO_2 determined by gravimetry; Rb, Sr, Sm, Nd determined by IDMS.

The **major elements** (Chapter 3) are the elements which predominate in any rock analysis. They are Si, Ti, Al, Fe, Mn, Mg, Ca, Na, K and P, and their concentrations are expressed as a weight per cent (wt %) of the oxide (Table 1.1). Major element determinations are usually made only for cations and it is assumed that they are accompanied by an appropriate amount of oxygen. Thus the sum of the major element oxides will total to about 100 % and the analysis total may be used as a rough guide to its reliability. Iron may be determined as FeO and Fe_2O_3, but is sometimes expressed as 'total Fe' and given as either $FeO_{(tot)}$ or $Fe_2O_{3(tot)}$.

Trace elements (Chapter 4) are defined as those elements which are present at less than the 0.1 % level and their concentrations are expressed in parts per million (ppm) or more rarely in parts per billion (ppb; 1 billion $= 10^9$) of the element (Table 1.1). Convention is not always followed however, and trace element concentrations exceeding the 0.1 % (1000 ppm) level are sometimes cited. The trace elements of importance in geochemistry are identified in Table 1.5 and shown in Figure 4.1.

Some elements behave as a major element in one group of rocks and as a trace element in another group of rocks. An example is the element K, which is a major constituent of rhyolites, making up more than 4 wt % of the rock and forming an essential structural part of minerals such as orthoclase and biotite. In some basalts, however, K concentrations are very low and there are no K-bearing phases. In this case K behaves as a trace element.

Volatiles such as H_2O, CO_2 and S are normally included in the major element analysis (Table 1.1). Water combined within the lattice of silicate minerals and released above 110 °C is described as H_2O+. Water present simply as dampness in the rock powder and driven off by heating below 110 °C is quoted as H_2O- and is not an important constituent of the rock. Sometimes the total volatile content of the rock is determined by ignition at 1000 °C and is expressed as 'loss on ignition' (Lechler and Desilets, 1987).

Isotopes are subdivided into radiogenic and stable isotopes. **Radiogenic isotopes** (Chapter 6) include those isotopes which decay spontaneously due to their natural radioactivity and those which are the final daughter products of such a decay scheme. They include the parent–daughter element pairs Rb–Sr, Sm–Nd, U–Pb, Th–Pb and K–Ar. They are expressed as ratios either in absolute terms $^{87}Sr/^{86}Sr$ (e.g.) or relative to a standard (the ε-notation) (Table 1.1).

Stable isotope studies in geology (Chapter 7) concentrate on the naturally occurring isotopes of light elements such as H, O, C and S which may be fractionated on the basis of mass differences between the isotopes of the element. For example, the isotope ^{18}O is 12.5 % heavier than the isotope ^{16}O and the two are fractionated during the evaporation of water. Stable isotopes contribute significantly to an understanding of fluid and volatile species in geology. They are expressed as ratios relative to a standard using the δ-notation (Table 1.1)

The major part of this book discusses the four main types of geochemical data outlined above and shows how they can be used to identify geochemical processes. In addition, Chapter 5 has been included to show the way in which trace and major element chemistry is used to determine the tectonic setting of some igneous and sedimentary rocks. Chapter 2 discusses some of the particular statistical problems which arise when analysing geochemical data-sets, and some recommendations are made about permissible and impermissible methods of data presentation.

In this introductory chapter we consider three topics: (1) the geochemical processes which are likely to be encountered in nature and their geochemical signatures; (2) the interaction between geological fieldwork and the interpretation of geochemical data; and (3) the different analytical methods currently in use in modern geochemistry.

1.2　Geological processes and their geochemical signatures

A major purpose of this text is to show how geochemical data can be used to identify geological processes. In this section the main geochemical signatures of igneous, sedimentary and metamorphic processes are briefly summarized and presented in graphical (Figures 1.1 to 1.3) and in tabular form (Tables 1.2 to 1.4). This brief survey is augmented by fuller discussions elsewhere in the text. Each of the Tables 1.2 to 1.4 lists the geological processes which may have a geochemical signature and identifies the sections in the book where the particular process is described and characterized using major or trace elements, and radiogenic or stable isotopes.

1.2.1　Processes which control the chemical composition of igneous rocks

The chemical composition and mineralogy of the source region exerts a fundamental control over the chemistry of magmatic rocks. The major and trace element composition of a melt is determined by the type of melting process and the degree of partial melting, although the composition of the melt can be substantially modified *en route* to the surface (Figure 1.1). The source region is best characterized by its radiogenic isotope composition because isotope ratios are not modified during partial melting and magma chamber processes. The composition of the source itself is a function of mixing processes in the source region. This is particularly pertinent to studies of the mantle, and in the last decade important advances have been made in understanding mantle dynamics through the isotopic study of mantle-derived oceanic basalts (see Section 6.3.6).

Most magmatic rocks are filtered through a magma chamber prior to their emplacement at or near the surface. Magma chamber processes frequently modify the chemical composition of the primary magma, produced by partial melting of the source, through fractional crystallization, magma mixing, contamination or a dynamic mixture of several of these processes. Resolving the chemical effects of these different processes requires the full range of geochemical tools — major and trace element studies coupled with the measurement of both radiogenic and stable isotope compositions. Excellent and detailed discussions of magma chamber processes are given by Hall (1987 — Chapter 7) and Wilson (1989 — Chapter 4).

Following emplacement or eruption, igneous rocks may be chemically modified, either by outgassing or by interaction with a fluid. The outgassing of igneous rocks chiefly affects the stable isotope chemistry whereas interaction with a fluid may affect all aspects of the rock chemistry. Ideally, igneous rocks selected for chemical analysis are completely fresh, but sometimes this cannot be achieved. For example,

Table 1.2   The geochemical signatures of igneous processes identified by section in subsequent chapters

	Major element signature	Trace element signature	Radiogenic isotope signature	Stable isotope signature
Near-surface processes				
Outgassing of melts				7.4.5
				7.5.2
				7.5.3
Interaction with groundwater				7.3.5
				7.3.3
Magma chamber processes				
Fractional crystallization	3.3.1	4.2.2		7.2.3
	3.3.3	4.3.3		7.5.3
	3.3.4	4.7.1		
		4.9.3		
		4.9.4		
Contamination		4.2.2	6.3.5	7.2.3
				7.5.3
Assimilation and fractional crystallization (AFC)	3.3.1	4.2.2	6.3.5	
		4.9.3		
Magma mixing		4.9.3	6.3.5	
Open system processes (the RTF magma chamber)		4.2.2		
		4.9.3		
Liquid immiscibility		4.5.2		
Source region processes				
Partial melting	3.3.1	4.2.2		
	3.4	4.3.3		
		4.9.3		
		4.9.4		
Zone refining		4.2.2		
Source mixing		4.9.3	6.3.5	7.2.3
Source character		4.7.1	6.3.2	7.2.3
			6.3.3	
			6.3.6	
Tectonic setting	5.2.2	5.2		
	5.2.3	5.3		
		5.5.2		

samples from the seafloor have most probably been subjected to weathering or even hydrothermal alteration by seawater. Many igneous plutonic bodies initiate, on emplacement, hydrothermal groundwater circulation in the surrounding country rocks, thus leading to the chemical alteration of the igneous pluton itself. Metamorphosed igneous rocks are also likely to be chemically modified by the interaction with a fluid phase, as is discussed below.

Table 1.3 The geochemical signatures of sedimentary processes, identified by section in subsequent chapters

	Major element signature	Trace element signature	Radiogenic isotope signature	Stable isotope signature
Provenance studies				
Tectonic setting of clastic sediments	5.4.1	4.7.2 5.4.2		
Source chemistry	5.4.1	4.3.3 4.4.2 5.4.2	6.2.3 6.3.4	
Source mixing	3.3.1			
Weathering	3.3.1	4.3.3		
Transport and erosion				
Sediment maturity	3.2.4	4.2.2 4.3.3		
River water chemistry		4.3.3		
Seawater chemistry			6.3.2	7.2.2 7.3.3 7.4.3
Meteoric water chemistry				7.3.3 7.4.2
Depositional processes				
Element residence times in seawater		4.3.3	6.3.2	
Precipitation of chemical sediments		4.3.3		7.5.2
Black-shale deposition				7.4.3
Sulphate reduction				7.5.2 7.5.4
Diagenetic processes		4.3.3 4.2.2		
Diagenetic temperatures				7.2.2 7.3.5
Pore water composition				7.3.3 7.4.2 7.3.5
Limestone diagenesis				7.4.2

1.2.2 Processes which control the chemical composition of sedimentary rocks

The chemical composition of the provenance is probably the major control on the chemistry of sedimentary rocks although this can be greatly modified by subsequent

Table 1.4   The geochemical signatures of metamorphic processes, identified by section in subsequent chapters

	Major element signature	*Trace element signature*	*Radiogenic isotope signature*	*Stable isotope signature*
The protolith				
Igneous	3.2.1			
	3.2.2			
	3.2.3			
Sedimentary	3.2.4			
Tectonic mixing	3.3.1			
Element mobility	3.3.1	4.2.2	6.3.1	
		4.4.1		
		4.8		
		4.9.3		
		5.1.2		
Diffusion in the solid state				
Blocking temperatures			6.2.3	
Thermometry				7.2.2
				7.4.6
Fluid movement				
H$_2$O				7.3.3
				7.3.4
				7.3.5
CO$_2$				7.4.5

processes (Figure 1.2). The composition of the provenance is a function of tectonic setting. Weathering conditions may leave their signature in the resultant sediment and major element studies of sedimentary rocks indicate that sometimes the former weathering conditions can be recognized from the chemistry of the sediments (Section 3.3.1). Significant chemical changes may also take place during transport: some trace elements become concentrated in the clay component and in the heavy mineral fraction whilst others are diluted in a quartz-rich coarse fraction. These processes are to a large extent also dependent upon the length of time spent between erosion and deposition.

Chemical changes during deposition will depend upon the depositional environment, which is chiefly controlled by subsidence rate. Chemical and biochemical processes controlling element solubilities in seawater, submarine weathering and redox conditions are also important for particular types of sediment. Post-depositional processes are best investigated using stable isotopes. The stable isotopes of oxygen and hydrogen are important tracers for different types of water, vital in the study of diagenetic fluids. Carbon and oxygen isotopes are used in the study of limestone diagenesis. The temperature-dependent fractionation of oxygen isotopes can be used to calculate the geothermal gradient during diagenesis and allows some control on the burial history of the rock.

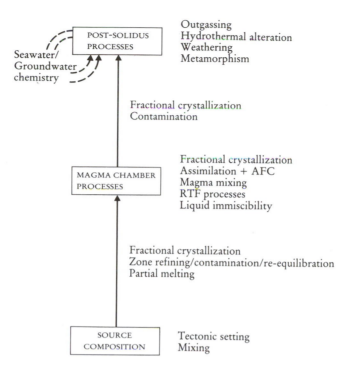

Figure 1.1   Flow diagram showing the principal processes which control the chemical composition of igneous rocks.

1.2.3   Processes which control the chemical composition of metamorphic rocks

The principal control on the chemical composition of a metamorphic rock is the composition of the pre-metamorphic protolith. Sometimes metamorphic recrystallization may be isochemical but most commonly there is a change in chemical composition (Figure 1.3). This is principally controlled by the movement of fluids and the thermal conditions during metamorphism. Metamorphism is frequently accompanied by deformation; particularly at high metamorphic grades, there may be tectonic interleaving of different protolith compositions which gives rise to a metamorphic rock of mixed parentage.

The ingress and expulsion of water during metamorphism, chiefly as a consequence of hydration and dehydration reactions, may give rise to changes in the chemical composition of the parent rock as a consequence of particular elements becoming mobile in the fluid. These processes are controlled by the composition of the fluid phase, its temperature and the ratio of metamorphic fluid to the host rock.

At high metamorphic grades and frequently in the presence of a hydrous fluid, melting may take place. The segregation and removal of this melt will clearly differentiate the parental rock into two compositionally distinct components — restite and melt. In this case, the precise nature of the chemical change is governed by the degree of melting and the melting process.

Chemical change in metamorphic rocks in the absence of a fluid phase is

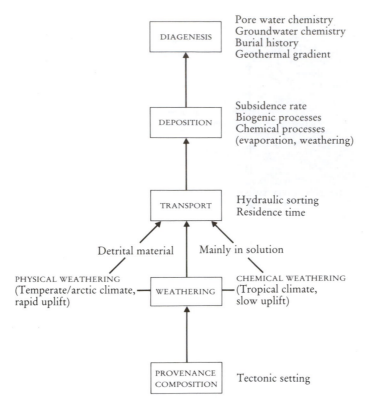

Figure 1.2   Flow diagram showing the principal processes which control the chemical composition of sedimentary rocks.

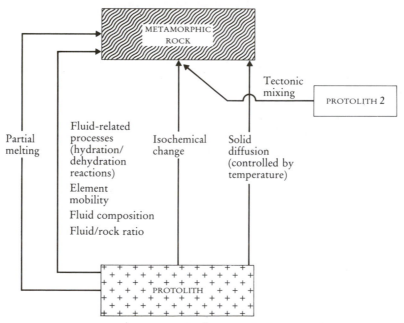

Figure 1.3   Flow diagram showing the principal processes which control the chemical composition of metamorphic rocks.

governed by diffusion of ions in the solid state. This is in response to changing mineral stabilities and metamorphic reactions — a function of the P–T conditions of metamorphism.

1.3 Geological controls on geochemical data

Geochemical investigations are most fruitful when a particular model or hypothesis is being tested. This ultimately hinges upon a clear understanding of the geological relationships. Thus, any successful geochemical investigation must be based upon a proper understanding of the geology of the area. It is not sufficient to carry out a 'smash and grab raid', returning to the laboratory with large numbers of samples, if the relationship between the samples is unknown and their relationship to the regional geology is unclear. It is normal to use the geology to interpet the geochemistry. Rarely is the converse true, for at best the results are ambiguous.

As an example, consider a migmatitic terrain in which there are several generations of melt produced from a number of possible sources. A regional study in which samples are collected on a grid pattern may have a statistically accurate feel and yet will provide limited information on the origin of the migmatite complex. What is required in such a study is the mapping of the age relationships between the units present, at the appropriate scale, followed by the careful sampling of each unit. This then allows chemical variations within the units to be investigated and models tested for the relationships between units. A fundamental thesis of this book is therefore that geochemical investigations must always be carried out in the light of a clear understanding of the geological relationships.

This approach leads naturally to the way in which geochemical data are presented. In the main this presentation is as bivariate (and trivariate) plots in which the variables are the geochemical data; a discussion of these plots forms the major part of this book. However, if the geology is also considered, then the additional variables of time and space may be plotted as well. Clearly, it can be informative on some occasions to examine chemical variations with time in a sedimentary pile. Furthermore, it is often valuable to examine the spatial distribution of geochemical data. This can vary over the entire range of geological investigations from the micro- to the global scale. Compare for example the 'map' of lead–isotopic analyses in a single galena grain (Shimizu and Hart, 1982) with the projection of isotopic anomalies in the Earth's mantle onto a world map, described by Hart (1984). Both are instructive.

1.4 Analytical methods in geochemistry

In this section the more widely used analytical methods are reviewed in order to provide a guide for those embarking on geochemical analysis. A summary of the techniques and their appropriateness to particular elements is given in Table 1.5. This survey leads into a discussion of the relative merits of the different analytical methods (Section 1.5). First, however, it is necessary to consider the criteria by which a particular analytical technique might be evaluated. Within the remit of this book, in which geochemical data are used to infer geochemical processes, it is the quality of the data which is important. Data quality may be measured in terms of their precision, accuracy and detection limit.

Precision refers to the repeatability of a measurement. It is a measure of the reproducibility of the method and is determined by making replicate measurements on the same sample. The limiting factor on precision is the counting statistics of the measuring device used. Precision can be defined by the coefficient of variation, which is 100 times the standard deviation divided by the mean (Till, 1974), also known as the relative standard deviation (Jarvis and Williams, 1989). A common practice, however, is to equate precision with one standard deviation from the mean (Norman *et al.*, 1989). It can be helpful to distinguish between precision during a given analysis session (repeatability) and precision over a period of days or weeks (reproducibility).

Accuracy is about getting the right answer. It is an estimate of how close our measured value is to the true value. Knowing the true value can be very difficult but it is normally done by reference to recommended values for international geochemical reference standards (see for example Govindaraju, 1984; Abbey, 1989). It is of course possible to obtain precise, but inaccurate, results. For most geological studies precision is more important than small differences in absolute concentration, for provided the data have all been obtained in the same laboratory the relative differences in rock chemistry can be used to infer geochemical processes.

The **detection limit** is the lowest concentration which can be 'seen' by a particular method and is a function of the level of background noise relative to an element signal (Norrish and Chappell, 1967).

The main analytical methods currently in use are briefly described below.

1.4.1   X-ray fluorescence (XRF)

X-ray fluorescence spectrometry (XRF) is currently the most widely used analytical technique in the determination of the major and trace element chemistry of rock samples. It is versatile and can analyse up to 80 elements over a wide range of sensitivities, detecting concentrations from 100 % down to a few parts per million. It is a rapid method and large numbers of precise analyses can be made in a relatively short space of time. The chief limitation is that elements lighter than Na (atomic number = 11) cannot be analysed by XRF. Good reviews of the XRF method are given by Norrish and Chappell (1977), Tertian and Claisse (1982), Williams (1987) and Ahmedali (1989).

X-ray fluorescence spectrometry is based upon the excitation of a sample by X-rays. A primary X-ray beam excites secondary X-rays (X-ray fluorescence) which have wavelengths characteristic of the elements present in the sample. The intensity of the secondary X-rays is used to determine the concentrations of the elements present by reference to calibration standards, with appropriate corrections being made for instrumental errors and the effects the composition of the sample has on its X-ray emission intensities. Alternatively, the X-rays may be detected without

Table 1.5 Elements readily analysed by XRF, INAA, IDMS, AAS, ICP and ICP–MS

Atomic Symbol Element No.			XRF	INAA	IDMS	AAS	ICP	ICP–MS
1	H	Hydrogen						
2	He	Helium						
3	Li	Lithium				x	x	x
4	Be	Beryllium				x		
5	B	Boron						x
6	C	Carbon						
7	N	Nitrogen						
8	O	Oxygen						
9	F	Fluorine						
10	Ne	Neon						
11	Na	Sodium	x			x	x	
12	Mg	Magnesium	x			x	x	
13	Al	Aluminium	x			x	x	
14	Si	Silicon	x			x	x	
15	P	Phosphorus	x				x	
16	S	Sulphur	x					
17	Cl	Chlorine	x					
18	Ar	Argon						
19	K	Potassium	x			x	x	
20	Ca	Calcium	x			x	x	
21	Sc	Scandium	x	x			x	x
22	Ti	Titanium	x			x	x	
23	V	Vanadium	x			x	x	
24	Cr	Chromium	x	x		x	x	
25	Mn	Manganese	x			x	x	
26	Fe	Iron	x			x	x	
27	Co	Cobalt	x	x		x	x	
28	Ni	Nickel	x	x		x	x	
29	Cu	Copper	x			x	x	
30	Zn	Zinc	x			x	x	
31	Ga	Gallium	x					
32	Ge	Germanium	x					
33	As	Arsenic	x					
34	Se	Selenium						
35	Br	Bromine						
36	Kr	Krypton						
37	Rb	Rubidium	x		x	x		x
38	Sr	Strontium	x		x	x	x	x
39	Y	Yttrium	x				x	x
40	Zr	Zirconium	x				x	x
41	Nb	Niobium	x				x	x
42	Mo	Molybdenum						
43	Tc	Technetium						
44	Ru	Ruthenium		x				
45	Rh	Rhodium						
46	Pd	Palladium		x				
47	Ag	Silver		x				
48	Cd	Cadmium						
49	In	Indium						
50	Sn	Tin	x					
51	Sb	Antimony						
52	Te	Tellurium						
53	I	Iodine						
54	Xe	Xenon						

Table 1.5   Continued

Atomic Symbol Element No.			XRF	INAA	IDMS	AAS	ICP	ICPMS
55	Cs	Caesium	x					x
56	Ba	Barium	x			x	x	x
57	La	Lanthanum	x	x	x		x	x
58	Ce	Cerium	x	x	x		x	x
59	Pr	Praseodymium					x	x
60	Nd	Neodymium	x	x	x		x	x
61	Pm	Promethium						
62	Sm	Samarium	x	x	x		x	x
63	Eu	Europium		x	x		x	x
64	Gd	Gadolinium		x	x		x	x
65	Tb	Terbium		x				x
66	Dy	Dysprosium		x	x		x	x
67	Ho	Holmium					x	x
68	Er	Erbium			x		x	x
69	Tm	Thulium		x				x
70	Yb	Ytterbium		x	x		x	x
71	Lu	Lutetium		x	x		x	x
72	Hf	Hafnium		x				x
73	Ta	Tantalum		x				x
74	W	Tungsten						
75	Re	Rhenium		x				
76	Os	Osmium		x				x
77	Ir	Iridium		x				
78	Pt	Platinum		x				
79	Au	Gold		x				
80	Hg	Mercury						
81	Tl	Thallium						
82	Pb	Lead	x		x	x		x
83	Bi	Bismuth						
84	Po	Polonium						
85	At	Astatine						
86	Rn	Radon						
87	Fr	Francium						
88	Ra	Radium						
89	Ac	Actinium						
90	Th	Thorium	x	x	x			x
91	Pa	Proactinium						
92	U	Uranium	x	x	x			x

being separated into different frequencies, using a detector which measures energy as well as intensity of the X-rays. This method, known as energy-dispersive X-ray fluorescence, is currently under investigation for routine trace element analysis (Potts *et al.*, 1990).

The typical XRF analysis of rock samples involves the preparation of the rock in two different forms — a pressed powder disc for trace element analysis (Leake *et al.*, 1969) and a glass bead made from the powdered sample fused with lithium metaborate or tetraborate for major element analysis (Norrish and Hutton, 1969; Claisse, 1989). The major elements are determined using one X-ray tube whereas trace elements are determined using one or more different tubes. X-ray tubes are delicate and tube changes are minimized to conserve their life, so data are normally obtained in batches over the space of several weeks or months.

1.4.2 Neutron activation analysis (INAA and RNAA)

Neutron activation analysis is a sensitive and versatile method of rock analysis, chiefly applicable to trace elements and capable of determining a large number of elements simultaneously without necessarily destroying the sample. There are two approaches. Instrumental neutron activation analysis (INAA) employs a powdered rock or mineral sample; radiochemical neutron activation analysis (RNAA) involves the chemical separation of selected elements. The range of elements analysed is given in Table 1.5 and the methods are described in detail by Muecke (1980).

In instrumental neutron activation analysis (INAA), about 100 mg of powdered rock or mineral sample is placed in a neutron flux in a neutron reactor together with standards. The sample and standards are irradiated for up to about 30 hours. The neutron flux gives rise to new, short-lived radioactive isotopes of the elements present which emit gamma radiations. Particular isotopes can be identified from the gamma radiations emitted and the intensities of these radiations are proportional to the amounts of the isotopes present. The gamma-ray spectrometry (the counting) is done at set intervals (several hours, several days and several weeks) after the irradiation in order to measure isotopes with different half-lives. Corrections are made for overlapping lines in the spectrum and concentrations are determined by comparison with the standards analysed at the same time. The method is particularly sensitive for the rare earth elements, the platinum group elements and a number of high field strength elements.

When elemental concentrations are below about 2 ppm, a chemical separation may be employed following the irradiation of the sample, but prior to counting. This approach, known as radiochemical neutron activation analysis (RNAA), clearly has the advantage of increased sensitivity.

1.4.3 Inductively coupled plasma emission spectrometry (ICP)

Inductively coupled plasma (ICP) emission spectrometry is a comparatively new technique with enormous potential in geochemistry. In principle the method is capable of measuring most elements in the periodic table with low detection limits and good precision over several orders of magnitude. Elements are measured simultaneously and a complete analysis can be made in the space of about two minutes, making it an extremely rapid analytical method. A full description of the method and its application is given by Walsh and Howie (1980) and Thompson and Walsh (1983).

ICP emission spectrometry is a 'flame' technique with a flame temperature in the range 6000–10 000 K. It is also a solution technique and standard silicate dissolution methods are employed. The sample solution is passed as an aerosol from a nebulizer into an argon plasma. The inductively coupled plasma is a stream of argon atoms, heated by the inductive heating of a radio-frequency coil and ignited by a high-frequency Tesla spark. The sample dissociates in the argon plasma and a large number of atomic and ionic spectral lines are excited. The spectral lines are detected by a range of photomultipliers, they are compared with calibration lines, and their intensities are converted into concentrations.

1.4.4 Atomic absorption spectrophotometry (AAS)

Atomic absorption spectrophotometry is based upon the observation that atoms of an element can absorb electromagnetic radiation. This occurs when the element is atomized and the wavelength of light absorbed is specific to each element. Thus the atomic absorption spectrophotometer comprises an atomizing device, a light source and a detector. A lowering of response in the detector during the atomization of a sample in a beam of light, as a consequence of atomic absorption, can be calibrated and is sensitive at the ppm level. The sample is prepared in solution and aspirated via a nebulizer and atomized in an acetylene–air or acetylene–nitrous oxide flame. The method is described in detail by Price (1972). There are two limitations to the routine use of AAS in silicate analysis. Firstly, the sample must be prepared as a solution, and secondly it is element-specific, i.e. only one element can usually be analysed at a time, although this latter limitation has in part been overcome by fitting instruments with multiple-turret lamp holders.

Table 1.5 lists the elements which can be determined by AAS with relative ease. It can be seen, for instance, that all the major elements, with the exception of P, can be measured and detection limits for Na, K, Mg and Ca are extremely low. There are methods in which all the major elements may be determined from one solution, although the cathode lamp has to be changed for each element. The trace elements Ba, Be, Co, Cr, Cu, Li, Ni, Pb, Rb, Sr, V and Zn are also readily determined.

AAS cannot compete with more rapid methods of silicate analysis such as XRF and ICP. Nevertheless, because AAS is comparatively cheap both in the capital outlay and in running costs, it most frequently finds its use in one of three specific applications.

(1) The determination of light elements such as Be and Li, with atomic numbers too low to be measured by XRF.
(2) Routine analysis of transition metals readily leached from soil or stream sediment. This application is commonly used in geochemical exploration.
(3) Non-routine trace element analysis using specialized applications of AAS in which, for example, the sample is atomized in a graphite furnace. This allows exceptionally low detection limits to be achieved for elements difficult to measure using other analytical techniques.

1.4.5 Mass spectrometry

Mass spectrometry in its various forms is the most effective method of measuring isotope ratios. It is normally preceded by the chemical separation of the element of interest. Charged ions are generated from the element to be analysed either by the bombardment of a gaseous sample with electrons (gas source) or by the volatilization of the sample on a glowing filament made of a high-melting-point metal (solid source). The ion beam is fired along a curved tube through a very powerful electromagnet which splits up the atoms according to their mass. A mass spectrum is produced in which the lighter ions are deflected with a smaller radius of curvature than heavy ions. The quantitative detection of the signal at two or more mass numbers allows an isotope ratio to be calculated. Gas source mass spectrometry is used in stable isotope studies and in argon isotope geology, whereas solid source mass spectrometry is used in other geochronological and isotope geology applications (Rb–Sr, Pb–Pb, U–Pb and Sm–Nd) and in trace element analysis by isotope dilution.

Isotope dilution mass spectrometry (IDMS)

Isotope dilution mass spectrometry is the most accurate and most sensitive of all trace element analytical techniques and is particularly suited to measuring very low concentrations. The method is described in some detail by Henderson and Pankhurst (1984) and depends upon the addition of an isotopic tracer or 'spike' to the sample. The spike contains a known concentration of a particular element whose isotopic composition is also known. If a known amount of spike and a known amount of sample are mixed, and the isotope ratio of the mixture determined, the concentration of the element in the sample can be calculated.

The method is particularly useful in determining the abundances of REE at low concentrations, although four of the REE (Pr, Tb, Ho and Tm) are mono-isotopic and cannot be analysed by this method. The main disadvantage is that even with automated mass spectrometry the method is time-consuming and expensive and so is normally reserved for measurements which can be used to calibrate other more rapid methods.

Inductively coupled plasma emission mass spectrometry (ICP–MS)

ICP–MS is a relatively new technique extending from the development of inductively coupled plasma emission spectrometry (Date and Jarvis, 1989). It is becoming increasingly accepted as a tool for trace element and isotopic analysis as a result of the very low detection limits and good accuracy and precision. It can be used for analysing a wide range of trace elements, in a single solution, using a small sample (Jenner *et al.*, 1990). Ions are extracted from the plasma through a pinhole-sized orifice into a pumped vacuum system and focused with an ion lens into a mass spectrometer.

Spark source mass spectro-metry (SSMS)

This is a less widely used analytical method in geochemistry but has been used in the analysis of trace elements. The method is described by Taylor and Gorton (1977) and its usefulness was debated by Kronberg *et al.* (1988) and Jochum and Hofman (1989). It is capable of the simultaneous determination of about 40 trace elements, has high sensitivity with detection limits of 1–10 ppb, requires small amounts of sample and has high precision and accuracy when the isotope dilution technique is used ($\pm$ 2–5 %). The sample is mixed with spiked graphite and briquetted into rod-shaped electrodes. A vacuum discharge is generated between the two sample electrodes. Elements are detected by mass on photoplates situated in the focal plane of the mass spectrometer. The mass spectra are analysed and ion intensities determined from line blackenings on the photoplate.

1.4.6 Electron microprobe analysis

The principles of electron microprobe analysis are very similar to those of X-ray fluorescence except that the sample is excited by a beam of electrons rather than an X-ray beam. Secondary X-rays are analysed according to their wavelength, the peak area counted relative to a standard and intensities converted into concentrations, making appropriate corrections for the matrix (Long, 1967). Energy-dispersive electron microprobe analysis utilizes an energy vs intensity spectrum (rather than wavelength vs intensity) and allows the simultaneous determination of the elements of interest. This mode of analysis is more rapid but less precise than the wavelength method.

Electron microprobe analysis is principally used for the major element analysis of minerals, although it is also used in the major element analysis of fused rock samples. The electron microprobe is not primarily a trace element instrument. Its chief merit is that it has excellent spatial resolution and commonly employs an electron beam of between 1 and 2 µm diameter. This means that extremely small sample areas can be analysed. The routine analysis of rock samples by the electron microprobe is restricted to the major element analysis of natural and synthetic glasses. A defocused electron beam is normally used in this application to minimize the problems of an inhomogeneous glass. The electron microprobe analysis of silicate glasses is of particular importance in the analysis of charges in experimental petrology, although less commonly fused discs of rock powder are analysed for major elements. Staudigel and Bryan (1981) have shown that this latter application gives results close to those for XRF analysis. Bender *et al.* (1984) cited the standard deviations on electron microprobe determinations of basalt glass compositions using a 35 µm beam as SiO_2 (0.30 %), Al_2O_3 (0.16 %), TiO_2 (0.04 %), FeO (0.15 %), MgO (0.05 %), CaO (0.20 %); Na_2O (0.05 %). Detection limits may be extended into the trace element range by using long counting times and precise background measurements (Merlet and Bodinier, 1990).

1.4.7 The ion microprobe

Ion microprobe technology was commercially developed in the late 1960s, but only in the last five to ten years has it had any impact upon geochemistry. The ion microprobe combines the analytical accuracy and precision of mass spectrometry with the very fine spatial resolution of the electron microprobe. It is currently used in the fields of geochronology, stable isotope geochemistry and trace element analysis, and in the study of element diffusion in minerals. Reviews of the method and its application are given by Reed (1989) and Hinton (1990). A finely focused beam of oxygen ions bombards an area of the sample (conventionally 20–30 µm in diameter) and causes secondary ions to be emitted. The ionization process, known as sputtering, drills a small hole in the surface of the sample. The secondary ion mass spectrum is analysed and used to determine the isotopic composition of the sample by secondary ion mass spectrometry (SIMS). There are, however, great complexities in relating the secondary ion spectrum to the composition of silicate materials, and much of the time between the original development of ion microprobe technology and its application in the earth sciences was spent resolving these problems.

1.5 Selecting an appropriate analytical technique

Choosing an analytical technique in geochemistry depends entirely upon the nature of the problem to be solved. It is important to know what elements are to be analysed, approximately what their concentrations are expected to be and how precise the results need to be. Additional considerations such as how many samples are to be analysed and the speed at which the analyses can be made may also be relevant.

In major element analysis the choice is between X-ray fluorescence and inductively coupled plasma emission spectroscopy. For XRF analysis the sample has to be prepared as a fused glass bead whereas in ICP analysis the sample has to be in solution. The ICP method is extremely fast, although the XRF method is more precise (Thompson and Walsh, 1983).

For trace element analysis there is much more choice and the following methods are available — XRF, INAA, RNAA, AAS, ICP, IDMS, SSMS with the possible addition of ICP–MS. Taking first the question of which elements are to be analysed, the XRF and the ICP methods are the most versatile, combining a wide range of elements with good precision and low detection limits. Elements lighter than Na, however, cannot be determined by XRF and either the AAS or ICP method must be used. Elements which are present in low concentrations require analytical methods with low detection limits such as INAA, RNAA, IDMS and SMSS. For the commonly sought rare earth element group, IDMS is the most precise although it is time-consuming and not all members of the group can be analysed by this method. RNAA also produces very precise results at low concentrations. Other methods currently in use require the separation of the REE by ion exchange methods prior to analysis, and both the ICP method (Walsh *et al.*, 1981; Zachmann, 1988; Roelandts, 1988) and XRF analysis (Robinson *et al.*, 1986) yield good results.

Isotope ratios are always analysed using a form of mass spectrometry.

1.6   Sources of error in geochemical analysis

Erroneous analytical results may arise for a variety of reasons and these are briefly described.

1.6.1   Contamination

Contamination during sample preparation can be a serious source of error in geochemical analysis. This is most likely to occur during crushing and grinding and may arise either as cross contamination from previously prepared samples or from the grinding apparatus itself. Cross contamination can be eliminated by careful cleaning and by precontaminating the apparatus with the sample to be crushed or ground. Contamination during grinding of the sample can only be controlled by the nature of the grinding surface. For the highest-precision analyses grinding should be carried out in agate, although this is delicate and expensive. Further, even agate may introduce occasional contamination (Jochum *et al.*, 1990). Tungsten carbide, a commonly used grinding material in either a shatter box or a ring mill, can introduce sizeable tungsten contamination, significant Co, Ta and Sc and trace levels of Nb (Nisbet *et al.*, 1979; Hickson and Juras, 1986; Norman *et al.*, 1989; Jochum *et al.*, 1990). In addition, the spectral lines for W overlap other elements and can spuriously enhance concentrations unless this is recognized. Chrome steel introduces sizeable amounts of Cr and Fe, moderate amounts of Mn and trace amounts of Dy, and high-carbon steel significant Fe, Cr, Cu, Mn, Zn and a trace of Ni (Hickson and Juras, 1986).

Other sources of contamination are in nature, when the sample is lightly coated with deposits from groundwater or seawater solutions. This may be remedied by leaching the rock chips after splitting but before powdering with 1M HCl for a few minutes. Contamination from impure reagents used in sample dissolution and preparation may also be important, even when using ultra-pure chemicals. A measurement of the level of contamination from this source can be made by analysing the reagents themselves in the dilutions used in sample preparation, and determining the composition of the 'blank'.

1.6.2   Calibration

All the methods of analysis described above, with the exception of some of the mass spectrometry applications, measure concentrations relative to a standard of known composition or to a calibration curve, drawn on the basis of standards of known composition. The standards used in the construction of calibration curves are either ultra-pure chemical reagents or, where matrix effects are important in some rock samples, well-analysed in-house samples and international reference samples (Govindaraju, 1984; Abbey, 1989). In either case the standards should be analysed using the most precise technique possible. Clearly the accuracy of the final analysis depends upon the accuracy of the standards used in calibration and systematic errors can easily be introduced.

1.6.3   Peak overlap

In most analytical techniques used in geochemistry there is little attempt to separate the element to be analysed from the rest of the rock or mineral sample. The only exception is in mass spectrometry. Thus there is the possibility of interference of spectral lines or peaks so that the value measured is spuriously high due to overlap from a subsidiary peak of another element present in the rock. The effect of these interferences must be calculated and removed.

1.6.4   Detecting errors in geochemical data

Errors in one's own data can be detected by running well-analysed in-house or international standards through the sample preparation and analytical system. Errors in published data are more difficult to spot unless the author has cited values for international reference standards.

| Chapter 2 | # Analysing geochemical data |

2.1 Introduction

Over the last 20 to 30 years there has developed a large body of literature on the statistical treatment of geochemical data. Some of this literature is in the form of warnings to the geochemical community that their practices are not sufficiently statistically rigorous. Other papers are concerned with improving the statistical techniques current amongst geochemists and they provide methods which are more appropriate to the peculiar properties of some geochemical data. Unfortunately, there is a tendency to leave the whole subject of the statistical treatment of geochemical data 'to the experts' and consequently a large part of this literature is ignored by most geochemists. This may in part be because the journals in which the papers are located are specific to mathematical geology. Another reason, however, is that mathematical geologists tend to write in order to communicate with other mathematicians rather than with geochemists. Whatever the reason, the net effect is that for a long time geo-statisticians have been advising geochemists in the practice of their art and yet for the most part their words have seemed irrelevant and have gone unheeded.

The purpose of this chapter, therefore, is to draw to the attention of geochemists some of the issues which our statistician colleagues have raised and to evaluate these issues in the context of presenting and interpreting geochemical data. This is not, therefore, a thorough review of statistics as applied to geochemistry, for that would require a book in itself (see for example Le Maitre, 1982; Rock, 1988a); rather, it is a discussion of some areas of statistics which directly impinge on the matters of this book. Brief definitions of the statistical terms used are given in Box 2.1.

The central problem which we have to address in considering the analysis of geochemical data is that, unfortunately for us as geochemists, our data are of a rather unusual kind. We express our compositions as parts of a whole, i.e. as percentages or as parts per million. These data are not amenable to the same type of analysis as are unconstrained data, and computer packages are not normally designed for the peculiarities of geochemical data. The particular nature of geochemical data raises all sorts of uncomfortable questions about the application of standard statistical techniques in geochemistry.

Box 2.1

The definition of some statistical terms.

Arithmetic mean $\bar{x}$	The arithmetic mean of a sample $\bar{x}$ (μ for the population) is the sum of the measurements divided by the number of measurements n: $\bar{x} = (x_1 + x_2 \ldots x_n) / n = \Sigma x/n$
Closed array	A closed array is a data array where the individual variables are not independent of each other but are related by, for example, being expressed as a percentage (Section 2.6).
Coefficient of variation C	The ratio of the standard deviation s to the mean $\bar{x}$.
Compositional data	Data which are expressed as part of a whole, such as percentages or parts per million. Compositional data form a closed array.
Correlation coefficient	Pearson's product-moment coefficient of linear correlation ρ measures the strength of the linear relationship between two variables x and y in a population. An estimation of the sample correlation coefficient r is given by $$r = \frac{\text{covariance } (x,y)}{\sqrt{[\text{variance}(x) \cdot \text{variance}(y)]}}$$ (see Eqns [2.1] and [2.2]). The Spearman rank coefficient of correlation r_s is calculated from the difference in rank order of the variables (see Eqn [2.3]).
Covariance	The product of the deviation from the mean for two variables x and y, averaged over the data-set $$S_{xy} = \frac{\Sigma(x - \bar{x})(y - \bar{y})}{n - 1}$$ Where all the covariances of a data-set are calculated they are repesented as a covariance matrix.
Degrees of freedom	The number of 'free' available observations (the sample size n) minus the number of parameters estimated from the sample.
Geometric mean, $\bar{x}_G$	The nth root of the product of positive values $X_1, X_2 \ldots X_n$: $\bar{x}_G = (X_1 \times X_2 \times X_3 \ldots X_n)^{1/n}$
Median	The median value divides the area under a distribution curve into two equal parts. An estimate of the median is the value in the sequence of individual values, ordered according to size, which divides the sequence in half. When the distribution is normal the median and the mean are the same.
Mode	For unimodal distributions the mode is the value of the measurement which has the greatest frequency. When there is a normal distribution the mode is the same as the mean.
Normal distribution	The normal or Gaussian distribution of samples is characterized by a symmetrical bell shape on a frequency diagram.

Box 2.1
(continued)

Null hypothesis	The hypothesis that two populations agree with regard to some parameter is called the null hypothesis. The null hypothesis is usually brought in to be rejected at a given level of probability.
Null value	The value assigned to a parameter in the null hypothesis.
Population	A set of measurements of a specified property of a group of objects. Normally only a sample of the population is studied. The symbols representing the population are represented by Greek letters whereas those relating to the sample are given in Roman lettering.
Regression	A measure of the intensity of the relationship between two variables. It is usually measured by fitting a straight line to the observations.
Robust test	A test is robust relative to a certain assumption if it provides sufficiently accurate results even when this assumption is violated.
Significance test	A measure of the probability level at which a null hypothesis is accepted or rejected — usually at the 5 % (0.05) or 1 % (0.01) level.
Spurious correlation	A correlation which appears to exist between ratios formed from variables where none exists between the original variables.
Standard deviation	The spread of values about the mean. It is calculated as the square root of the variance: $$s = \sqrt{\Sigma(x - \bar{x})^2 / (n - 1)}$$ In a normally distributed set of numbers 68.26 % of them will lie within one standard deviation of the mean and 95.46 % will lie within two standard deviations from the mean. The standard deviation of the population is σ.
Variance, s^2	A measure of the deviation of individual values about the mean: $$s^2 = \Sigma(x - \bar{x})^2 / (n - 1)$$ The variance is the square of the standard deviation.

2.2 Averages

Geochemists frequently use 'average values' in order to present their data in summary form. In addition, it is sometimes useful in comparative studies to work with the composition of average rocks. For example, average values of mid-ocean ridge basalts and of chondritic meteorites are frequently used as normalizing values in trace element studies. Averages are also used in analytical geochemistry where a signal may be measured a number of times and then an average taken of the measurements.

Rock (1987a, 1988b) has drawn attention to the way in which geochemists calculate average values and has shown that the methods most frequently used are in fact quite inadequate. The most commonly used summary statistic is the arithmetic mean, with the standard deviation sometimes quoted as a measure of the spread of the data. Less frequently used are the geometric mean, the median and the mode. These terms are defined in Box 2.1.

The arithmetic mean is an inappropriate choice as a method of averaging geochemical data for two principal reasons. Firstly, it assumes that the sample population is either normally or log-normally distributed. For many geological data-sets this is not the case, nor is there any *a priori* reason why it should be so. The normality of geochemical data, therefore, should never be assumed. Secondly, many geochemical data-sets contain outliers, single values completely outside the range of the other measured values. Outliers occur as a result of contamination and by the introduction of samples from other populations. They cause extreme distortion to an arithmetic mean if not screened out.

Similar problems occur with estimates of the standard deviation and with the geometric mean; Rock (1988b) proposes that robust estimates of average values should be used instead of the mean. A robust statistical test is one which works even when the original assumptions of the test are not strictly fulfilled. In the case of average values, robust estimates have the advantage of being able to reject outliers and work well with non-normal small data-sets. Robust estimates of average values are calculated using the computer program of Rock (1987a).

This is all rather bad news for the geochemist, who has become used to a number of standard statistical devices available in computer packages and even on pocket calculators. Rock (1988b) emphasizes, however, that in the statistical literature, of the various methods used to assess average values, the 'mean and standard deviation consistently performed the worst'. An example from Rock (1988b — Table 3) makes the point. The value quoted by Nakamura (1974) for La in chondritic meteorites is 0.329. However, the arithmetic mean of 53 samples is 0.365 whilst the geometric mean is 0.329. Twenty robust estimates range from 0.322 to 0.34, although of the three robust methods recommended by Rock, values are restricted to the range 0.322 to 0.329. The message seems to be that robust estimates are to be preferred. They also have the advantage that the calculation of several robust estimates allows outliers to be recognized from the inconsistencies between calculated values.

2.3 Correlation

One of the most important questions asked by a geochemist when inspecting tabulated geochemical data is to enquire what are the associations between the listed oxides and elements. For example, in the list of analyses of tonalitic and trondhjemitic gneisses in Table 2.2, do the oxides CaO and Al_2O_3 vary together? Is there a linear relationship between K_2O and Na_2O? This type of question is traditionally answered by using the statistical technique of correlation.

2.3.1   The correlation coefficient

Correlation may be defined as a measure of the strength of association between two variables measured on a number of individuals, and is quantified using the Pearson product–moment coefficient of linear correlation, usually known as the correlation coefficient. Thus the calculation of the correlation coefficients between CaO and Al_2O_3 and K_2O and Na_2O can provide an answer to the questions asked above.

When, as is normal in geochemistry, only a sample of the total population is measured, the sample correlation coefficient (r) may be calculated from the expression

$$r = \frac{\text{covariance } (x,y)}{\sqrt{[\text{variance } (x) \times \text{variance } (y)]}} \qquad\qquad [2.1]$$

where there are n values of variable x $(x_1 \ldots x_n)$ and of variable y $(y_1 \ldots y_n)$.

An easier form for computation is

$$r = \frac{\text{CSCP}}{\sqrt{(\text{CSSX . CSSY})}} \qquad\qquad [2.2]$$

where CSCP (corrected sum of cross products) $= \Sigma(xy) - \Sigma(x).\Sigma(y)/n$
       CSSX (corrected sum of squares for x) $= \Sigma(x^2) - \Sigma(x).\Sigma(x)/n$
       CSSY (corrected sum of squares for y) $= \Sigma(y^2) - \Sigma(y).\Sigma(y)/n$

Values of r vary from -1 to $+1$. When $r = +1$ then there is perfect sympathy between x and y and there is a perfect linear relationship. When $r = -1$ there is perfect antipathy between x and y. If $r = 0$ then there is no relationship between x and y at all. The value of r^2 is also useful, for it is a measure of the fraction of the total variance of x and y that is explained by the linear relationship. For instance, if the correlation coefficient $r = 0.90$, then $r^2 = 0.81$; that is, 81 % of the total variance is explained by the linear relationship.

2.3.2   The significance of the correlation coefficient (r)

The sample correlation coefficient (r) is an estimate of the population correlation coefficient (ρ), i.e. the correlation that exists in the total population of which only a sample has been measured.

It is important to know whether a calculated value for r represents a statistically significant relationship between x and y. That is, does the relationship observed in the sample hold for the population? The probability that this is the case may be estimated for different levels of significance, usually at the 5 % (or 0.05) level or the 1 % (0.01) level. (These values may also be expressed as confidence limits, in this case 95 % or 99 % respectively.) Estimates of this sort are normally made by reference to a table of values for r (Table 2.1). For a given number of degrees of freedom (number of samples minus 2, in this case), values for r are tabulated for different significance levels. The values represent the minimum values for rejecting the null hypothesis that the correlation coefficient of the population is zero $(\rho = 0)$ at the given level of significance. Two sets of tables are given depending upon whether the sign of the correlation coefficient is important. The one-sided test may be used when the alternative to the null hypothesis $(\rho = 0)$ is either $\rho > 0$ or $\rho < 0$.

Table 2.1   Values for the correlation coefficient (r) above which it is considered statistically significant (at the 1 % or 5 % level) for a given number of degrees of freedom (DF)*

DF	Two-sided test		One-sided test		DF	Two-sided test		One-sided test	
	5 %	1 %	5 %	1 %		5 %	1 %	5 %	1 %
1	0.997	0.999	0.988	0.999	25	0.381	0.487	0.323	0.445
2	0.950	0.990	0.900	0.980	30	0.349	0.449	0.296	0.409
3	0.878	0.959	0.805	0.934	35	0.325	0.418	0.275	0.381
4	0.811	0.917	0.729	0.882	40	0.304	0.393	0.257	0.358
5	0.754	0.875	0.669	0.833	50	0.273	0.354	0.231	0.322
6	0.707	0.834	0.621	0.789	60	0.250	0.325	0.211	0.295
7	0.666	0.798	0.582	0.750	70	0.232	0.302	0.195	0.274
8	0.632	0.765	0.549	0.715	80	0.217	0.283	0.183	0.257
9	0.602	0.735	0.521	0.685	90	0.205	0.267	0.173	0.242
10	0.576	0.708	0.497	0.658	100	0.195	0.254	0.164	0.230
12	0.532	0.661	0.457	0.612	150	0.159	0.208	0.134	0.189
14	0.497	0.623	0.426	0.574	200	0.138	0.181	0.116	0.164
16	0.468	0.590	0.400	0.543	300	0.113	0.148	0.095	0.134
18	0.444	0.561	0.378	0.516	400	0.098	0.128	0.082	0.116
20	0.423	0.537	0.360	0.492	500	0.088	0.115	0.074	0.104

* Data from Sachs (1984).

The two-sided test is used when $\rho \neq 0$. For example, the data-set in Table 2.2 contains 31 samples and the calculated correlation coefficient between CaO and Al_2O_3 is 0.568 (Table 2.3a). Inspection of tabulated values for r (one-sided test) shows that at the 5 % significance level and 29 degrees of freedom ($n-2$) the tabulated value for r is 0.301. Since the calculated value (0.568) is greater than the tabulated value (0.301), the correlation coefficient in the sample is statistically significant at the 5 % level. That is, there is 95 % chance that the relationship observed in the sample also applies to the population. Hence, the null hypothesis that $\rho = 0$ is rejected.

2.3.3   Assumptions in the calculation of the product–moment coefficient of correlation

The Pearson product-moment coefficient of linear correlation is based upon the following assumptions:

(1)   The units of measurement are equidistant for both variables.
(2)   There is a linear relationship between the variables.
(3)   Both variables should be normally or nearly normally distributed.

Assumption (3) is frequently regarded as an important prerequisite for linear correlation (e.g. Till, 1974). However, this is not always practised (Sachs, 1984) for strictly it is the variation of y from the estimated value of y for each value of x that must be normally distibuted and rarely is the sample population large enough for this criterion to be satisfactorily tested. Nevertheless, when testing for the significance of r, the data should be normally distributed.

Table 2.2   Major element data for tonalitic and trondhjemitic gneisses from the north marginal zone, Limpopo belt, Zimbabwe

Rock no.	1	2	3	4	5	6	7	8	9	10	11	12	13	14	15	16
SiO_2	61.50	62.15	62.58	62.59	62.82	63.19	63.62	63.71	66.67	67.18	67.31	67.63	67.68	67.89	68.00	68.55
TiO_2	0.61	0.75	0.56	0.58	0.61	0.82	0.61	0.66	0.72	0.77	0.31	0.47	0.41	0.37	0.78	0.47
Al_2O_3	15.88	18.35	18.10	16.02	17.46	16.66	16.87	15.81	15.41	16.08	18.37	15.47	14.72	15.72	11.70	16.20
Fe_2O_3	7.96	4.69	5.34	6.64	5.96	6.16	5.22	5.53	5.61	4.87	2.77	4.44	3.99	2.45	7.50	3.73
MnO	0.15	0.05	0.09	0.12	0.08	0.10	0.08	0.07	0.09	0.07	0.03	0.08	0.07	0.03	0.10	0.03
MgO	3.60	1.61	1.71	2.56	2.36	1.98	1.82	2.49	1.32	1.24	0.93	1.60	1.16	0.64	1.94	1.13
CaO	4.96	4.01	4.38	5.50	5.70	5.21	4.24	5.01	4.79	3.99	4.23	4.38	4.51	2.66	5.33	4.24
Na_2O	4.42	5.57	6.01	4.79	4.19	4.77	4.94	3.32	4.03	4.73	5.75	4.22	4.01	4.86	3.43	4.47
K_2O	0.82	1.89	1.30	1.22	0.87	1.28	1.70	1.99	1.05	1.44	1.21	1.02	1.06	3.18	0.54	1.31
P_2O_5	0.28	0.19	0.20	0.15	0.22	0.27	0.17	0.19	0.17	0.30	0.10	0.09	0.09	0.10	0.16	0.13
Sum	100.20	99.30	100.30	100.20	100.30	100.40	99.30	98.80	99.90	100.70	101.00	99.40	97.70	97.90	99.50	100.30

Rock No.	17	18	19	20	21	22	23	24	25	26	27	28	29	30	31
SiO_2	70.05	70.13	70.91	71.11	71.34	71.47	71.87	72.81	73.68	74.08	74.25	75.62	75.45	77.88	78.44
TiO_2	0.32	0.34	0.39	0.28	0.26	0.25	0.37	0.39	0.33	0.15	0.18	0.13	0.12	0.19	0.14
Al_2O_3	15.93	15.49	15.11	16.83	15.68	14.82	15.48	14.15	14.33	14.63	13.96	13.32	13.80	11.04	11.04
Fe_2O_3	2.88	3.52	3.88	1.92	2.64	3.04	2.42	3.16	2.28	1.49	1.53	1.95	1.40	3.08	3.14
MnO	0.03	0.05	0.09	0.00	0.05	0.04	0.02	0.06	0.02	0.04	0.03	0.05	0.04	0.04	0.05
MgO	0.82	1.36	0.92	0.35	0.68	0.63	0.45	0.68	0.83	0.39	0.32	0.20	0.06	0.01	0.01
CaO	3.11	3.49	3.52	2.94	3.51	3.22	3.26	3.41	2.87	2.38	2.17	1.39	1.09	1.36	1.04
Na_2O	5.06	4.65	4.80	5.69	4.77	5.16	5.00	3.90	4.57	4.50	4.15	4.23	4.76	4.78	4.71
K_2O	1.95	1.42	1.39	1.60	1.46	1.04	1.32	1.75	2.83	2.97	2.76	2.98	3.25	1.23	1.75
P_2O_5	0.12	0.12	0.15	0.03	0.11	0.04	0.14	0.10	0.07	0.06	0.05	0.04	0.04	0.01	0.07
Sum	100.30	100.60	101.20	100.80	100.50	99.70	100.30	100.40	101.80	100.70	99.40	99.90	100.00	99.60	100.40

2.3.4   Spearman rank correlation

Sometimes geochemical data cannot strictly be used in product-moment correlation of the type described above for they do not fulfil the requisite conditions. For example, some populations are not normally distributed and others include outliers. An alternative, therefore, to Pearson's product-moment coefficient of linear correlation is the Spearman rank coefficent of correlation, usually designated r_s. This type of correlation is applicable to major or trace element data measured on a ranking scale rather than the equidistant scale used in Pearson's product-moment correlation. The Spearman rank correlation coefficient is calculated as follows:

$$r_s = 1 - \left[\frac{6\Sigma D^2}{n(n^2 - 1)} \right] \qquad [2.3]$$

where D is the difference in ranking between the x-values and y-values and n is the number of pairs. In this case the only assumptions are that x and y are continuous random variables, which are at least ranked and are independent paired observations. If the rank orders are the same then $D = 0$ and $r_s = +1.0$. If the rank orders are the reverse of each other then $r_s = -1.0$. The significance of r_s may be assessed using significance tables for the Spearman rank coefficient of correlation (Table 2.4) in a similar way to that described for product-moment correlation in Section 2.3.2 above. Table 2.3(b) shows the Spearman rank coefficients of correlation for the major element data of Table 2.2. In this instance, the calculated values do not differ greatly from the Pearson product-moment coefficient of correlation

Table 2.3 Correlation matrices for the data in Table 2.2

(a) *Pearson product-moment coefficient of correlation*

	SiO_2	TiO_2	Al_2O_3	Fe_2O_3	MnO	MgO	CaO	Na_2O	K_2O	P_2O_5	Sum
SiO_2	1.0000										
TiO_2	−0.8394	1.0000									
Al_2O_3	−0.7468	0.4438	1.0000								
Fe_2O_3	−0.7840	0.8593	0.2267	1.0000							
MnO	−0.6233	0.6410	0.0915	0.8798	1.0000						
MgO	−0.8837	0.7838	0.4543	0.9041	0.8021	1.0000					
CaO	−0.8862	0.8437	0.5680	0.8225	0.6415	0.8617	1.0000				
Na_2O	−0.0690	−0.1688	0.4938	−0.2563	−0.3027	−0.2165	−0.1805	1.0000			
K_2O	0.4580	−0.5244	−0.1708	−0.6742	−0.5094	−0.5140	−0.6727	0.0173	1.0000		
P_2O_5	−0.7919	0.8639	0.5049	0.7858	0.6756	0.7790	0.7309	−0.0588	−0.4404	1.0000	
Sum	0.2071	−0.1397	0.0942	−0.1360	−0.1008	−0.1100	−0.0864	0.3163	−0.0431	0.0585	1.0000

(b) *Spearman rank coefficient of correlation*

	SiO_2	TiO_2	Al_2O_3	Fe_2O_3	MnO	MgO	CaO	Na_2O	K_2O	P_2O_5	Sum
SiO_2	1.0000										
TiO_2	−0.8405	1.0000									
Al_2O_3	−0.7801	0.5131	1.0000								
Fe_2O_3	−0.7964	0.8750	0.3700	1.0000							
MnO	−0.6082	0.6609	0.1170	0.8586	1.0000						
MgO	−0.9074	0.8731	0.5818	0.8810	0.7221	1.0000					
CaO	−0.8558	0.8592	0.5336	0.8768	0.7228	0.9326	1.0000				
Na_2O	−0.0958	−0.1798	0.4866	−0.2289	−0.3114	−0.1697	−0.2297	1.0000			
K_2O	0.4704	−0.4789	−0.1977	−0.6624	−0.5273	−0.5205	−0.6851	0.0889	1.0000		
P_2O_5	−0.8260	0.8746	0.6046	0.8048	0.6457	0.8333	0.7702	−0.0450	−0.3561	1.0000	
Sum	0.2153	−0.2059	0.1232	−0.2346	−0.1896	−0.1628	−0.1707	0.2522	0.0643	−0.0060	1.0000

The particular advantages of the Spearman rank correlation coefficient are: (1) they alone are applicable to ranked data; and (2) they are superior to the product-moment correlation coefficient when applied to populations that are not normally distributed and/or include outliers. A further advantage is that the Spearman rank correlation coefficient (r_s) is speedy to calculate and may be used as a quick approximation for the product-moment correlation coefficient (r).

2.3.5 Correlation matrices

Frequently a geochemical data-set will have as many as 30 variables. This means that there are 435 possible scatter diagrams that can be drawn for this one data-set. It has been traditional, therefore, to calculate the correlation coefficient for each pair of variables and present the data as a matrix (see Table 2.3). The correlation matrix may be used as an end in itself from which significant correlations may be identified, but it may also be the prelude to more sophisticated statistical techniques. The calculation of a correlation matrix will frequently be the initial step in the examination of a geochemical data-set. It is traditionally used to identify the most highly correlated element pairs prior to plotting the data on conventional scatter diagrams (see Section 3.3.2). Examples of the use of the Pearson product-moment coefficient of correlation are found in Beach and Tarney (1978), where

Table 2.4

Significance tables for the Spearman rank correlation coefficient for sample size n, at the 10 %, 1 % and 0.1 % significance levels (one-sided test) and the 20 %, 2 % and 0.2 % significance levels (two-sided test)*

	One-sided test		
n	10 %	1 %	0.1 %
4	1.000		
5	0.800	1.000	
6	0.657	0.943	
7	0.571	0.893	1.000
8	0.524	0.833	0.952
9	0.483	0.783	0.917
10	0.455	0.745	0.879
12	0.406	0.687	0.818
14	0.367	0.626	0.771
16	0.341	0.582	0.729
18	0.317	0.550	0.695
20	0.299	0.520	0.662
25	0.265	0.466	0.598
30	0.240	0.425	0.549
35	0.222	0.394	0.510
40	0.207	0.368	0.479
45	0.194	0.347	0.453
50	0.184	0.329	0.430
60	0.168	0.300	0.394
70	0.155	0.278	0.365
80	0.145	0.260	0.342
90	0.136	0.245	0.323
100	0.129	0.233	0.307
	20 %	2 %	0.2 %
	Two-sided test		

* Data from Sachs (1984).

correlation matrices are used to study the problem of element mobility during the retrogression of high-grade gneisses, and in Weaver *et al.* (1981), in a study of the trace element chemistry of a highly metamorphosed anorthosite.

It must be noted, however, that whilst the correlation matrix is a very useful device and correlation coefficients are useful statistical descriptors, we must be cautious about their use in geochemistry because of the very special nature of geochemical data (see Section 2.6). As I shall show more fully below, the value of a correlation coefficient may not be in doubt but the petrogenetic meaning ascribed to that value is open to interpretation.

2.3.6 Correlation coefficient patterns

Cox and Clifford (1982) have proposed a way of presenting correlation coefficient data for a suite of rocks in a diagrammatic form. Their method, which is purely descriptive, uses the Pearson product-moment coefficient of correlation and is an attempt to utilize and display graphically the large amount of information contained in a correlation matrix, without resorting to plotting the enormous number of

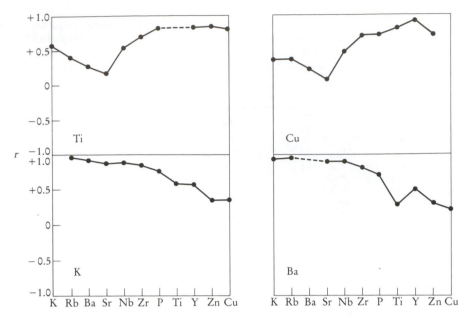

Figure 2.1   Correlation coefficient patterns for Ti, Cu, K and Ba for the Rooi Rand dolerites. Data from the correlation matrix of Cox and Clifford (1982 — Table 4). The similarity in the patterns for K and Ba implies some geochemical coherence.

possible two-element variation diagrams. They have selected a range of geologically important trace elements which they have arranged in a somewhat arbitrary order and plotted them as the x-axis of a bivariate graph. The correlation coefficient for a given element from −1 to +1 is plotted as the y-axis and the positions of each of the plotted points are joined to form a correlation coefficient pattern (Figure 2.1).

The underlying assumption of the correlation coefficient diagram is that for a suite of lavas undergoing a single-stage process such as crystal fractionation, the interelement correlation coefficients will be +1, −1 or 0. In this case values close to zero and zero values may have significance in the sense that where a correlation might be expected and none is found some process must have operated to produce this result. Usually, however, the evolution of an igneous liquid is not a single-stage process and a number of different processes are acting together. The greater the number of competing petrological processes, the greater the scatter is likely to be on a two-element variation diagram and the smaller the numerical value of the correlation coefficient. Thus correlation coeffcient patterns can be used in two ways: (1) for patterns for the same element, but for different rock suites, contrasting sets of processes may be identified; (2) for different elements in the same rock suite, their contrasting roles in the petrogenetic processes may be identified. This is illustrated in Figure 2.1.

2.4   Regression

Often in geochemistry the strength of an association, as defined by the correlation coefficient, is sufficient information from which to draw petrological conclusions.

Sometimes, however, it is also useful to quantify that association. This is traditionally done using regression analysis. For example, in the instance of the association between CaO and Al_2O_3 in the tonalites and trondhjemites of Table 2.2 the question 'If the CaO concentration were 3.5 wt %, what would be the concentration of Al_2O_3?' can be answered by calculating the regression equation for the variables CaO and Al_2O_3.

The quantification of an association is carried out by fitting a straight line through the data and finding the equation of that line. The equation for a straight line relating variables x and y is

$$y = a + bx \qquad\qquad [2.4]$$

The constant a is the value of y given by the straight line at $x = 0$. The constant b is the slope of the line and shows the number of units increase (or decrease) in y that accompanies an increase in one unit of x. The constants a and b are determined by fitting the straight line to the data. The relation above is ideal and does not allow for any deviation from the line. However, in reality this is not the case for most observations are made with some error; so often the data form a cloud of points to which a straight line must be fitted. It is this which introduces some uncertainty to line-fitting procedures and has resulted in a number of alternative approaches. Regression analysis is the subject of a number of statistical texts (e.g. Draper and Smith, 1981) and a useful review of fitting procedures in the earth sciences is given by Troutman and Williams (1987). Below some of the more popular forms of regression are described.

2.4.1   Ordinary least squares regression

Ordinary least squares regression is traditionally one of the most commonly used line-fitting techniques in geochemistry because it is relatively simple to use and because computer software with which to perform the calculations is generally readily available. Unfortunately, it is often not appropriate.

The least squares best-fit line is constructed so that the sum of the squares of the vertical deviations about the line is a minimum. In this case the variable x is the independent (non-random) variable and is assumed to have a very small error; y, on the other hand, is the dependent variable (the random variable), with errors an order of magnitude or more greater than the errors on x, and is to be determined from values of x. In this case we say that y is regressed on x (Figure 2.2a). It is possible to regress x on y and in this case the best-fit line minimizes the sum of the squares of the horizontal deviations about the line (Figure 2.2b). Thus there are two possible regression lines for the same data, a rather unsatisfactory situation for physical scientists who prefer a unique line. The two lines intersect at the mean of the sample (Figure 2.2c) and approach each other as the value of the correlation coefficient (r) increases until they coincide at $r = 1$.

In the case of ordinary least squares regression, where y is regressed on x, the value of the intercept, a, may be computed from:

$$a = \bar{y} - b\bar{x} \qquad\qquad [2.5]$$

where $\bar{x}$ and $\bar{y}$ are the mean values for variables x and y and b is the slope of the line. The slope b is computed from

$$b = r(S_y/S_x) \qquad\qquad [2.6]$$

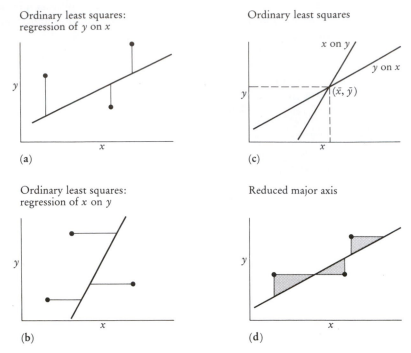

Figure 2.2   Three types of regression line discussed in the text. (a) Ordinary least squares regression of *y* on *x*; in this case the vertical distance between the point and the line is minimized. (b) Ordinary least squares regression of *x* on *y*; the horizontal distance between the point and the line is minimized. (c) Both ordinary least squares lines pass through the means $(\bar{x}, \bar{y})$, the centroid of the data. (d) Reduced major axis regression; the line is fitted to minimize the area of the shaded triangles.

where *r* is the product-moment correlation coefficient and S_x and S_y are the standard deviations of the samples of *x*- and *y*-values.

Confidence intervals on the slope of the line *b* and the intercept *a* for a given significance level may be computed from the equations given by Till (1974 — p.97). Thus confidence intervals on values of *y* for a number of values of *x* may be used to draw a confidence band on the regression line. This confidence band will be wider at the ends of the fitted line because there are more points near the mean values.

Despite its great usage in geochemistry, ordinary least squares regression has a number of disadvantages. Firstly, the method yields two different lines, neither of which may represent the actual relation between the variables. Secondly, it is assumed that the errors on the independent variable are small, whereas errors on the dependent variable are much larger. Generally in geochemistry it is meaningless to define one variable as the dependent variable and the other as the independent variable, for both will have been determined by the same technique and are subject to the same errors. The treatment of the errors in the measurement of *x* and *y* has given rise to a number of alternatives to least squares regression. The third disadvantage is that least squares regression is chiefly used in a predictive sense — variable *y* is estimated from variable *x*. In geochemistry, however, regression is more commonly used to confirm the strength of association between variables and to calculate the slope and intercept of a line. Williams (1983) gives an interesting discussion of the misuse of ordinary least squares regression in the earth sciences.

Thus, whilst an ordinary least squares line may be calculated as a first step in the regression analysis, an elaboration of or an alternative to ordinary least squares regression is needed for most geochemical applications. Some of these alternatives are reviewed below.

2.4.2 Reduced major axis regression

Reduced major axis regression is a more appropriate form of regression analysis for geochemistry than the more popular ordinary least squares regression. The method (Kermack and Haldane, 1950) is based upon minimizing the areas of the triangles between points and the best-fit line (Figure 2.2d).

The slope b of the reduced major axis line is given by

$$b = \pm (S_y/S_x) \tag{2.7}$$

where S_x and S_y are the standard deviations of sample values x and y and the sign is taken from the correlation coefficient. Thus, unlike ordinary least squares and least normal squares regression, the slope of the reduced major axis line is independent of the correlation coefficient r. The intercept a is taken from Eqn [2.5] above. Butler (1986) comments that there is some difficulty in estimating the confidence limits for the reduced major axis line but follows Moran (1971) in using the 95 % confidence bounds of the two ordinary least squares lines as the 95 % confidence limits. Till (1974 — p.102) calculates confidence intervals in terms of standard deviations.

In Figure 2.3 a comparison of regression lines is shown. Using the variables Fe_2O_3 and CaO from Table 2.2 the ordinary least squares (regressing both x on y and y on x), and the reduced major axis methods are used to fit straight lines to the data. The equation for each line is given.

2.4.3 Weighted least squares regression

A more specialized treatment of data may require weighted least squares regression. This is necessary when some data points are less reliable than others and so are more subject to error. In this case different weights may be ascribed to each data point before performing the line fitting. The weighting must be assigned by the researcher and is normally achieved by first calculating an ordinary least squares line and then investigating the residuals — the differences between the data and the ordinary least squares line.

The weighted least squares technique is the most commonly employed method of constructing an isochron in geochronology (York, 1967, 1969) although in detail the different isotopic methods require slightly different approaches. For example, Brooks et al. (1972) showed that for Rb/Sr geochronology the errors in the isotope ratios are normally distributed and for $^{86}Sr/^{87}Sr$ ratios less than 1.0 (the usual situation in whole-rock analysis) the errors are not correlated. In Pb isotope geology, however, the errors between the lead isotope ratios are highly correlated and require a slightly different treatment (see York, 1969).

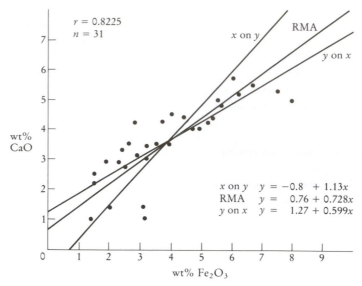

Figure 2.3   Three different regression lines drawn for the same data with their regression equations (data taken from Table 2.2). The regression lines are: ordinary least squares regression of x on y (x on y) — slope and intercept calculated from Eqns [2.5] and [2.6]; reduced major axis (RMA) — slope and intercept calculated from Eqns [2.7] and [2.5]; ordinary least squares regression of y on x (y on x) — slope and intercept calculated from Eqns [2.5] and [2.6].

2.4.4   Robust regression

Robust regression is another more specialized line-fitting technique and is another form of weighted least squares regression. Robust linear regression does not allow a single data point such as an outlier to have a disproportionate influence on the computed value of the slope and intercept. This is important, for ordinary least squares estimates can be seriously distorted by the influence of one or two outlying values. For this reason the data should be inspected for outliers. Outlying observations should be examined to see if they are in error, although no point should be discarded simply because it is an outlier. Inspection for outliers may be carried out visually using a scatter plot or with an exploratory data analysis computer program such as that described by Rock (1987a).

In robust regression, outlying values are downweighted. Zhou (1987) gives an example of the use of this technique in geochemical exploration where outliers (often anomalies and the object of such an exercise) may hamper their own identification by distorting the results of statistical analysis.

2.4.5   Some problems with traditional approaches to correlation and regression

At the beginning of this section we asked questions about the association between pairs of elements. Typical questions are about the strength of association between oxides, for example, to what degree the oxides CaO and Al_2O_3 are associated in the data of Table 2.2. A more disturbing question, and one that is not usually asked, is to what extent is the association between CaO and Al_2O_3 controlled by the other associations in the data-set. For example, does the fact that CaO correlates well with SiO_2 affect in any way its correlation with Al_2O_3?

Traditionally geochemists have looked at the relationships between pairs of elements in isolation from the other members of the data-set by plotting a large number of bivariate variation diagrams or by constructing a correlation matrix of the type described above. Yet the nature of geochemical data is that they are multivariate, i.e. simultaneous measurements have been made on more than one variable. In other words, geochemists have tended to use a bivariate approach to a multivariate problem. This is not to say that bivariate analysis of the data is totally useless, and parameters such as the correlation coefficient can be used as sample descriptors. More appropriate techniques are therefore those of multivariate analysis, and many of the methods are described in some detail for the petrologist by Le Maitre (1982).

This approach is not described in this text for there is a still more fundamental problem, and one which is not resolved directly by the application of multivariate techniques. This is the problem caused by the summation of major element analyses to 100 % — the constant sum problem. The statistical difficulties resulting from this feature of geochemical data are formidable and are discussed below in Section 2.6.

2.5   Ratio correlation

One specialized application of correlation and regression is in ratio correlation. The correlation of ratios can lead the user into a great deal of trouble and should normally be avoided. The exception is in geochronology, and this is discussed in Section 2.5.3 below. The dangers of ratio correlation in geochemistry have been documented by Butler (1982, 1986) and Rollinson and Roberts (1986) and are the subject of a text by Chayes (1971). A summary of the arguments is presented below.

Given a set of variables $X_1, X_2, X_3 \ldots$ which show no correlation, ratios formed from these pairs which have parts in common such as X_1/X_2 vs X_3/X_2, X_1/X_2 vs X_1/X_3 or X_1 vs X_1/X_2 will be highly correlated. This was first recognized by Pearson (1896) in the context of simple anatomical measurements and brought to the attention of geologists by Chayes (1949). For the case where X_1/X_2 is plotted against X_3/X_2 Pearson (1896) showed that a first-order approximation for the correlation coefficient r is given by the expression

$$r = \left[\frac{r_{13}C_1C_3 - r_{12}C_1C_2 - r_{23}C_2C_3 + C_2^2}{\sqrt{(C_1^2 + C_2^2 - 2r_{12}C_1C_2)}.\sqrt{(C_3^2 + C_2^2 - 2r_{23}C_3C_2)}} \right] \qquad [2.8]$$

where r_{12} is the correlation coefficient between variables X_1 and X_2 and C_3 is the coefficient of variation (the standard deviation divided by the mean) of variable X_3, etc. This expression holds for small values of C (< 0.3) and when the relative variance of X_2 is not large and when the absolute measurements are normally

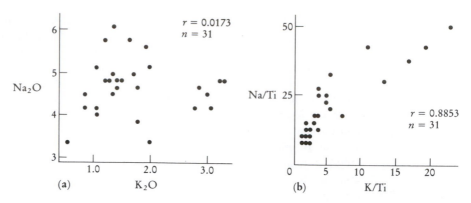

Figure 2.4   (a) Bivariate plot of Na_2O vs K_2O from the data presented in Table 2.2 — the correlation coefficient of 0.0173 is not significant at 99.9 %. (b) Bivariate plot of the molecular proportions of Na/Ti vs K/Ti for the same data illustrating the effects of spurious correlation ($r = 0.8853$).

distributed. The more general form of this equation for X_1/X_2 vs X_3/X_4 is given by Chayes (1971 — p.11).

If the variables X_1, X_2, X_3 are uncorrelated (i.e. $r_{12} = r_{23} = r_{13} = 0$) and the coefficients of variation are all the same (i.e. $C_1 = C_2 = C_3$) then the expression reduces to 0.5. Thus even though the variables X_1, X_2, X_3 are uncorrelated, the correlation coefficient between the ratios X_1/X_2 and X_3/X_2 is 0.5. In the case where X_1, X_2 and X_3 are uncorrelated, C_1 and C_3 are equal and C_2 is three times their value, then the expression reduces to 0.9. These correlation coefficients are spurious correlations for they appear to indicate a correlation where none exists between the original variables. This is illustrated in Figure 2.4, where Na_2O concentrations are plotted against K_2O concentrations (Figure 2.4a — data taken from Table 2.2). The points appear to be randomly scattered (although it might be possible to recognize two populations of data points) and they have a very low coefficient of correlation. If the weight per cent concentrations are converted to molecular proportions and Na and K ratioed to Ti, it can be seen (Figure 2.4b) that the points are highly correlated, with a correlation coefficient of 0.8853.

In the light of these observations Butler (1986) argued that in the case of ratio correlation the assessment of the strength of a linear association cannot be tested in the usual way, against a null value of zero. Rather, the null value must be the value computed for the spurious correlation (i.e. r in Eqn [2.8] above) and will therefore vary for every diagram plotted. An even more complex null hypothesis proposed by Nicholls (1988) is that the correlation coefficient of the data-set is compared with that for a set of random numbers with a similar range of values, means and variances as the data-set under investigation. This is not, however, a fruitful approach.

2.5.1   An example of the improper use of ratio correlation — Pearce element ratio diagrams

An example of the misuse of ratio correlation can be seen in the molecular proportion diagrams of T.H. Pearce, more recently called 'Pearce element ratio diagrams' or 'Pearce diagrams'. These diagrams require the plotting of ratios of oxides recast as cation quantities on an $X-Y$ graph. The ratios have an element in common, usually a common denominator (Pearce, 1968, 1970). Pearce diagrams were originally developed to avoid the effects of closure inherent in plotting percentages, the conventional method of displaying major element geochemical data. This is not a trivial problem and is discussed in some detail in Section 2.6 below. More recently users of Pearce diagrams have claimed that they can be used to discriminate between rival petrological hypotheses and identify geological process (Russell *et al.*, 1990; Ernst *et al.*, 1988).

Pearce element ratios have enjoyed a limited but varied use in petrology, although it should be stressed that the spurious nature of ratio plots makes the conclusions drawn from these diagrams at least doubtful and more commonly plain wrong for it is impossible to disentangle any geological effect from the spurious correlation effect. The most popular applications have been in identifying the fractionating phase(s) in igneous suites and in identifying mobile elements in altered volcanic rocks, particularly in altered basalts and komatiites. They have also been used in petrogenetic studies to aid in the identification of mantle phases influencing the chemistry of partial melts and in identifying participating phases in crystal fractionation. A variant of this latter use has been to investigate mixing between crystals and melt using a Pearce element ratio diagram *because* these diagrams generate a straight line. Pearce element ratio diagrams have also been used in mineralogy to identify atomic substitutions in mineral lattices.

The underlying assumption of Pearce diagrams is that in a system undergoing change any element which remains constant during the geological process under investigation may be used as a standard against which the change (in the other elements) may be measured. In the case of Pearce diagrams the constant element, or conserved element as it is often called, becomes the ratioing element and it is chosen because of its assumed constancy in the system under investigation. However, regardless of the geological logic which may underlie these diagrams, the fact remains that they are ratio plots and are subject to the peculiarities of data presented as ratios. Thus one cannot escape from the problem of spurious correlation.

A second property of Pearce diagrams which is claimed by their adherents is that the slope of a trend on a ratio plot is of significance in discriminating between rival hypotheses. For example, it is claimed that it is possible to discriminate between olivine and orthopyroxene fractionation in a basaltic magma from the slope of the regression line on a ratio plot. In other words, the slope is a function of the stoichiometry of the mass transfer process (Russell *et al.*, 1990). Thus different slopes identify different mass transfer processes. However, this argument too is flawed, for regression lines drawn through the data have incorrect slopes. This is because, in the case of ordinary least squares regression, the slope of the line is directly related to the correlation coefficient (Eqn [2.6]), which in this case is spurious. The slopes cannot therefore be used with any confidence to infer petrological processes.

Some authors have sought to circumvent the problems of ratio correlation by transforming their data into logarithmic form. Unfortunately this also does not solve any problems for the problems are preserved even as log-ratios (see Kenny, 1982; Rollinson and Roberts, 1986).

It is the variability with regard to its own mean, in the common element in Pearce element ratio diagrams, that gives rise to the potentially spurious nature of such diagrams. Spurious correlations arise primarily as a function of the means, standard deviations and correlation coefficients of the unratioed elemental data set. Small variations in the concentration of the ratioing element give rise to large and similar variations in both ratios of the Pearce plot. These elements often have a large coefficient of variation contributing to a high value for r, the spurious correlation coefficient.

2.5.2   Application to trace element diagrams

A number of elemental plots of trace elements are presented as ratio plots of the form X_1/X_2 vs X_3/X_2, X_1 vs X_1/X_2 or X_1 vs X_2/X_1 and are all subject to the constraints of ratio correlation discussed above. In some cases the trace element diagrams are designed only for classification purposes but, where linear trends are important for petrogenetic interpretation, then the problem of spurious correlation applies. In this case the plots should be considered carefully and ideas tested on alternative plots before any petrological conclusions are drawn from the data.

2.5.3   Ratio correlation in isotope geology

Ratios with a common denominator are the staple diet of much of geochronology and isotope geology and the statistical validity of Rb–Sr isochron diagrams was questioned by Chayes (1977) and discussed more fully by Butler (1982) and Dodson (1982). Butler (1982) pointed out that in the case of Rb–Sr isochron diagrams where the isotope ratio $^{87}Sr/^{86}Sr$ is plotted against $^{87}Rb/^{86}Sr$, the presence of a common denominator (^{86}Sr) 'should raise the suspicion that some or all of the observed variation on the scatter diagram may be due to the effects of having formed ratios with a common denominator'. Dodson (1982) responded to this argument by showing that isotopic ratios such as $^{87}Sr/^{86}Sr$ are never calculated from independent measurements (unlike ratios formed from major element oxide pairs). Rather, they are a directly measured property of the element under consideration, unrelated to the amount sampled, and can only be altered in a limited number of ways, the most important of which is radioactive decay. Dodson proposed the null hypothesis for isotope geochemistry that 'variations in the measured isotopic composition of an element are unrelated to its concentration or to any other petrochemical property of the materials sampled'. He showed that if the null hypothesis is true then the expected value of the ratio correlation coefficient is zero and that isochron diagrams are not subject to the common denominator effect.

A related problem in isotope geology is the correlation between the ratio $^{87}Sr/^{86}Sr$ and ^{86}Sr concentrations. In this case (provided there is no analytical error) the correlation is most likely to result from the random mixing of two uniform sources of Sr with different isotopic compositions) although there is also the possibility that the observed correlations have arisen as a consequence of the common denominator effect resulting from ratio correlation.

2.6   The constant sum problem

Geochemists are used to expressing the major element compositions of rocks and minerals as percentages, so that the sum of the major elements will always be about 100 %. This has to be, simply because there is no other way of presenting a chemical analysis in a form which can be compared with other analyses. This standard form, which is universally used by geochemists for both rock and mineral analyses, is a source of much grief to statisticians, who for 30 years have been informing geochemists that they are working in a minefield of spurious correlations and that their interpretation of major and trace element chemistry is potentially unsound. Data expressed as part of a whole (percentages or parts per million) are described as **compositional data**.

In brief, the problem is as follows. Percentages are highly complex ratios containing variables in their denominators which represent all the constituents being examined. Thus, components of percentage data are not free to vary independently. As the proportion of one component increases, the proportion of one or more other components must decrease. For instance, the first sample in Table 2.2 contains 61.5 % silica and so the value for the second variable (TiO_2) is not free to take any value but must be restrained to equal to or less than (100 − 61.5) %. The next oxide (for example Al_2O_3) is further restrained, and so on. This has a number of effects on the data. Firstly, it introduces a negative bias into correlations. This may be illustrated as follows. If one component has a significantly higher value than all the other components — such as SiO_2 — then bivariate graphs of SiO_2 and the other components will show a marked negative tendency. This attribute of percentage data is fully explored by Chayes (1960) and is illustrated in Figure 3.15. Secondly, the summing of analyses to 100 % forces a correlation between components of the data-set. The problem has been illustrated by Meisch (1969), who demonstrates that an apparently significant correlation between two oxides may have arisen through the closure process from an actual correlation between two other oxides. These induced correlations may or may not conceal important geological correlations. Stated in another way, the null value for the correlation coefficient for variables from a percentage array is not the usual null value of zero. The problem for geochemists is that what the null value should be in these circumstances is not known. In terms of correlation theory the usual criterion of unrelatedness or independence does not hold, nor are the variances and covariances of the data-set independent.

A third consequence of percentage formation is that subcompositions, frequently used in variation diagrams such as the AFM diagram, do not reflect the variations present in the 'parent' data-set. Aitchison (1986 — Table 3.1) shows that the correlation coefficient between pairs of variables varies substantially in subsets of a data-set and that there is no apparent pattern to the changes. In addition, subcompositions have variances which show different rank orderings from those in the parent data-set. For example, in the subset of data AFM (Na_2O+K_2O, $FeO_{(total)}$, MgO) the variances may be A > F > M, but in the parent data-set the variances may be F > A > M.

A normal percentage array of major element data is described by Chayes (1960) as 'closed' (Aitchison, 1986 and elsewhere, uses the term 'composition') and a closed array or data-set has inherently the possibility of non-zero correlations. The implication of a closed array is that there exists behind it an open array ('basis' in the nomenclature of Aitchison) from which these data have been derived. This open array is not subject to the constant sum effect and would be much easier to treat statistically. However, the problem in geochemistry is that we have no access to the open array and often we cannot envisage what it is like.

2.6.1   The consequences of closure

Correlating compositional data

The principal consequence of closure for geochemistry is that correlation, frequently used to examine major and trace element interrelationships, can produced misleading results. Consider the data presented in Table 2.2. Here there is a strong inverse correlation between the wt % concentrations of SiO_2 and MgO. That there is such an association is beyond doubt and the strength of the association is shown by the high correlation coefficient (−0.8837). However, the cause of the association and hence its meaning is not at all clear because we are dealing with percentage data in which one constituent makes up more than half of the composition and the other variable makes up a proportion of the remainder. There is inevitably therefore a forced association between the two. This problem can be demonstrated for major element plots, particularly those involving silica. However, it applies more generally to major and trace element data (for both are expressed as fractions of a whole), for forced correlations are a property of compositional data (Butler, 1986; Skala 1979).

The means of compositional data-sets

Woronow (1990) and Woronow and Love (1990) have argued that a second consequence of closure is that the means of percentage (and ppm) data have no interpretative value. They show that an infinite number of possible changes can explain an observed change in the mean of compositional data. Thus it is impossible to produce a meaningful statistical test to evaluate the similarities between means of compositional data. This observation has important implications for approaches such as discriminant analysis, frequently used in geochemistry (see Section 2.9).

Invalid escape routes

The problem of closure is not removed when the data is transformed into cations (Butler, 1981) nor is it removed when the problem is presented in graphical rather than statistical form. Furthermore, closure remains even when the constituents of an analysis do not sum exactly to 100 %, through analytical error or incomplete analysis, for there is always the category of 'others' which can be used to produce a sum of 100 %.

2.6.2   Aitchison's solution to the constant sum effect

Aitchison has addressed the constant sum effect in a series of detailed papers (Aitchison, 1981, 1982, 1984) and in a substantial text (Aitchison, 1986) in which he illustrates his work with examples drawn from a number of disciplines including geochemistry. Aitchison's fundamental premise is that 'the study of compositions is essentially concerned with the *relative* magnitudes of the ingredients rather than their absolute values (this) leads naturally to a conclusion that we should think in terms of ratios' (Aitchison, 1986 — p.65). This frees percentage data from their restricted region (a 'simplex' in the terminology of Aitchison) to spread more freely though sample space. Thus when formulating questions about associations between variables in a geochemical data-set our thinking should be based upon ratios rather than upon percentages as has traditionally been the case. Aitchison goes on to point out that handling the variances and covariances of ratios is difficult and that a mathematically simpler approach is to take logarithms of these values. Thus he proposes that compositional data should be expressed as the covariances of (natural) log-ratios of the variables rather than the raw percentages. The calculation of log-ratios may seem tedious and unnecessary but it has the consequence of freeing sample values from a restricted range to vary between $\pm \infty$.

At first sight there appears to be some similarity between the method of Aitchison quoted with approval here and the ratio correlation technique heavily criticized in an earlier section of this chapter. The two are the same inasmuch as both require the taking of ratios, but there the similarity ceases. At the heart of Aitchison's argument is the observation that you cannot correlate ratios, and yet this is an essential part of ratio correlation. Rather, Aitchison's method involves not the correlation of log-ratios themselves but the structure of a log-ratio covariance matrix.

In his 1986 text Aitchison proves (for the mathematically literate reader) that the covariance structure of log-ratios is superior to the covariance structure of a percentage array (the 'crude' covariance structure, as it is termed in his text). The covariance structure of log-ratios is free from the problems of the negative bias and of subcompositions which bedevil percentage data. In detail he shows that there are three ways in which the compositional covariance structure can be specified. Each is illustrated in Table 2.5. Firstly, it can be presented as a **variation matrix** in which the log-ratio variances are plotted for every variable ratioed to every other variable. This matrix provides a measure of the relative variation of every pair of variables and can be used in a descriptive sense to identify relationships within the data array and in a comparative mode between data arrays.

A second approach is to ratio every variable against a common divisor. The covariances of these log-ratios are presented as a log-ratio covariance matrix. (The correlation coefficient between two variables x and y is the covariance normalized to the square root of the product of their variances — Eqn [2.1]). The choice of variable as the divisor is immaterial because it is the structure of the matrix which is of importance rather than the individual values of the covariances. Nevertheless, this does give rise to a large number of different solutions. This form of covariance structure is used by Aitchison (1986 — Chapter 7) in a wide variety of statistical tests on log-ratio data which include tests of normality and regression. An example of its use in testing the independence of subcompositions in pollen analysis is given by Woronow and Butler (1986).

The final form of compositional covariance structure is centred log-ratio covariance matrix. In this case the single divisor of the log-ratio covariance matrix is replaced by the geometric mean of all the components. This form is used with

Table 2.5   The forms of data presentation proposed by Aitchison (1986)

(a) *Variation matrix [where Al/Si, etc., means the variance of $\log(Al_2O_3/SiO_2)$]*

	SiO_2	Al_2O_3	TiO_2	Fe_2O_3	MnO	MgO	CaO ...
SiO_2	0	Al/Si	Ti/Si	Fe/Si	Mn/Si	Mg/Si	Ca/Si
Al_2O_3		0	Ti/Al	Fe/Al	Mn/Al	Mg/Al	Ca/Al
TiO_2			0	Fe/Ti	Mn/Ti	Mg/Ti	Ca/Ti
Fe_2O_3				0	Mn/Fe	Mg/Fe	Ca/Fe
MnO					0	Mg/Mn	Ca/Mn
MgO						0	Ca/Mg
CaO							0
"							
"							
"							

(b) *Log-ratio covariance matrix [where Al,Si, etc., means the covariance of the log-ratios $(Al_2O_3/TiO_2, SiO_2/TiO_2)$]*

	SiO_2/TiO_2	Al_2O_3/TiO_2	Fe_2O_3/TiO_2	MnO/TiO_2	MgO/TiO_2	CaO/TiO_2. . .
SiO_2/TiO_2	Si,Si	Al,Si	Fe,Si	Mn,Si	Mg,Si	Ca,Si
Al_2O_3/TiO_2		Al,Al	Fe,Al	Mn,Al	Mg,Al	Ca,Al
Fe_2O_3/TiO_2			Fe,Fe	Mn,Fe	Mg,Fe	Ca,Fe
MnO/TiO_2				Mn,Mn	Mg,Mn	Ca,Mn
MgO/TiO_2					Mg,Mg	Ca,Mg
CaO/TiO_2						Ca,Ca
"						
"						
"						

(c) *Centred log-ratio covariance matrix (where Al,Si, etc., means the covariance of the log-ratios $Al_2O_3/g, SiO_2/g$ and where g is the geometric mean)*

	SiO_2/g	Al_2O_3/g	TiO_2/g	Fe_2O_3/g	MnO/g	MgO/g	CaO/g. . .
SiO_2/g	Si,Si	Al,Si	Ti,Si	Fe,Si	Mn,Si	Mg,Si	Ca,Si
Al_2O_3/g		Al,Al	Ti,Al	Fe,Al	Mn,Al	Mg,Al	Ca,Al
TiO_2/g			Ti,Ti	Fe,Ti	Mn,Ti	Mg,Ti	Ca,Ti
Fe_2O_3/g				Fe,Fe	Mn,Fe	Mg,Fe	Ca,Fe
MnO/g					Mn,Mn	Mg,Mn	Ca,Mn
MgO/g						Mg,Mg	Ca,Mg
CaO/g							Ca,Ca
"							
"							
"							

simulated data by Butler and Woronow (1986 — Appendix 2) and has the conceptual advantage for the geochemist over the log-ratio covariance form that no one element is singled out as the divisor. It is used by Aitchison in principal component analysis. Each of the three forms outlined is the same in the sense that they each determine the same covariance structure, but in a particular form of compositional analysis one may be easier to use than another.

A few multivariate procedures such as regression and principal component analysis survive the log-ratio transformation in a recognizable form.

An example — basalts from Kilauea Iki lava lake, Hawaii

Aitchison's method was tested with a suite of basalts from Kilauea Iki lava lake, Hawaii (Rollinson, 1992). These were chosen for their apparent geological simplicity. The chemical variablity in these basalts appears to be the product of only one process — the fractionation of magnesian olivine (Fo_{85}) from a single batch of magma (Richter and Moore, 1966). Percentage data presented in Table 3.3 were recalculated as log-ratios using TiO_2 as the common divisor and the variation matrix, the covariance matrix and the centred covariance matrix were calculated. The conventional interpretation of these data using a correlation matrix indicates strong associations between almost all element pairs except for those involving Mn. Data inspection showed that:

1. The variation matrix shows that there is the greatest relative variation between MgO (indicative of olivine — the fractionating phase) and the elements excluded from olivine and concentrated in the melt — K, Ti, P, Na, Ca and Al.

2. The log-ratio covariance matrix shows that the greatest covariances are between the prime constituents of olivine (Mg–Fe–Si and Mn), the fractionating phase. Na, K and P, which show very small covariances, are elements excluded from olivine and are concentrated in the melt.

3. The centred log-ratio covariance matrix shows negative covariances between Mg–K, Mg–Ti, Mg–P, Mg–Na and Mg–Ca, emphasizing the strong antipathy between Mg (in olivine) and the elements K, Ti, P, Na and Ca which are concentrated in the melt. The positive covariance between Fe and Mg is reflected in the strong association between Mg and Fe in the mineral olivine. In contrast the elements with very small covariances Si, Al, Fe, Mn and Ca show no strong association with most other elements.

The interpretation of log-ratios

From the study of Kilauea basalts (Rollinson, 1992) it can be concluded that Aitchison's log-ratio method yields results which are geologically reasonable. Log-ratios and log-ratio matrices may be interpreted as follows.

1. High values in a variation matrix will identify the element pairs which show the greatest variability. In igneous rocks this may be between a crystallizing mineral and the melt or between two or more crystallizing minerals.

2. High positive values in the two covariance matrices indicate strong associations between the elements and this is interpreted to mean that they coexist in the same mineral.

3. Large negative values in the two covariance matrices tend to confirm the variability indicated in the variation matrix.

4. Total variance may turn out to be an important indicator of the processes operating in a suite of rocks.

A final caveat should be added, namely that the methodology outlined here is very sensitive to non-normal, small data-sets.

2.7   The interpretation of trends on triangular diagrams

Triangular diagrams may be seen as a special case of the closed array or constant sum problem (Chayes, 1971 — Chapter 4), for in preparing a triangular diagram a subcomposition comprising three components is selected from data in the form of either an open or a closed array and is reformed as percentages of their sum. That is, the data are formed into a closed array. This then introduces the problems of induced correlations already discussed above. Therefore, the interpretation of trends on triangular diagrams must be carried out with great care.

Butler has analysed the summary statistics of $CaO–Na_2O–K_2O$ and A–F–M diagrams for 114 igneous rocks from the Big Bend National Park area, Texas (Butler, 1979 — Tables 1 and 2) and showed that the process of recasting the variables as percentages may dramatically change the statistical properties of the data. It is possible that in the process of percentage formation the rank order of the means, variances and correlation coefficients may change so that, for example, the variable with the smallest variance in the initial data-set may have the largest variance in the ternary data-set. He concludes that 'given the fact that major reversals of variance can occur simply as the result of ternary percentage formation it should be reasonable to expect that at least part of any trend is artificial'.

This returns us to a now-familiar petrological problem — that of disaggregating the statistical and petrological controls on a given trend, in this case as projected in a triangular diagram. Butler (1979) recommends that it is unprofitable to base genetic interpretations on ternary trends alone.

2.8   Principal component analysis

Principal component analysis is a technique for reducing the number of variables (separate oxides and trace elements) with which the investigator has to grapple. It has the advantage that a large number of variables may be reduced to a few uncorrelated variables, so that a variation diagram (see Section 3.3) may contain information about a large number of variables instead of the usual two or three. On the other hand, there are two obvious disadvantages to this approach. One is that complex plotting parameters on variation diagrams are difficult to comprehend; secondly, one of the principal arts of geochemical detective work is to identify the roles that different elements play in elucidating geochemical processes.

The method is well described by Le Maitre (1982). An original set of variables is transformed into a new set of variables called **principal component coordinates**. The new variables are simply a more convenient way of expressing the same results and provide the optimum way of viewing the data. If only a few principal coordinates account for a high proportion of the variance then the remainder may be discarded, and this reduces the number of variables under consideration.

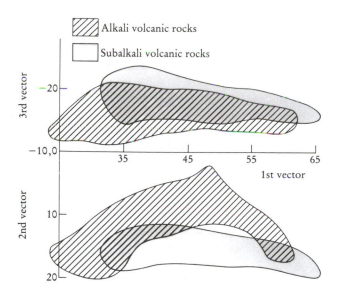

Figure 2.5   First, second and third eigenvectors from principal component analysis plotted in two dimensions to illustrate chemical differences between alkali and subalkali rock series (from Le Maitre, 1968).

The method defines a new set of orthogonal axes called **eigenvectors** or **latent vectors**. For instance, the first eigenvector is the direction of maximum spread of the data in terms of n-dimensional space. It is a 'best fit' line in n-dimensional space and the original data can be projected onto this vector using the first set of principal component coordinates. The variance of these coordinates is the first **eigenvalue** or **latent root**, and is a measure of the spread in the direction of the first eigenvector. For example, eigenvector 1 may be expressed as

Eigenvector $1 = x_1SiO_2 + x_2TiO_2 + x_3Al_2O_3 \ldots$

where x_1, x_2, x_3 etc., define the principal component coordinates. The method then defines a second eigenvector which has maximum spread at right angles to the first eigenvector, and so on. The eigenvalues are used to measure the proportion of data used in each eigenvector. By definition, the first eigenvector will contain the most information and succeeding eigenvectors will contain progressively less information. Thus it is often the case that the majority of information is contained in the first two or three eigenvectors. Eigenvectors and eigenvalues may be calculated either from a covariance matrix (where the variables are measured in the same units, e.g. wt% or ppm) or from a correlation matrix where the variables are expressed in different units. The choice of method is discussed by Le Maitre (1982 — p.110).

The chief use of principal component analysis is in the production of variation diagrams and this is illustrated by Le Maitre (1968) where he describes the chemical variation within and between alkali and subalkali basalts. He reduces nine variable major element analyses to three eigenvectors which define more than 90% of the analyses. The three eigenvectors (or latent vectors as they are called here) are plotted as a two-dimensional graph (Figure 2.5).

Aitchison (1984 and 1986 — Chapter 8) describes how log-ratio data may be used in principal component analysis in preference to percentage data.

2.9   Discriminant analysis

Discriminant analysis is a powerful technique in classifying samples into predefined groups on the basis of multiple variables. In geochemistry, discriminant analysis has been applied particularly fruitfully in the investigation of relationships between the major and trace element chemistry and the tectonic setting of volcanic and sedimentary rocks. This short section, therefore, serves as a theoretical introduction to the discussion of petrological discrimination diagrams which is given in Chapter 5.

In discriminant analysis a set of standard samples are nominated as belonging to two or more groups. From the distributions of these groups it is possible to calculate one or more linear functions of the variables measured which will achieve the greatest possible discrimination between the groups.

The functions have the form

$$F_i = a_i \, x_1 + b_i \, x_2 + c_i \, x_3 \ldots p_i \, x_p \qquad [2.9]$$

where $x_1, x_2 \ldots x_p$ are the discriminating variables (major elements or trace elements), $a_i, b_i \ldots p_i$ are the discriminating function coefficients and F_i is the discriminant score. The magnitudes of the discriminating function coefficients associated with the variables show the relative importance of the variables in separating the groups along the discriminant function.

2.9.1   An example from igneous petrology

An excellent example of the use of discriminant analysis in igneous petrology is found in the work of J.A. Pearce (1976), who employed this technique in an attempt to classify basalts on the basis of their major element chemistry (see also Section 5.2.2). The study is based upon a collection of recent basalts taken from six different tectonic environments — ocean-floor basalts, island-arc tholeiites, calc–alkaline basalts, shoshonites, ocean-island basalts and continental basalts. The objective of the study was to see if there is a relationship between major element chemistry and tectonic setting.

The initial part of the investigation was an analysis of the within-group and between-group variation, for in this way the parameters which are most likely to contribute to the separation of groups can be identified and those which are likely to be least effective can be discarded. This is then followed by the quantitative step — the discriminant analysis, which calculates the characteristics of the data-set which contribute most to the separation of the groups. These characteristics are expressed as the following parameters (see Table 2.6).

(1)   Eigenvectors, i.e. the coefficients of the discriminant function equations (see Eqn [2.9]).
(2)   An eigenvalue for each discriminant function. This shows the contribution made by the function to the total discriminating power. In the case of F1 it can be seen from Table 2.6 that it contributes to 49.7 % of the discrimination and that F1 and F2 together contribute to 76.1 % of the total discrimination.

Table 2.6   Results of discriminant analysis for basaltic rocks*

	F1	F2	F3	F4	F5
		Eigenvectors			
SiO_2	+0.0088	−0.0130	−0.0221	+0.0036	+0.0212
TiO_2	−0.0774	−0.0185	−0.0532	−0.0326	+0.0042
Al_2O_3	+0.0102	−0.0129	−0.0361	−0.0096	−0.0071
FeO	+0.0066	−0.0134	−0.0016	+0.0088	+0.0141
MgO	−0.0017	−0.0300	−0.0310	+0.0277	−0.0017
CaO	−0.0143	−0.0204	−0.0237	+0.0321	+0.0153
Na_2O	−0.0155	−0.0481	−0.0614	+0.0140	+0.0701
K_2O	−0.0007	+0.0715	−0.0289	+0.0899	+0.0075
Eigenvalues	2.58	1.37	0.65	0.5	0.09
Percentage of trace	49.7	26.4	12.4	9.7	1.7
Cumulative percentage	49.7	76.1	88.5	98.2	99.9
		Scaled eigenvectors			
SiO_2	+0.34	−0.51	−0.86	+0.14	+0.83
TiO_2	−0.85	−0.20	−0.59	−0.36	+0.05
Al_2O_3	+0.32	−0.40	−1.12	−0.30	−0.22
FeO	+0.18	−0.37	−0.04	+0.24	+0.24
MgO	−0.04	−0.74	−0.76	+0.68	−0.04
CaO	−0.29	−0.41	−0.48	+0.65	+0.31
Na_2O	−0.17	−0.54	−0.69	+0.16	+0.79
K_2O	−0.01	+0.70	−0.28	+0.88	+0.07

* From Pearce (1976).

(3)   A set of scaled eigenvectors. These show the relative contributions of each variable to the discriminant function. In the case of F1 the variables TiO_2 (−0.85) and SiO_2 (+0.34) show the largest scores and will dominate this particular discriminant function.

A convenient way of visually examining the group separation may be obtained by plotting the discriminating functions F1 and F2 as the axes of an *x–y* graph (Figure 5.19). Individual analyses are plotted as their F1 and F2 discriminant function scores. The only disadvantage of this plot is that the discriminating functions are less easy to visualize than the original oxide variables. The value of a discriminant function diagram is measured by its success rate in correctly classifying the data.

A numerical procedure may also be used for classifying the groups. This utilizes only a part of the data as a 'training set' for which the discriminating functions are derived. The remainder of the data is used as a 'testing set' with which the calculated functions may be optimized so as to minimize the number of misclassifications.

2.9.2   Other applications of discriminant analysis

Chayes and Velde (1965) used discriminant analysis to subdivide basaltic lavas found in ocean islands from those of island arcs on the basis of their major element chemistry. This approach was extended by Pearce and Cann (1971, 1973), who used discriminant analysis to subdivide basaltic rocks according to their tectonic setting on the basis of their trace element compositions. They used the elements Ti, Zr and Y, generally thought to be immobile during submarine weathering and greenschist facies metamorphism, and showed that it is possible to differentiate between four groups of basaltic rocks, namely those which have formed on the ocean floor, in volcanic arcs, in ocean islands and within continents. Their work had a wide impact: because of the choice of immobile elements, it had applicability to altered and metamorphosed basalts and held the potential for the identification of the tectonic setting of ancient basalts. This work was followed by a number of other workers who produced similar 'discrimination diagrams' for basalts based on trace element concentrations. These diagrams are presented in Chapter 5 and are discussed there in more detail.

The work of Pearce and Cann (1971, 1973) was criticized by Butler and Woronow (1986) for their use of a ternary diagram. This is because the formation of ternary percentages will induce closure into the data so that an unknown amount of the depicted variability is an artifact of closure. Instead they propose a diagram based upon principal component analysis in which the first two principal components (see Section 2.8) are used as the axes of a 'discrimination' diagram (Figure 5.4)

Discrimination diagrams have also been applied to the environment of deposition of sedimentary rocks. Potter *et al.* (1965) showed that a discriminant function based upon the trace elements B and V could be used to distinguish between freshwater and marine argillaceous sediments. More recently tectonic discrimination diagrams have also been extended to sedimentary rocks. Bhatia (1983) and Roser and Korsch (1988) have produced discriminant function diagrams which allow the identification of the provenance of sandstones according to their plate tectonic setting.

2.10   Whither geochemical data analysis?

The particular nature of geochemical data — compositional data — has been the subject of interest for more than 30 years and yet only recently have geochemists had a tool with which to handle their data correctly. At the same time, however, those past 30 years have seen an explosion in the generation of geochemical data and a huge edifice of interpretive petrology has been built — upon inferences that might be wrong. As a 'worst possible scenario', consider the exchange between Rock (1988b, 1989) and Aitchison (1989). If the means of compositional data are an invalid measure of location, then not only are many widely quoted summary statistics in error but the manner by which the analyses were obtained is also in error. For Rock (1989) points out that 'all machine-based analytical data are already dependent upon pre-existing location estimates for standard rocks'. In other words, all our XRF analyses might be slightly wrong. The magnitude of possible error is mind-blowing. Clearly the pressing need is for a suite of readily accessible computer packages which can handle compositional data.

There is however some hope, for not all uses of geochemical data are in error and that is what this book is about. Most of the examples of the misuse of geochemical data relate to major element studies, but petrological models increasingly depend upon trace element and isotopic studies. The ideas developed in this chapter will recur through the following pages — sometimes as warnings, sometimes to disallow what has previously been said. Most importantly, though, the approach must be that of 'hypothesis testing'. Diagrams must be constructed with a clear, testable model in mind so that hypotheses can be refuted or verified.

Chapter 3

Using major element data

3.1 Introduction

This chapter will examine the ways in which major element data are used in geochemistry. The discussion will be restricted to the ten elements traditionally listed as oxides in a major element chemical analysis — Si, Ti, Al, Fe, Mn, Mg, Ca, Na, K and P. Geochemists make use of major element data in three principal ways — in rock classification, in the construction of variation diagrams and as a means of comparison with experimentally determined rock compositions, whose conditions of formation are known. Each of these uses will be discussed in a separate section of this chapter. In addition, major elements are used, often together with trace elements, in the identification of the original tectonic setting of igneous and some sedimentary rocks. This topic will be discussed in Chapter 5.

The application of major element chemistry to rock classification and nomenclature is widely used in igneous petrology but is also useful for some sedimentary rocks. The second use of major element data, in the construction of variation diagrams, displays the data as bivariate or trivariate plots, i.e. on either an x–y graph or on a triangular graph. This type of diagram is used to show the interrelationship between elements in the data-set and it is from these relationships that geochemical processes may be inferred. Variation diagrams are not restricted in their use to major elements and will be discussed again in later sections of this book. The third use of major element data — the plotting of the chemical composition of an igneous rock onto a phase diagram — assumes that the chemistry of the rock is unchanged from that of the original igneous melt. In this case the comparison of rock compositions with experimentally determined phase boundaries for melts of similar composition under a range of physical conditions may allow inferences to be made about the conditions of melting and/or the subsequent crystallization history of the melt.

3.2 Rock classification

With the advent of automated XRF analysis, most geochemical investigations produce a large volume of major element data. Thus, increasingly it is both useful and in some cases necessary to attempt to classify rocks on the basis of their chemical composition. This section reviews the classification schemes in current use and outlines the rock types for which they may be specifically suited. A summary is given in Box 3.1. The criteria employed in the evaluation of a classification scheme are that it should be easy to use and widely applicable, its logical basis should be readily understood and that as far as possible it accurately should reflect the existing nomenclature, based upon mineralogical criteria.

Box 3.1

Summary of chemical classification schemes described in Section 3.2

Igneous rocks
3.2.1 Oxide–oxide plots
    The total alkalis–silica diagram (TAS)
    — for volcanic rocks
    — for plutonic rocks
    — for discriminating between alkaline and subalkaline rock series
    Subdivision of subalkaline volcanic rocks using K_2O vs SiO_2
3.2.2 Norm-based classifications
    Basalt classification using the Ne–Di–Ol–Hy–Q diagram
    Granite classification using the Ab–An–Or diagram
    Volcanic and plutonic rock classification using Q'(F')–ANOR
3.2.3 Cation classifications
    Volcanic and plutonic rocks using R1 and R2
    Komatiitic, tholeiitic and calc–alkaline volcanic rocks using the Jensen plot

Sedimentary rocks
3.2.4 Arenite/wacke
    Mudrocks

3.2.1 Classifying igneous rocks using oxide–oxide plots

Bivariate oxide–oxide major element plots are probably the most straightforward way in which to classify igneous rocks, although to date these methods are most suitable for volcanic rocks.

The total alkalis–silica diagram (TAS)   The total alkalis–silica diagram is one of the most useful classification schemes available for volcanic rocks. Chemical data — the sum of the Na_2O and K_2O content (total alkalis, TA) and the SiO_2 content (S) — are taken directly from a rock analysis as wt % oxides and plotted onto the classification diagram.

The usefulness of the TAS diagram was demonstrated by Cox *et al.* (1979), who showed that there are sound theoretical reasons for choosing SiO_2 and $Na_2O + K_2O$ as a basis for the classification of volcanic rocks. The current version of the diagram (Figure 3.1, after Le Maitre *et al.*, 1989) was constructed from a data-set of 24 000 analyses of fresh volcanic rocks carrying the names used in their original classification. The field boundaries are defined according to current usage with the minimum of overlap between adjoining fields. The coordinates of the field boundaries are shown in Figure 3.2.

The TAS diagram divides rocks into ultrabasic, basic, intermediate and acid on the basis of their silica content (following the usage of Peccerillo and Taylor, 1976). The nomenclature is based upon a system of root names with additional qualifiers to be used when necessary. For example, the root name basalt may be qualified to

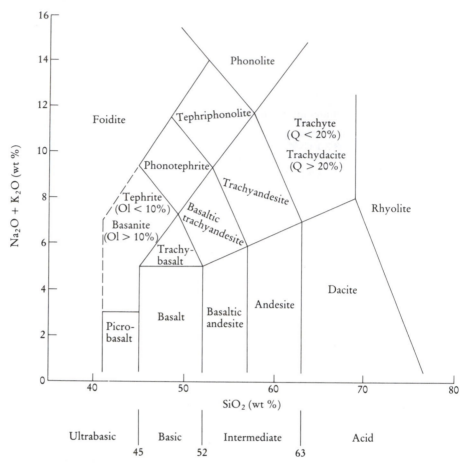

Figure 3.1   The chemical classification and nomenclature of volcanic rocks using the total alkalis versus silica (TAS) diagram of Le Maitre *et al.* (1989). Q = normative quartz; Ol = normative olivine.

alkali basalt or subalkali basalt. Some rock names cannot be allocated until a norm calculation has been performed (see Section 3.2.2 for norm calculations). For example, a tephrite contains less than 10 % normative olivine, whereas a basanite contains more than 10 % normative olivine.

(a) Using TAS with volcanic rocks   The TAS classification scheme is intended for the more common, fresh volcanic rocks. It is inappropriate for potash-rich rocks and highly magnesian rocks and should not be normally used with weathered, altered or metamorphosed volcanic rocks because the alkalis are likely to be mobilized. Rocks showing obvious signs of crystal fractionation should also be avoided. Analyses should be recalculated to 100 % on an H_2O- and CO_2-free basis.

(b) A TAS diagram for plutonic rocks   Wilson (1989) uses TAS to give a preliminary classification of plutonic igneous rocks (Figure 3.3). This diagram is of great practical use for there is no other simple classification of plutonic rocks. Unfortunately, however, the boundaries are largely based upon an earlier version of

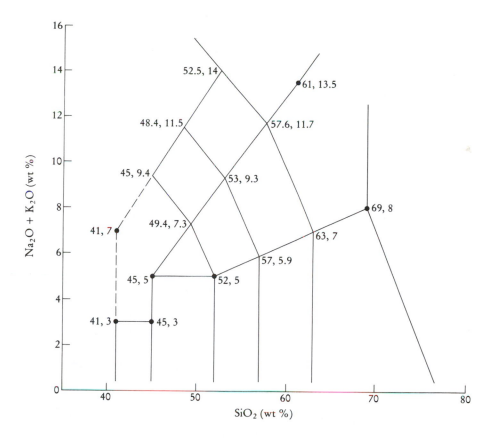

Figure 3.2   Plotting coordinates expressed as SiO_2, (Na_2O+K_2O) for the field boundaries of the total alkalis versus silica diagram of Le Maitre *et al.* (1989).

TAS devised for volcanic rocks by Cox *et al.* (1979) and therefore are not consistent with the boundaries of the TAS diagram for volcanic rocks given in Figure 3.1.

(c) Discrimination between the alkaline and subalkaline rock series using TAS Volcanic rocks may be subdivided into two major magma series — the alkaline and subalkaline (originally termed tholeiitic) series — on a total alkalis–silica diagram. MacDonald and Katsura (1964) and MacDonald (1968), using data from basalts in Hawaii, were amongst the first to publish boundary lines on a TAS diagram which allowed a distinction to be made between the alkaline series and the tholeiitic series of basalts. Similar diagrams are given by Kuno (1966), working with Tertiary volcanic rocks from Eastern Asia, and Irvine and Baragar (1971).

Different authors locate the boundary line in slightly different places on the TAS diagram and Rickwood (1989) has produced a compilation of the published lines and their plotting coordinates. They are plotted onto the TAS classification diagram of Le Maitre *et al.* (1989) in Figure 3.4 and their coordinates are given in the figure caption.

The K_2O vs SiO_2 diagram for the subdivision of the subalkaline series   Volcanic rocks of the subalkaline series have been further subdivided on the basis of their concentrations of K_2O and SiO_2 (Peccerillo and Taylor, 1976). Le Maitre *et al.* (1989) propose a division of subalkaline rocks into low-K, medium-K and high-K types and suggest that these terms may be used to qualify the names basalt,

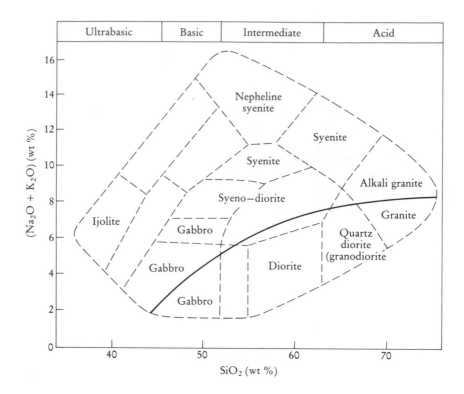

Figure 3.3   The chemcial classification and nomenclature of plutonic rocks using the total alkalis versus silica (TAS) diagram of Cox *et al.* (1979) adapted by Wilson (1989) for plutonic rocks. The curved solid line subdivides the alkalic from subalkalic rocks.

basaltic andesite, andesite, dacite and rhyolite. This nomenclature broadly coincides with the terms low-K (tholeiite) series, calc–alkaline series and high-K (calc–alkaline) in the synthesis of Rickwood (1989), and with the terms low-K subalkalic basalt, subalkalic basalt and alkalic basalt used by Middlemost (1975). A compilation of curves from Rickwood (1989) and Le Maitre *et al.* (1989) is given in Figure 3.5.

3.2.2   Classifying igneous rocks using the norm

The norm calculation is a way of working out the mineralogy of a rock from a chemical analysis, and in the context of rock classification makes possible a pseudo-mineralogical classification. The CIPW norm is the most commonly used calculation scheme, devised at the beginning of this century by three petrologists named Cross, Iddings and Pirrson and a geochemist named Washington — hence the acronym, CIPW.

The norm of a rock may be substantially different from the observed mineralogy — the mode. The calculations assume that the magma is anhydrous: thus minerals such as biotite or hornblende are not permitted. Further simplifying assumptions are that no account is taken of the minor solid solution of elements such as Ti and Al in ferromagnesian minerals and that the Fe/Mg ratio of all ferromagnesian minerals is assumed to be the same. The normative mineralogy is based entirely

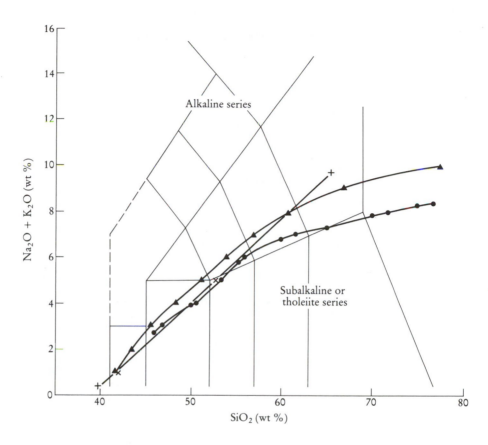

Figure 3.4   The subdivision of volcanic rocks into alkaline and subalkaline (tholeiitic) on a total alkalis vs silica diagram. The boundaries and plotting coordinates (SiO₂, total alkalis) are as follows: ×, MacDonald and Katsura (1964), straight line from 41.75,1.0 to 52.5,5.0. +, MacDonald (1968), straight line from 39.8,0.35 to 65.5,9.7. ●, Kuno (1966) curved line 45.85,2.75; 46.85,3.0; 50.0,3.9; 50.3,4.0; 53.1,5.0; 55.0,5.8; 55.6,6.0; 60.0,6.8; 61.5,7.0; 65.0,7.35; 70.0,7.85; 71.6,8.0; 75.0,8.3; 76.4,8.4. ▲, Irvine and Baragar (1971), curved line 39.2,0.0; 40.0,0.4; 43.2,2.0; 45.0,2.8; 48.0,4.0; 50.0,4.75; 53.7,6.0; 55.0,6.4; 60.0,8.0; 65.0,8.8; 77.4,10.0]. From the compilation of Rickwood (1989).

upon chemistry; thus fine-grained, coarse-grained and metamorphosed igneous rocks all with the same chemistry will have the same normative composition.

   The rules for the norm calculation are given by Cross *et al.* (1903) and a simplified version appropriate to most common rocks is given in Box 3.2 (after Kelsey, 1965, and Cox *et al.*, 1979). The calculations are normally made by computer; there are various published programs for the calculation of the CIPW norm either as part of a larger suite of programs (Till, 1977; Verma, 1986) or as a specific program (Glazner, 1984; Fears, 1985). In the CIPW norm calculation, the rock chemistry is converted to molecular proportions (by dividing the wt % oxides by their molecular weight; see Table 3.1, columns 1–3) and at the end of the calculation the proportions of normative minerals are recast as wt % by multiplying by the molecular weight. This is known as the **wt % norm** (Table 3.1, column 7).

Box 3.2

Rules for the CIPW norm calculation after Kelsey (1965) and Cox _et al._ (1979)

It is helpful to use a standard form when calculating a norm by hand. This may be done by laying out the oxides (with their molecular weights and therefore the molecular proportions) as columns along the top of the page and the more common normative minerals as rows along the left-hand margin. The 'boxes' of the table are filled as the calculation is made and can be used to keep a check on when an element is totally used up. Constants used in the CIPW norm calculation are listed below.

Oxide	Molecular weight	Normative mineral		Formula	Molecular weight
SiO_2	60	Q	Quartz	S	60
TiO_2	80	Or	Orthoclase	KAS_6	556
Al_2O_3	102	Ab	Albite	NAS_6	524
Fe_2O_3	160	An	Anorthite	CAS_2	278
FeO	72	Lc	Leucite	KAS_4	436
MnO	71	Ne	Nepheline	NAS_2	284
MgO	40	C	Corundum	A	102
CaO	56	Ac	Acmite	$NF^{3+}S_4$	462
Na_2O	62	Di { Wo	Wollastonite	CS	116
K_2O	94	En	Enstatite	MS	100
P_2O_5	142	Fs	Ferrosilite	$F^{2+}S$	132
		Wo	Wollastonite	CS	116
		Hy { En	Enstatite	MS	100
		Fs	Ferrosilite	$F^{2+}S$	132
		Ol { Fo	Forsterite	M_2S	140
		Fa	Fayalite	$F_2^{2+}S$	204
		Mt	Magnetite	$F^{2+}F^{3+}$	232
		He	Hematite	F^{3+}	160
		Il	Ilmenite	$F^{2+}T$	152
		Ap	Apatite	$C_{3.33}P$	310

Method:

(1) Calculate molecular proportions of the oxides by dividing by their molecular weights.
(2) Add MnO to FeO.
(3) Allocate CaO equal to $3.33 \times P_2O_5$ to apatite.
(4) If FeO > TiO_2: allocate FeO equal to the amount of TiO_2 present to ilmenite. If FeO < TiO_2: an excess of TiO_2 is provisionally made into sphene, using an equal amount of CaO (although, only after CaO has been allocated to anorthite). If there is still an excess of TiO_2 it is allocated to rutile.
(5) Provisionally allocate Al_2O_3 equal to K_2O for orthoclase.
(6) Provisionally allocate to any excess Al_2O_3 equal Na_2O for albite. If there is insufficient Al_2O_3 go to step 10.
(7) Any excess of Al_2O_3 over Na_2O + K_2O is matched with an equal amount of CaO for anorthite
(8) If there is an excess of Al_2O_3 over CaO it is allocated to corundum.
(9) An excess of CaO over Al_2O_3 is used for diopside and wollastonite.
(10) An excess of Na_2O over Al_2O_3 is used in acmite; there is no anorthite in the norm. Allocate Fe_2O_3 equal to the excess Na_2O for acmite.

Box 3.2
(continued)

(11) If $Fe_2O_3 > Na_2O$, allocate an equal amount of FeO for magnetite.

(12) If Fe_2O_3 is still in excess, it is calculated as hematite.

(13) Sum MgO + remaining FeO. Calculate their relative proportions.

(14) Any CaO unused after anorthite (step 7) is allocated to diopside using an equal amount of FeO + MgO (allotted in proportion to that determined in step 13).

(15) Excess CaO is provisionally allocated to wollastonite.

(16) Excess MgO + FeO is provisionally allocated to hypersthene.

(17) Allocate SiO_2 to sphene, acmite, provisional orthoclase, albite and anorthite, diopside, wollastonite and hypersthene in the proportions of the formulae above.

(18) An excess of SiO_2 is calculated as quartz.

(19) If there is insufficient SiO_2 at step 17 the SiO_2 allocated to hypersthene is omitted from the sum of SiO_2 used. If at this stage there is an excess of SiO_2, that remaining is allocated between hypersthene and olivine using the equations:

$$x = 2S - M$$
$$y = M - x$$

where x is the number of hypersthene molecules and y the number of olivine molecules, M the value of available MgO + FeO and S the amount of available SiO_2. If there is insufficient SiO_2 to match half the amount of MgO + FeO, then MgO + FeO is made into olivine (rather than hypersthene).

(20) If there is still a deficiency of SiO_2 in step 19, SiO_2 allocated to sphene is subtracted from the total in step 17 and CaO and TiO_2 are calculated as perovskite.

(21) If there is still a deficiency in SiO_2, the total in step 17 is calculated substituting perovskite for sphene and olivine for hypersthene. Albite is omitted and Na is distributed between albite and nepheline according to the rules

$$x = (S - 2N)/4$$
$$y = N - x$$

where x is the number of albite molecules, y the number of nepheline molecules, N the amount of available Na_2O and S the amount of available SiO_2.

(22) If there is not enough SiO_2 at step 21 to equal twice the Na_2O, all the Na_2O is made into nepheline and K_2O is distributed between leucite and orthoclase, although now the total available SiO_2 is the total from step 17 when perovskite, olivine and nepheline are made and orthoclase is omitted. In this case

$$x = (S - 4K)/2$$
$$y = (K - x)$$

where x is the number of orthoclase molecules and y the number of leucite molecules, K the available K_2O and S the available SiO_2.

(23) If there is still a deficiency in SiO_2, the CaO of wollastonite and diopside is distributed between these two minerals and calcium orthosilicate and the silica allocated accordingly.

(24) Finally the weight percentages of the normative minerals are calculated by multiplying the oxide amounts by the molecular weight of the minerals. An oxide is selected which appears as unity in the formulae cited above and that value (the molecular proportion) is multiplied by the molecular weights given above. This gives the normative constituents in weight per cent terms. These should sum to approximately the same percentage as the original analysis.

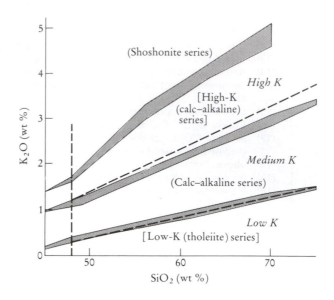

Figure 3.5 The subdivision of subalkalic rocks using the K_2O vs silica diagram. The diagram shows the subdivisions of Le Maitre *et al.* (1989) (broken lines with nomenclature in italics) and of Rickwood (1989) (nomenclature in parentheses). The shaded bands are the fields in which fall the boundary lines of Peccerillo and Taylor (1976), Ewart (1982), Innocenti *et al.* (1982), Carr (1985) and Middlemost (1985) as summarized by Rickwood (1989). The plotting parameters are as follows (SiO_2, K_2O): Le Maitre *et al.* (1989) (broken lines): high-K/medium-K boundary 48.0,1.2; 68.0,3.1; medium-K/low-K boundary 48.0,0.3; 68.0,1.2; vertical boundary at 48 % SiO_2. Rickwood (1989) band between shoshonitic series and high-K series 45.0,1.38; 48.0,1.7; 56.0,3.3; 63.0,4.2; 70.0,5.1; and 45.0,1.37; 48.0,1.6; 56.0,2.98; 63.0,3.87; 70.0,4.61; band between high-K and calc–alkaline series 45.0,0.98; 49.0,1.28; 52.0,1.5; 63.0,2.48; 70.0,3.1; 75.0,3.43; and 45.0,0.92; 49.0,1.1; 52.0,1.35; 63.0,2.32; 70.0,2.86; 75.0,3.25; band between calc–alkaline and low-K series 45.0,0.2; 48.0,0.41; 61.0,0.97; 70.0,1.38; 75.0,1.51; and 45.0,0.15; 48.0,0.3; 61.0,0.8; 70.0,1.23; 75.0,1.44.

Table 3.1   Comparison of CIPW and cation norms for an average tonalite*

	1	2	3	4	5	6		7	8
	Wt % oxide of rock	Mol. wt	Molecular proportions	Number of cations	Cationic proportions	Cation %		CIPW norm	Cation norm
SiO_2	61.52	60.09	1.0238	1.00	1.0238	58.03	Q	15.94	15.12
TiO_2	0.73	79.90	0.0091	1.00	0.0091	0.52	Or	12.23	12.45
Al_2O_3	16.48	101.96	0.1616	2.00	0.3233	18.32	Ab	30.71	33.2
Fe_2O_3	1.83	159.69	0.0115	2.00	0.0229	1.30	An	22.52	22.98
FeO	3.82	71.85	0.0532	1.00	0.0532	3.01	Di	2.17	2.18
MnO	0.08	70.94	0.0011	1.00	0.0011	0.06	Hy	10.32	10.47
MgO	2.80	40.30	0.0695	1.00	0.0695	3.94	Mt	2.64	1.95
CaO	5.42	56.08	0.0966	1.00	0.0966	5.48	Il	1.38	1.04
Na_2O	3.63	61.98	0.0586	2.00	0.1171	6.64	Ap	0.56	0.53
K_2O	2.07	94.20	0.0220	2.00	0.0439	2.49			
P_2O_5	0.25	141.95	0.0018	2.00	0.0035	0.20	Total	98.47	99.92
Total	98.63				1.7641	99.99	An	34 %	33 %
							Ab	47 %	48 %
							Or	19 %	18 %

* Analysis from Le Maitre (1976).

Cation norms   Barth (1952) proposed an alternative calculation based upon molecular proportions and cations. This type of norm calculation has become variously known as the **Barth–Niggli norm** (after its proponents) or the **cation norm** or the **molecular norm**. In this case the calculations are made using the equivalent weights of the oxides, i.e. the molecular weight when one cation is present. In the case of oxides such as CaO or TiO_2 the equivalent weight is the same as the molecular weight but in the case of Al_2O_3 or Na_2O the equivalent weight is half the molecular weight.

The calculation of a cation norm is illustrated in Table 3.1. The wt % oxide values (column 1) are divided by their equivalent weights (divide by column 2 and multiply by column 4), converted into cation proportions (column 5) and then converted into cation % (column 6). Molecules are then constructed according to the standard CIPW rules (column 8), although the proportions of the components in which cations are allocated is different from the CIPW norm. In the case of the CIPW norm the proportions of components allocated to albite are Na/Al/Si = 1:1:6 (on the basis of the combined oxygen), whereas in the case of the cation norm the allocation is 1:1:3 (on the basis of the cation proportions). The cation norm is not recalculated on a wt % basis; rather the result is expressed as a molecular percentage (mol %). One merit of the cation norm is that the proportions of opaque minerals are closer to their volume percentages as seen in thin section.

Norm calcula-   A particular difficulty with the norm calculation is that it is very sensitive to the
tions and the   oxidation state of the iron. This is more of a problem in mafic rocks where the Fe
oxidation state of   content is much higher than that of most felsic rocks. Middlemost (1989) has
iron   proposed a range of oxidation ratios (Fe_2O_3/FeO) for use with volcanic rocks drawn from the geological literature. His data are presented by rock type on a TAS diagram in Figure 3.6.

Basalt classifica-   Thompson (1984) proposed a classification scheme for basalts based upon their
tion using the   CIPW normative proportions of nepheline (and other feldspathoids), olivine,
Ne–Di–Ol–Hy–Q   diopside, hypersthene and quartz. The diagram is shown in Figure 3.7 and is an
diagram of   expanded version of a portion of the Yoder–Tilley low-pressure basalt tetrahedron
Thompson   illustrated in Figure 3.26(a). The three equilateral triangles of this diagram
(1984)   Ne–Ol–Di, Ol–Di–Hy and Di–Hy–Q represent basaltic and related rocks which are respectively undersaturated, saturated and oversaturated with silica (Figure 3.7). Thus silica–undersaturated basalts (alkali basalts) are characterized by normative olivine and nepheline, silica–saturated basalts (olivine tholeiites) are characterized by normative hypersthene and olivine, and silica–oversaturated basalts (quartz tholeiites) are characterized by normative quartz and hypersthene. Silica saturation is particularly important in basaltic magmas, because in dry magmas this single parameter determines the crystallization sequence of minerals and direction of evolution during fractional crystallization.

Normative compositions [calculated assuming FeO/(FeO + Fe_2O_3) = 0.85, or Fe_2O_3/FeO = 0.18] are projected onto one of the three triangles by summing the three relevant normative parameters (calculated as the wt % norm) and calculating the value of each as a percentage of their sum. The calculation and plotting procedure for triangular diagrams is given in Section 3.3.2. This diagram should not be used for highly evolved magmas and is best applied to basalts which have more than 6 % MgO. A disadvantage of this classification is that it uses only about half of the calculated norm and so is not fully representative of the rock. It is also very sensitive to small errors in Na_2O and so is inappropriate for altered rocks.

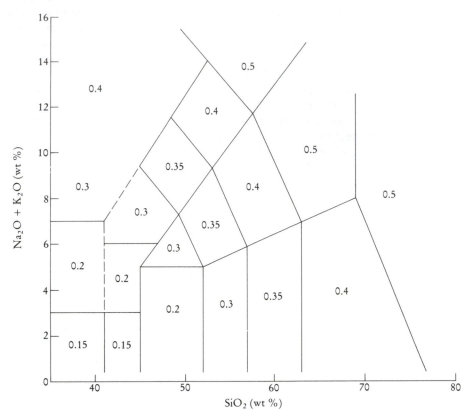

Figure 3.6   The total-alkalis vs silica diagram (TAS) of Le Maitre *et al.* (1989) showing the different Fe$_2$O$_3$/FeO ratios recommended by Middlemost (1989) for volcanic rocks, plotted according to rock type as defined in the TAS classification shown in Figure 3.1.

Granite classifi-
cation using the
Ab–An–Or
diagram of
O'Connor
(1965)

The Ab–An–Or classification diagram of O'Connor (1965) can be applied to felsic rocks with more than 10 % normative quartz. The diagram (Figure 3.8) is based entirely upon the normative feldspar composition recast to 100 % and represents a projection from quartz onto the feldspar face of the 'granite' tetrahedron Q–Ab–An–Or. The Ab–An–Or diagram has been chiefly used to classify plutonic rocks although O'Connor intended his classification to be used with both volcanic and plutonic felsic rocks. The feldspar compositions in this classification are calculated using the Barth–Niggli molecular norm, although it is not clear whether this procedure has been followed by all users. However, the plotting parameters in the Ab–An–Or projection are within 2 % of each other in either calculation scheme. This is illustrated in Table 3.1, where the An–Ab–Or plotting parameters for a tonalite are calculated using both the CIPW and the molecular norms. The field boundaries of the diagram were empirically defined from a data-set of 125 plutonic rocks for which there were both normative and modal data. Barker (1979) has proposed a modification to the diagram, slightly expanding the field of trondhjemite at the expense of tonalite (Figure 3.8). This modification is widely used.

The O'Connor (1965) Ab–An–Or diagram as modified by Barker (1979) provides a convenient way of classifying felsic plutonic rocks on the basis of their major

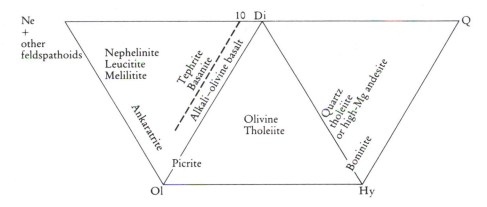

Figure 3.7   The classification of basalts and related basic and ultrabasic magmatic rocks according to their CIPW normative composition expressed as Ne–Ol–Di, Ol–Di–Hy or Di–Hy–Q (after Thompson, 1984).

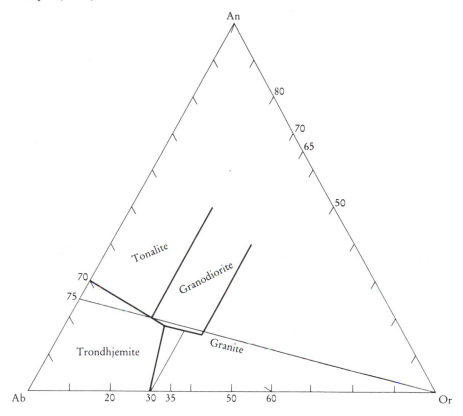

Figure 3.8   The classification of 'granitic' rocks according to their molecular normative An–Ab–Or composition after Barker (1979) (heavy lines). The original fields of O'Connor (1965) are shown in faint lines.

element chemistry. It has the advantage of field boundaries which are clear and easy to reproduce and which effectively separate tonalites, trondhjemites, granites and granodiorites from each other. The normative calculation provides a more accurate estimate of feldspar compositions than a modal classification, for it reflects any solid solution in the feldspars. The diagram can also be used (with caution) with deformed and metamorphosed granitic rocks and permits an estimate of their original magma type.

Streckeisen and Le Maitre (1979) proposed a classification diagram for volcanic and plutonic rocks (Figure 3.9), based upon their normative composition, which was intended to mirror the Streckeisen QAPF classification. The norm calculations are made using the Barth–Niggli molecular norm. The diagram has rectangular coordinates of which the *y*-axis reflects the degree of silica saturation and is either a measure of the quartz content [Q' = Q/(Q + Or + Ab + An)] or the feldspathoid content [F' = (Ne + Lc + Kp)/(Ne + Lc + Kp + Or +Ab + An)]. The *x*-axis reflects changing feldspar composition [ANOR = 100 × An/(Or + An)]. The exclusion of albite from the feldspar axis avoids the difficult task of allocating it to either plagioclase or alkali feldspar. The field boundaries were defined for the plutonic rocks empirically using a data-bank of 15 000 analyses of rocks, for many of which the mineralogical composition was also known.

Unfortunately this diagram does not work very well. Streckeisen and Le Maitre (1979) show contoured plots for a number of different rock types in which the distribution of the rock type is plotted on the Q' (F')–ANOR diagram. Few rocks

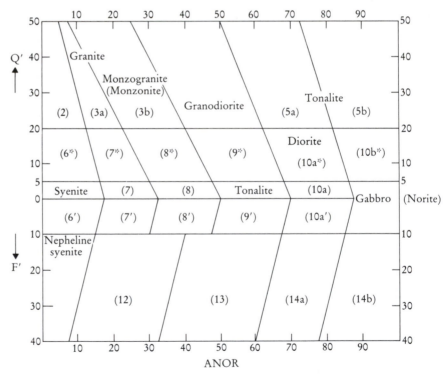

Figure 3.9   The classification of plutonic rocks using their molecular normative compositions (after Streckeisen and Le Maitre, 1979). The numbered fields correspond to those in the modal QAPF diagram of Streckeisen (1976). The named rock types are plotted at the position of their highest concentration in the data file used by Streckeisen and Le Maitre. Q' = Q/(Q + Or + Ab + An); F' = (Ne + Lc + Kp)/(Ne + Lc + Kp + Or + Ab + An); ANOR = 100 × An/(Or + An).

are strictly confined to their allotted fields and there is a significant overlap between fields in a number of cases, notably for tonalite. As a consequence the diagram has not been widely used.

3.2.3   Classifying igneous rocks using cations

In order to avoid the criticism that wt % oxide data do not faithfully reflect the cation distribution of a sample, a number of authors prefer to recalculate the rock composition as cations. This is the same as the initial stage of the cation norm calculation. The wt % of the oxide is divided by the equivalent weight of the oxide set to one cation. It is often expressed slightly differently — the wt % oxide is divided by the molecular weight of the oxide and multiplied by the number of cations in the formula unit. Thus the wt % SiO_2 is divided by 60.09. However, the wt % Al_2O_3 is divided by 101.96 and then multiplied by 2. In some cases the cationic proportions are multiplied by 1000 and described as millications. A worked example is given in Table 3.2.

The R1–R2 diagram of De la Roche et al. (1980)

De la Roche and LeTerrier (1973) and De la Roche *et al.* (1980) proposed a classification scheme for volcanic and plutonic igneous rocks based upon their cation proportions, expressed as millications. This diagram is most useful for plutonic rocks. The results are plotted on an x–y bivariate graph using the plotting parameters R1 and R2. R1 is plotted on the x–axis and is defined as:

$$R1 = [4Si - 11(Na + K) - 2(Fe + Ti)]$$

Table 3.2   Cation calculation for an average tonalite*

	1	2	3	4	5	6	7
	Wt % oxide of rock	Mol. wt	No. of cations	Cationic Propor- tions	Milli- cations	Cation %	Mol %
SiO_2	61.52	60.09	1.00	1.0238	1023.80	58.03	67.86
TiO_2	0.73	79.9	1.00	0.0091	9.14	0.52	0.61
Al_2O_3	16.48	101.96	2.00	0.3233	323.26	18.32	10.71
Fe_2O_3	1.83	159.69	2.00	0.0229	22.92	1.30	0.76
FeO	3.82	71.85	1.00	0.0532	53.17	3.01	3.52
MnO	0.08	70.94	1.00	0.0011	1.13	0.06	0.07
MgO	2.80	40.3	1.00	0.0695	69.48	3.94	4.61
CaO	5.42	56.08	1.00	0.0966	96.65	5.48	6.41
Na_2O	3.63	61.98	2.00	0.1171	117.13	6.64	3.88
K_2O	2.07	94.2	2.00	0.0439	43.95	2.49	1.46
P_2O_5	0.25	141.95	2.00	0.0035	3.52	0.20	0.12
			Sum	1.7644		99.99	100.01

$R1 = 4Si - 11(Na + K) - 2(Fe + Ti) = 2152.83$
$R2 = 6Ca + 2Mg + Al = 1042.11$

* Analysis from Le Maitre (1976).

Fe represents total iron. R2 is plotted along the *y*-axis and is defined as:

R2 = (Al + 2Mg + 6Ca)

The diagram for plutonic rocks is given in Figure 3.10 and an example of the calculation of R1 and R2 is given in Table 3.2. The advantages of this classification scheme are that:

(1) the entire major element chemistry of the rock is used in the classification;
(2) the scheme is sufficiently general to apply to all types of igneous rock;
(3) mineral compositions can also be plotted on the diagram, allowing a broad comparison between modal and chemical data; and
(4) the degree of silica saturation and changing feldspar compositions can be shown.

The authors claim that the R1–R2 diagram is simpler to use than a norm-based classification.

The problem with this particular classification diagram is that it is difficult to understand and difficult to use. The parameters R1 and R2 have no immediate meaning, making the diagram difficult to understand at first sight. In addition, the field boundaries are curvilinear and so are difficult to reproduce.

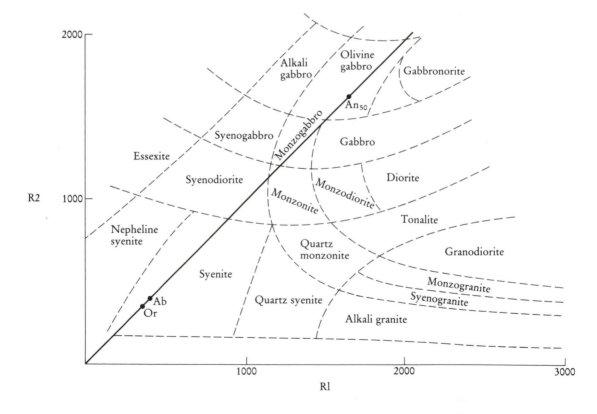

Figure 3.10   The classification of plutonic rocks using the parameters R1 and R2 (after de la Roche *et al.*, 1980), calculated from millication proportions. R1 = 4Si − 11(Na + K) − 2(Fe + Ti); R2 = 6Ca + 2Mg + Al.

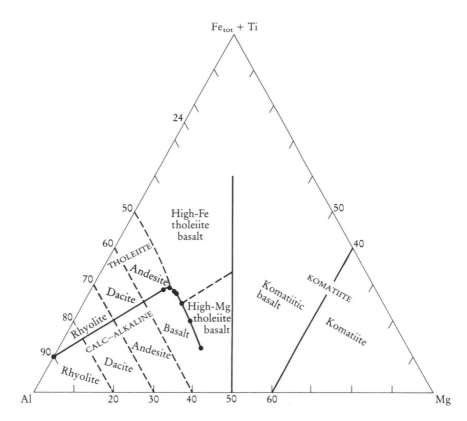

Figure 3.11 The classification of volcanic rocks according to their cation percentages of Al, (Fe$_{[total]}$ + Ti) and Mg showing the tholeiite, calc–alkaline and komatiite fields. In addition to the coordinates shown on the diagram, the boundary between the tholeiitic and calc–alkaline fields is defined by the coordinates (Al, Fe + Ti, Mg): 90,10,0; 53.5,28.5,18; 52.5,29,18.5; 51.5,29,19.5 50.5,27.5,22; 50.3,25,24.7; 50.8,20,29.2; 51.5,12.5,36 (Rickwood, 1989; corrected).

The Jensen cation plot (Jensen, 1976)

The Jensen cation plot is a classification scheme for subalkaline volcanic rocks and is particularly useful for komatiites. It is based upon the proportions of the cations (Fe^{2+} + Fe^{3+} + Ti), Al and Mg recalculated to 100 % and plotted on a triangular diagram (Figure 3.11). The elements were selected for their variability within subalkaline rocks, for the way in which they vary in inverse proportion to each other and for their stability under low grades of metamorphism. Thus this classification scheme can be used successfully with metamorphosed volcanic rocks which have suffered mild metasomatic loss of alkalis. This is a distinct advantage over other classification schemes for volcanic rocks.

The main importance of this diagram, however, is that it shows komatiites clearly as a separate field from basalts and from calc–alkaline rocks, and so is useful for Archean metavolcanics. The original diagram of Jensen (1976) was slightly modified by Jensen and Pyke (1982), who moved the komatiitic basalt/komatiite field boundary to a lower Mg value. This is the version presented here (Figure 3.11). The plotting parameters of the field boundaries are taken from Rickwood (1989).

3.2.4   The chemical classification of sedimentary rocks

The geochemical classification of sedimentary rocks is not as developed as for igneous rocks and most systems of sedimentary rock classification utilize features which can be observed in hand specimen or in thin section, such as grain size and the mineralogy of the particles and matrix.

Arenite/wacke   Unlike many igneous rocks it is difficult to find a simple relationship between the mineralogy of sandstones and their chemical composition. For this reason the geochemical classification of sandstones does not mimic the conventional mineralogical classification of sandstones based upon quartz–feldspar–lithic fragments. Rather it differentiates between mature and immature sediments.

The most commonly used geochemical criteria of sediment maturity are the SiO_2 content and the SiO_2/Al_2O_3 ratio (Potter, 1978), reflecting the abundance of quartz and the clay and feldspar content. Another useful index of chemical maturity is the alkali content ($Na_2O + K_2O$), also a measure of the feldspar content. Using an index of chemical maturity and the Na_2O/K_2O ratio, Pettijohn *et al.* (1972) proposed a classification for terrigenous sands based upon a plot of $\log(Na_2O/K_2O)$ vs $\log(SiO_2/Al_2O_3)$. Although they state that their diagram is 'not particularly useful for naming purposes but simply shows some of the relationships between elemental composition, mineralogy and rock type', it is widely used. The Pettijohn *et al.* diagram as modified by Herron (1988) is presented in Figure 3.12. A similar diagram is the Na_2O/K_2O vs SiO_2 diagram used by Middleton (1960) to subdivide greywackes, and adapted more recently by Roser and Korsch (1986) to determine the tectonic setting of sandstone–mudstone suites. However, such diagrams must be applied with caution, for Na and K are likely to be mobilized during diagenesis and metamorphism.

Herron (1988) modified the diagram of Pettijohn *et al.* using $\log(Fe_2O_3/K_2O)$ along the *y*-axis instead of $\log(Na_2O/K_2O)$ (Figure 3.13). The ratio $Fe_2O_3[total]/K_2O$ allows arkoses to be more successfully classified and is also a measure of mineral stability, for ferromagnesian minerals tend to be amongst the less stable

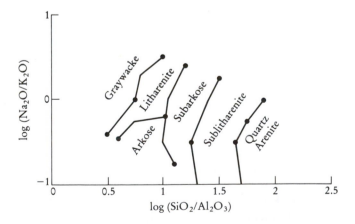

Figure 3.12   The classification of terrigenous sandstones using $\log(Na_2O/K_2O)$ vs $\log(SiO_2/Al_2O_3)$ from Pettijohn *et al.* (1972) with the boundaries redrawn by Herron (1988).

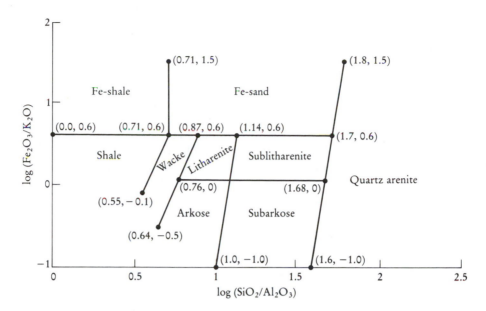

Figure 3.13   The classification of terrigenous sandstones and shales using $\log(Fe_2O_3/K_2O)$ vs $\log(SiO_2/Al_2O_3)$ (after Herron, 1988). The numbers shown in parentheses are the plotting coordinates for the field boundaries expressed as $[\log(SiO_2/Al_2O_3), \log(Fe_2O_3/K_2O)]$.

minerals during weathering. A third axis (not shown) plots total Ca — a measure of the calcium carbonate content of the rock cement. Shale, which is not considered in the Pettijohn *et al.* scheme, is identified on the basis of a very low SiO_2/Al_2O_3 ratio. A further advantage of the Herron classification is that it can be used to identify shales, sandstones, arkoses and carbonate rocks *in situ* from geochemical well logs, using neutron activation and gamma-ray tools (Herron and Herron, 1990).

Blatt *et al.* (1972) devised a classification based upon the factor analysis of the chemical composition of clastic sediments. Their scheme plots Na_2O, total Fe (as Fe_2O_3) + MgO and K_2O at the apices of a triangular diagram and shows fields for greywacke, lithic sandstone and arkose. In a study of recent big-river sediments, however, Potter (1978) found that he could not duplicate the fields of Blatt *et al.* (1972) and suggested that the chemical composition of ancient sediments is strongly influenced by diagenesis. Blatt *et al.* (1972) also suggested that greywacke and arkose could be subdivided on the basis of their FeO content on an Fe_2O_3-FeO plot. Again this is not mirrored in recent big-river sediments (Potter, 1978) and Fe^{2+}/Fe^{3+} must be strongly controlled by the oxidation state of the sediment.

Mudrocks   There is no widely recognized chemical classification of mudrocks. This is surprising since they have a much more variable chemistry than sandstones and so readily lend themselves to chemical classification. Englund and Jorgensen (1973) proposed a classification of mudrocks on the basis of their (K_2O + Na_2O + CaO) vs (MgO + FeO) vs Al_2O_3 contents expressed on a triangular diagram. Wronkiewicz and Condie (1987) use a similar diagram based upon $Fe_2O_{3[total]}$ vs K_2O vs Al_2O_3 to classify the Witwatersrand shales of South Africa. They also used an expanded version of the $\log(Na_2O/K_2O)$ vs $\log(SiO_2/Al_2O_3)$ plot of Pettijohn *et al.* (1972) on which they plot the fields of Archaean greenstone belt shales and Phanerozoic shales. Bjorlykke (1974) plotted the $(Al_2O_3+K_2O)/(MgO+Na_2O)$ content of shales and used it as an indicator of volcanic-arc provenance.

3.2.5   Discussion

For volcanic igneous rocks there is a variety of chemical classification schemes which work and which are simple to use, the best of which is the TAS diagram. Plutonic rocks are more problematic. An adequate classification scheme exists for granitic rocks in the O'Connor diagram but there is no simple and widely accepted classification for all plutonic rocks. Probably, the most comprehensive is the R1–R2 diagram of De la Roche *et al.* (1980) although this diagram is difficult both to understand and to apply.

3.3   Variation diagrams

A table of geochemical data from a particular igneous province, metamorphic terrain or sedimentary succession may at first sight show an almost incomprehensible variation in the concentration of individual elements. Given that the samples are likely to be geologically related, a major task for the geochemist is to devise a way in which the variation between individual rocks may be simplified and condensed so that relationships between the individual rocks may be identified. The device which is most commonly used and has proved invaluable in the examination of geochemical data is the variation diagram. This is a bivariate graph or scattergram on which two selected variables are plotted. Diagrams of this type were popularized as long ago as 1909 by Alfred Harker in his *Natural history of igneous rocks*, and one particular type of variation diagram, in which SiO_2 is plotted along the *x*-axis, has become known as the 'Harker diagram'.

An illustration of the usefulness of variation diagrams can be seen from a comparison of the data in Table 3.3 and the variation diagrams plotted for the same data (Figure 3.14). It is clear that the variation diagrams have condensed and rationalized a large volume of numerical information and show qualitatively that there is an excellent correlation (either positive or negative) between each of the major elements displayed and SiO_2. Traditionally this strong geochemical coherence between the major elements has been used to suggest that there is an underlying process which will explain the relationships between the major elements.

3.3.1   Recognizing geochemical processes on a major element variation diagram

Most trends on variation diagrams are the result of mixing. In igneous rocks the mixing may be that of two magmas, the addition and/or subtraction of solid phases during contamination or fractional crystallization, or mixing due to the addition of

Table 3.3   Chemical analyses of rocks from Kilauea Iki lava lake, Hawaii*

	1	2	3	4	5	6	7	8	9	10	11	12	13	14	15	16	17
SiO_2	48.29	48.83	45.61	45.50	49.27	46.53	48.12	47.93	46.96	49.16	48.41	47.90	48.45	48.98	48.74	49.61	49.20
TiO_2	2.33	2.47	1.70	1.54	3.30	1.99	2.34	2.32	2.01	2.73	2.47	2.24	2.35	2.48	2.44	3.03	2.50
Al_2O_3	11.48	12.38	8.33	8.17	12.10	9.49	11.43	11.18	9.90	12.54	11.80	11.17	11.64	12.05	11.60	12.91	12.32
Fe_2O_3	1.59	2.15	2.12	1.60	1.77	2.16	2.26	2.46	2.13	1.83	2.81	2.41	1.04	1.39	1.38	1.60	1.26
FeO	10.03	9.41	10.02	10.44	9.89	9.79	9.46	9.36	9.72	10.02	8.91	9.36	10.37	10.17	10.18	9.68	10.13
MnO	0.18	0.17	0.17	0.17	0.17	0.18	0.18	0.18	0.18	0.18	0.18	0.18	0.18	0.18	0.18	0.17	0.18
MgO	13.58	11.08	23.06	23.87	10.46	19.28	13.65	14.33	18.31	10.05	12.52	14.64	13.23	11.18	12.35	8.84	10.51
CaO	9.85	10.64	6.98	6.79	9.65	8.18	9.87	9.64	8.58	10.55	10.18	9.58	10.13	10.83	10.45	10.96	11.05
Na_2O	1.90	2.02	1.33	1.28	2.25	1.54	1.89	1.86	1.58	2.09	1.93	1.82	1.89	1.73	1.67	2.24	2.02
K_2O	0.44	0.47	0.32	0.31	0.65	0.38	0.46	0.45	0.37	0.56	0.48	0.41	0.45	0.80	0.79	0.55	0.48
P_2O_5	0.23	0.24	0.16	0.15	0.30	0.18	0.22	0.21	0.19	0.26	0.23	0.21	0.23	0.24	0.23	0.27	0.23
H_2O^+	0.05	0.00	0.00	0.00	0.00	0.08	0.03	0.01	0.00	0.06	0.08	0.00	0.09	0.02	0.04	0.02	0.04
H_2O^-	0.05	0.03	0.04	0.04	0.03	0.04	0.05	0.04	0.00	0.02	0.02	0.02	0.00	0.00	0.01	0.01	0.02
CO_2	0.01	0.00	0.00	0.00	0.00	0.11	0.04	0.02	0.00	0.00	0.00	0.01	0.00	0.01	0.01	0.01	0.01
Total	100.01	99.89	99.84	99.86	99.84	99.93	100.00	99.99	99.93	100.05	100.02	99.95	100.05	100.06	100.07	99.90	99.95

* From Richter and Moore (1966), courtesy of the US Geological Survey.

melt increments during partial melting. In sedimentary rocks trends on a variation diagram will also result from mixing, but in this case the mixing of chemically distinct components which contribute to the composition of the sediment. In metamorphic rocks, trends on a variation diagram will usually reflect the processes in the igneous or sedimentary precursor, masked to some degree by specific metamorphic processes such as metasomatism. In some instances, however, deformation may 'smear' together more than one rock type, giving rise to a mixing line of metamorphic origin.

Below we consider some of the more important mixing processes.

Fractional crystallization   Fractional crystallization is a major process in the evolution of many igneous rocks, and is frequently the cause of trends seen on variation diagrams for igneous rocks. The fractionating mineral assemblage is normally indicated by the phenocrysts present. A test of crystal fractionation may be made by accurately determining the composition of the phenocrysts using the electron microprobe and then plotting the compositions on the same graph as the rock analyses. If trends on a variation diagram are controlled by phenocryst compositions then it may be possible to infer that the rock chemistry is controlled by crystal fractionation. It should be noted, however, that fractional crystallization may also take place at depth and in this case the fractionating phases may not be represented in the phenocryst assemblage.

The importance of fractional crystallization was expounded at length by Bowen (1928) in his book *The evolution of the igneous rocks*; he argued that geochemical trends for volcanic rocks represent a 'liquid line of descent'. This is the path taken by residual liquids as they evolve through the differential withdrawal of minerals from the magma. The ideas of Bowen now need to be qualified in the light of modern findings in the following ways: (1) trends identical to those produced by crystal fractionation can also be produced by partial melting; (2) only phenocryst-poor or aphyric volcanic rocks will give a true indication of the liquid path; (3) rarely does a suite of volcanic rocks showing a progressive chemical change erupt as a time sequence. Thus even a highly correlated trend for phenocryst-free volcanic

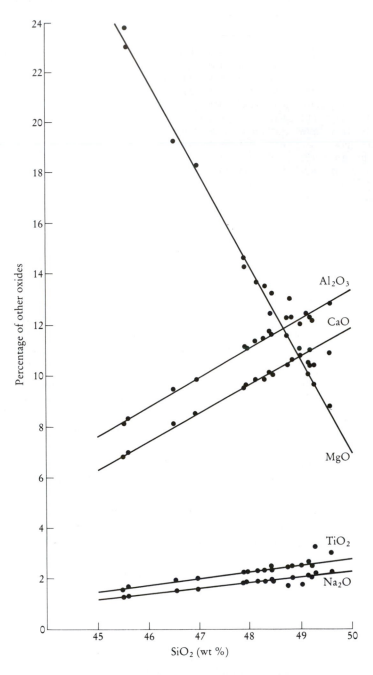

Figure 3.14 Bivariate plots of the oxides Al$_2$O$_3$, CaO, MgO, TiO$_2$, Na$_2$O vs SiO$_2$ in basaltic lavas from Kilauea Iki lava lake from the 1959–1960 eruption of Kilauea volcano, Hawaii (from Richter and Moore, 1966). The data are given in Table 3.3.

rocks on a variation diagram for a single volcano is unlikely to represent a liquid line of descent. Rather it is an approximation to a liquid line of descent of a bundle of similar, overlapping, subparallel lines of descent. Such lavas are not related to a single parental magma but rather to a series of similar and related magmas (Cox *et al.*, 1979).

Assimilation and fractional crystallization

If phenocryst compositions cannot explain trends in a rock series and a fractional crystallization model does not appear to work, it is instructive to consider the possibility of simultaneous assimilation of the country rock and fractional crystallization. This process, often abreviated to AFC, was first proposed by Bowen (1928), who argued that the latent heat of crystallization during fractional crystallization can provide sufficient thermal energy to consume the wall-rock. Anderson and Cullers (1987) argued for an AFC model to explain the major element chemistry of a Proterozoic tonalite–trondhjemite suite hosted by Archean gneisses. The suite showed marked chemical variability but trends on Harker diagrams were not compatible with any simple fractionation scheme based upon the chemistry of minerals present, or once present, in the original melt. Their calculations showed, however, that if in addition to crystal fractionation the melt was contaminated with a small amount (*ca* 7 %) of the enclosing Archean gneiss, the observed trends were duplicated.

O'Hara (1980) has argued that contamination can result in the 'decoupling' of the major and trace element or isotope chemistry and is not always demonstrable from the major element data. For example, contamination of a basalt precipitating olivine, clinopyroxene and plagioclase will result in increased precipitation of fractionating minerals but may cause only a minor change in composition of the liquid as measured, for example, in its silica content. Trace element levels and isotope ratios, however, will be changed and provide a better means of recognizing assimilation.

Partial melting

Progressive fractional melting will show a trend on a variation diagram which is controlled by the chemistry of the solid phases being added to the melt. However, this can be very difficult to distinguish from a fractional crystallization trend on a major element variation diagram, for both processes represent crystal–liquid equilibria involving almost identical liquids and identical crystals. One situation in which progressive partial melting and fractional crystallization may be differentiated is if the two processes take place under different physical conditions. For example, if partial melting takes place at great depth in the mantle and fractional crystallization is a crustal phenomenon, then some of the phases involved in partial melting will be different from those involved in fractional crystallization.

Mixing lines in sedimentary rocks

Trends on variation diagrams for sedimentary rocks may result from the mixing of the different ingredients which constitute the sediment. There are a number of examples of this effect in the literature. Bhatia (1983), in a study of turbidite sandstones from eastern Australia, shows Harker diagrams in which there is a change in mineralogical maturity, i.e. an increase in quartz coupled with a decrease in the proportions of lithic fragments and feldspar (Figure 3.15). Argast and Donnelly (1987) show how strongly correlated trends may result from two-component (quartz–illite, quartz/feldspar–illite) mixing and curvilinear or scattered trends result from three-component mixing (quartz–illite–calcite).

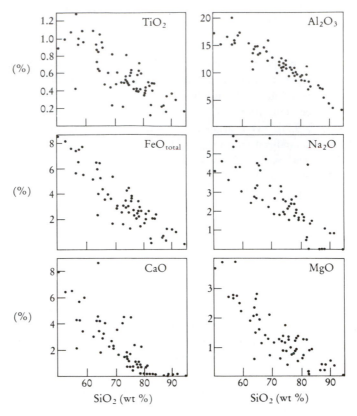

Figure 3.15   Harker variation diagrams for quartz-rich sandstone suites from eastern Australia (after Bhatia, 1983). The increase in SiO_2 reflects an increased mineralogical maturity, i.e. a greater quartz content and a smaller proportion of detrital grains.

The identification of former weathering conditions from sedimentary rocks

A good measure of the degree of chemical weathering can be obtained from the chemical index of alteration (CIA; Nesbitt and Young, 1982).

$$CIA = [Al_2O_3/(Al_2O_3 + CaO* + Na_2O + K_2O)]$$

In addition, weathering trends can be displayed on a $(CaO* + Na_2O) - Al_2O_3 - K_2O$ triangular plot (Nesbitt and Young, 1984, 1989). On a diagram of this type the initial stages of weathering form a trend parallel to the $(CaO + Na_2O) - Al_2O_3$ side of the diagram, whereas advanced weathering shows a marked loss in K_2O as compositions move towards the Al_2O_3 apex (Figure 3.16). The trends follow mixing lines representing the removal of alkalis and Ca in solution during the breakdown of first plagioclase and then potassium feldspar and ferromagnesian silicates.

The CIA and trends on triangular plots have been used in two different ways. Firstly, chemical changes in a recent weathering profile such as that illustrated in Figure 3.16 are used as a template against which the chemical history of an ancient profile can be read. Deviations from such trends can be used to infer chemical changes resulting from diagenesis or metasomatism (Nesbitt and Young, 1984, 1989). The second application is to mudstones. The major and trace element chemistry of modern muds reflects the degree of weathering in their source (Nesbitt

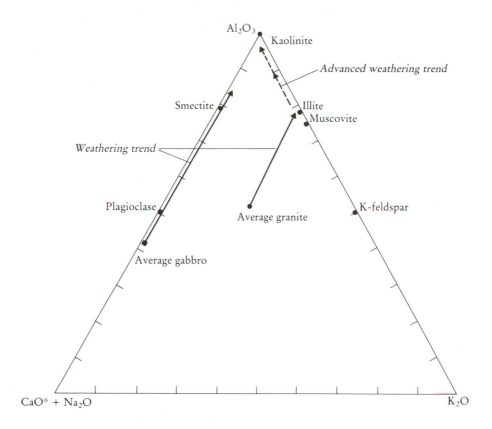

Figure 3.16 The $(Na_2O + CaO) - Al_2O_3 - K_2O$ diagram of Nesbitt and Young (1984, 1989) showing the weathering trends for average granite and average gabbro. The advanced weathering trend for granite is also shown. Compositions are plotted as molar proportions and the compositions of plagioclase, K-feldspar, muscovite and kaolinite are shown. CaO* represents the CaO associated with the silicate fraction of the sample.

et al., 1990); thus the chemical composition of ancient muds may be used in a similar way to make inferences about past weathering conditions.

Mixing in metamorphic rocks Banded gneisses from the Archaean Lewisian complex of northwest Scotland show linear trends on major element variation diagrams. There are two possible explanations of such trends. On the one hand the banded gneisses may be the product of the tectonic mixing of mafic and felsic end-members in the gneiss suite, in which case trends on the variation diagrams also reflect this mixing (Tarney, 1976). Alternatively, the trend could be pre-metamorphic and magmatic in origin and indicate the approximate igneous composition of the gneiss suite.

Element mobility Element mobility describes the chemical changes which take place in rock after its formation, usually through interaction with a fluid. Most commonly, element mobility will take place during weathering, diagenesis and metamorphism or through interaction with a hydrothermal fluid. In metamorphic rocks, element mobility may also take place as a result of solid-state diffusion and melt generation. Here, however, we are chiefly concerned with fluid-controlled element mobility on the scale of several centimetres or more. The mobility of major elements is controlled by three main factors — the stability and composition of the minerals in the unaltered rock, the stability and composition of the minerals in the alteration product and the composition, temperature and volume of the fluid phase.

Element mobility may be detected from mineralogical phase and compositional changes that have taken place in a rock as a result of metamorphism or hydrothermal activity and from the mineral assemblages present in associated veins. Scattered trends on variation diagrams are also a useful indicator, although chemical alteration can sometimes produce systematic changes which may mimic other mixing processes such as crystal fractionation. These apparent trends may result from volume changes arising from the removal or addition of a single component of the rock. Variation diagrams which can be used to identify element mobility are discussed in Section 4.9.3.

Table 3.4 summarizes the main mobile elements in a range of common rock

Table 3.4   Major element mobility in common rock types under a variety of hydrothermal conditions

Rock type		Si	Ti	Al	Fe	Mn	Mg	Ca	Na	K	P	Reference
Komatiite		×						×	×	×		Arndt (1983)
Basalt	Hydrothermal alteration	−			−	−	+	−	−	−		Mottl (1983)
Basalt	Hydrothermal alteration	+	−		−	−	−	−	+			MacGeehan and MacLean (1980)
Basalt	Submarine weathering	−			+		−	−	−	+		Pearce (1976)
Basalt	Weathering	−					−	−	−	−		Pearce (1976)
Basalt	Greenschist facies metamorphism				×		×		×	×		Pearce (1976), Gelinas *et al.* (1982)
Basalt	Amphibolite facies metamorphism	×						×	×			Rollinson (1983)
Granite	Weathering				×		×	×	−	×		Nesbitt and Young (1989)
Granite	Contact metamorphism			+	−		−	−	−	+		Vernon *et al.* (1987)
'Granite'	Granulite facies metamorphism						−			×		Allen *et al.* (1985)
Calcareous sediments	Medium grade metamorphism								−	−		Ferry (1983)
Calcareous sediments	Contact metamorphism	×			×		×					Burcher-Nurminen (1981)
Sandstone–clay	Diagenesis	×			×		×	×		×		Boles and Franks (1979)

Key: ×, element mobile; −, element depleted; +, element enriched.

types under a variety of hydrothermal conditions. Basaltic rocks are well documented and most studies show that Ti, Al and P are generally immobile whereas Ca and Na are almost always mobilized. Similarly, in granites Ti, Al and P are generally immobile. Sediments are not so well studied although chemical changes during the diagenesis of a sandstone–clay sequence are well described and reflect the breakdown, with progressive burial of orthoclase and plagioclase and the conversion of smectite to illite. At higher metamorphic grades little is known except for a consensus that Al is immobile (Ferry, 1983).

Artificial trends Sometimes trends on a variation diagram are artificially produced by the numerical processes used in plotting the data and do not automatically signify geochemical relationships. This is well documented by Chayes (1960) and Aitchison (1986), who have shown that correlations in compositional data can be forced as a result of the unit sum constraint (see Section 2.6). The most helpful way to circumvent this problem is to examine trends on variation diagrams in the light of a specific hypothesis to be tested. The closeness of fit between the model and the data can then be used to evaluate the hypothesis.

3.3.2 Selecting a variation diagram

The two main types of variation diagram currently used by geochemists are considered in this section — bivariate plots and triangular variation diagrams.

Bivariate plots The principal aim of a bivariate plot, such as that illustrated in Figure 3.14, is to show variation between samples and to identify trends. Hence the element plotted along the *x*-axis of the diagram should be selected either to show the maximum variability between samples or to illustrate a particular geochemical process. Normally the oxide which shows the greatest range in the data-set would be selected; in many cases this would be SiO_2, but in basic igneous rocks it might be MgO and in clay-bearing sediments Al_2O_3.

In a reconnaisance geochemical study of a problem it might be necessary to prepare a very large number of variation diagrams in order to delimit the possible geological processes operating. In this case the initial screening of the data is best done by computer (see for example Barnes, 1988). If a correlation matrix is used, it is important to remember that good correlations may arise through a cluster of data points and a single outlier. Similarly, poor correlations can arise if the data-set contains several populations, each with a different trend.

More normally, and more fruitfully, however, most geochemical investigations are designed to solve a particular problem and to test a hypothesis — usually formulated from geological or other geochemical data. In this case the plotting parameter for a variation diagram should be selected as far as possible with the process to be tested in mind. For example, if in the case of igneous rocks a crystal fractionation mechanism is envisaged, then an element should be selected which is contained in the fractionating mineral and which will be enriched or depleted in the melt.

(a) Harker diagrams — bivariate plots using SiO_2 along the x-axis Variation diagrams in which oxides are plotted against SiO_2 are often called **Harker**

diagrams. They are the oldest form of variation diagram and are one of the most frequently used means of displaying major element data (see Figures 3.14, 3.15). SiO_2 is commonly chosen as the plotting parameter for many igneous rock series and for suites of sedimentary rocks with a variable quartz content because it is the major constituent of the rock and shows greater variability than any of the other oxides. However, the very fact that SiO_2 is the most abundant oxide means that there are a number of inherent problems of which the user must be aware. These are: (1) a negative tendency (see Figure 3.15), (2) spurious correlations and (3) a reduced scatter of values as SiO_2 increases (see the Al_2O_3–SiO_2 plot in Figure 3.15). These problems are fully discussed in Section 2.6.

(b) Bivariate plots which use MgO on the x-*axis*   One of the most commonly used alternatives to the Harker diagram is the MgO plot. This is most appropriate for rock series which include abundant mafic members, for in this case the range of SiO_2 concentrations may be small. MgO, on the other hand, is an important component of the solid phases in equilibrium with mafic melts and shows a great deal of variation either as a consequence of the breakdown of magnesian phases during partial melting or their removal during fractional crystallization.

(c) Bivariate plots using cations   It is sometimes simpler to display mineral compositions on a variation diagram if major element chemical data are plotted as cation %, that is the wt % oxide value divided by the molecular weight and multiplied by the number of cations in the oxide formula and then recast to 100 % (see Table 3.2, columns 1–4 and 6); see for example Francis (1985). An identical calculation with the result expressed as mol % cations instead of cation % is used by Hanson and Langmuir (1978) in their MgO–FeO cation diagram (see Section 3.3.4). Roedder and Emslie (1970) in a similar diagram use mol % of MgO and FeO (see Table 3.2, column 7).

(d) Bivariate plots using the magnesium number   The older geochemical literature carries a large number of examples of complex, multi-element plotting parameters which were used as a measure of fractionation during the evolution of an igneous sequence. These are rather complicated to use and difficult to interpret and so have fallen into disuse. However, one which is useful, and so survives, is the **magnesium–iron ratio**, or **magnesium number** as it is sometimes called. The magnesium–iron ratio is particularly useful as an index of crystal fractionation in basaltic liquids (see Oskarsson *et al.*, 1982; Wilkinson, 1982) for here the Mg–Fe ratio changes markedly in the early stages of crystallization as a result of the higher Mg–Fe ratio of the liquidus ferromagnesian minerals than their host melts. The magnesium–iron ratio is expressed either in wt % form as $100[MgO/(MgO + FeO)]$ or $100[MgO/(MgO + FeO + Fe_2O_3)]$ or as an atomic fraction as $100[Mg^{2+}/(Mg^{2+} + Fe^{2+})]$. The inverse of this ratio is also used as a measure of iron enrichment.

Triangular   Triangular variation diagrams are used when it is necessary to show simultaneous
variation   change between three variables. However, this practice is not recommended and
diagrams   bivariate plots are to be preferred, since both the computation of the plotting parameters and the interpretation of the resultant trends raise a number of important problems (see Section 2.7).

The plotting procedure for triangular diagrams is illustrated in Figure 3.17. This is most conveniently done by microcomputer and Topley and Burwell (1984) give an example of a versatile interactive program written in BASIC.

(a) The AFM diagram The AFM diagram (Figure 3.18) is the most popular of triangular variation diagrams and takes its name from the oxides plotted at its apices — Alkalis ($Na_2O + K_2O$), Fe oxides ($FeO + Fe_2O_3$) and MgO. The igneous AFM diagram should not be confused with the metamorphic diagram of the same name which is used to show changing mineral compositions in Al_2O_3–FeO–MgO space. The plotting parameters are calculated by summing the oxides ($Na_2O + K_2O$) + [($FeO + Fe_2O_3$) recalculated as FeO] + MgO and then recalculating each as a percentage of the sum. There is some ambiguity over the way in which the Fe-oxides should be treated and the following alternatives are in current use:

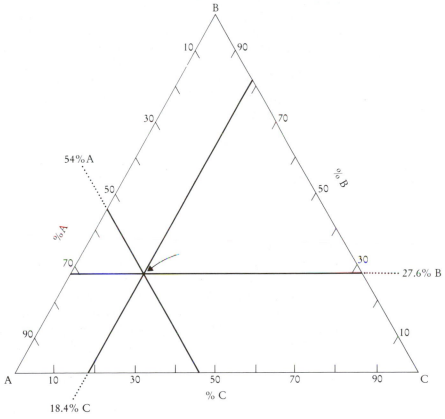

Figure 3.17 Construction lines for plotting the point A = 54 %, B = 27.6 % and C = 18.4 % on a triangular diagram. The values are plotted as follows: variable B is 100 % at the top of the plotting triangle and 0 % along the base of the triangle and so, counting upwards from the base (the concentrations are given on the right-hand side of the triangle), the horizontal line representing 27.6 % is located. In a similar way the line representing 54 % A is located, parallel to the right-hand side of the triangle. The point at which the two lines intersect is the plotting position. To check that it has been accurately located, the line for variable C should pass through the intersection of the two other lines.

$F = (FeO + 0.8998Fe_2O_3)$, i.e. all Fe_2O_3 converted to FeO
is the same as
$F = $ total $(FeO + Fe_2O_3)$ expressed as FeO i.e. $FeO_{[total]}$
but different from
$F = (FeO + Fe_2O_3)$ expressed as raw wt %

In addition some authors include MnO with the Fe-oxides. Rickwood (1989) has shown that the differences are minor and are unlikely to result in serious misplotting; nevertheless it is recommended that a standard procedure is adopted and that F is calculated as total Fe, i.e. $(FeO + Fe_2O_3)$ recast as FeO. This then accommodates XRF analytical data in which the separate oxidation states of iron cannot be determined.

Most authors use oxide wt % when plotting data on an AFM diagram but in a few cases atomic proportions are used and it is not always clear which method has been adopted. The shape of the trend is similar in each case but the position of the atomic proportions plot is shifted away from the Fe apex relative to the position of the oxide plot for the same data (see Barker, 1978).

The AFM diagram is most commonly used to distinguish between tholeiitic and calc–alkaline differentiation trends in the subalkaline magma series. Kuno (1968) and Irvine and Baragar (1971) present dividing lines separating the rocks of the calc–alkaline series and rocks of the tholeiite series (Figure 3.18). Kuno's boundary line yields a smaller area for the tholeiitic suite. Both authors use wt % plots in which F is calculated as $(FeO + Fe_2O_3)$ expressed as FeO. The coordinates for the boundary lines are given in the caption to Figure 3.18 (Rickwood, 1989).

Examples of the trends characteristic of the tholeiitic and calc–alkaline rock series are also plotted in Figure 3.18. The tholeiitic trend is illustrated by Thingmuli volcano in Iceland and the calc–alkaline trend is for the average compositions of the Cascades lavas (Carmichael, 1964).

(b) Problems in the use of the AFM diagram   It is important to note that the AFM diagram is limited in the extent to which petrogenetic information may be extracted (Wright, 1974). This is chiefly a function of the trivariate plotting procedure, which does not use absolute values and only a part of the rock chemistry. In most rocks the A–F–M parameters make up less than 50 % of the oxide weight percentages and cannot therefore fully represent the rock chemistry. In addition, when plotting a rock series different proportions of each rock are normalized to 100 %. This distorts the plotted values. For example, in a series of volcanic rocks with a compositional range from basalt to dacite about 40 % of the basalt is used when plotting onto an AFM diagram whereas only about 15 % of the dacite is used. A further problem with the trivariate plotting procedure has been noted by Butler (1979), who argues that not only do trends on an AFM plot lead to non-quantitative expressions of mineralogical control but that the trends themselves could be an artefact of ternary percentage formation (see Section 2.7).

Barker (1978) advocated plotting mineral compositions on an AFM diagram in addition to rock compositions to assess mineralogical control of magmatic processes. However, this approach can only be semiquantitative since the lever rule (see below), which works well in bivariate plots, cannot be applied because of the disparate proportions of the compositions before projection. Thus AFM diagrams cannot be used in petrogenetic studies to extract quantitative information about

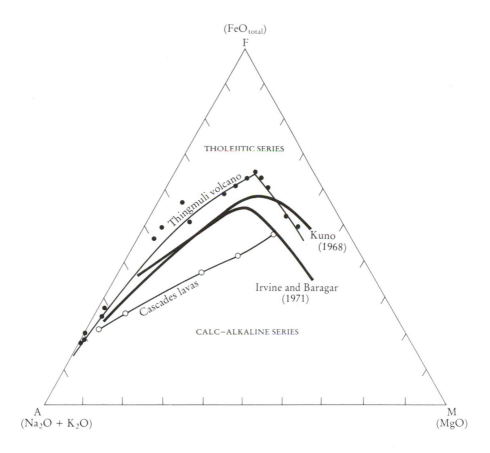

Figure 3.18   An AFM diagram showing the boundary between the calc–alkaline field and the tholeiitic field after Kuno (1968) and Irvine and Baragar (1971) (heavy lines). Also shown are lava compositions and trends (faint lines) for a typical tholeiitic sequence (Thingmuli volcano, Iceland — shown as filled circles — from Carmichael, 1964) and a typical calc–alkaline trend (the average composition of Cascades lavas — shown as open rings — from Carmichael, 1964). The coordinates for points on the boundary lines of Kuno (1968) are A,F,M : 72.0,24.0,4.0; 50.0,39.5,10.5; 34.5,50.0,15.5; 21.5,57.0,21.5; 16.5,58.0,25.5; 12.5,55.5,32.0; 9.5,50.5,40.0; and for Irvine and Baragar (1971), A,F,M: 58.8,36.2,5.0; 47.6,42.4,10.0; 29.6,52.6,17.8; 25.4,54.6,20.0; 21.4,54.6,24.0; 19.4,52.8,27.8; 18.9,51.1,30.0; 16.6,43.4,40.0; 15.0,35.0,50.0 (from Rickwood, 1989).

processes. This must be done using bivariate oxide diagrams. The main usefulness of AFM diagrams, therefore, is to show trends which can be used to identify rock series as illustrated above.

Finally, however, it is worth noting that trends can be generated on an AFM diagram which have no geological meaning at all. Le Maitre (1976) plotted 26 000 samples of unrelated igneous rocks collected from around the world onto an AFM diagram and showed that they defined a marked calc–alkaline trend. He concluded that unrelated analyses taken at random can define trends on an AFM diagram and urged caution in interpreting such trends. A closer inspection of his data shows, however, that the samples are not entirely random since about 85 % of the rocks were collected from continents. Maybe these data have something to say about the composition and origin of the continental crust.

3.3.3   Interpreting trends on variation diagrams

It has been shown above that there are a variety of processes which can produce similar-looking trends on major element variation diagrams. It is important therefore to discover the extent to which these several processes might be distinguished from one another and identified.

Extract calculations   One approach is to try to calculate the composition of the materials added to or subtracted from a magma and to quantify the amount of material involved. This may be done using an extract calculation, a device described in some detail by Cox *et al.* (1979).

The method is illustrated in Figure 3.19(a), in which the chemical compositions (expressed in terms of variables A and B) of both minerals and rocks are plotted on the same variation diagram. Mineral X crystallized from liquid L_1 and the residual liquid follows the path to L_2. The distance from L_1 to L_2 will depend upon the amount of crystallization of mineral X.

This may be quantified as follows:

The amount of L_2 is proportional to the distance $X–L_1$
The amount of X is proportional to the distance $L_1–L_2$

Thus:

The percentage of $L_2 = 100 \times XL_1/XL_2$
Percentage of $X = 100 \times L_1L_2/XL_2$

This relationship is known as **the lever rule**.

If there are two or more minerals crystallizing simultaneously from liquid L_1 in such proportions that their average composition is C [Figure 3.19(b) and (c)], then the liquid path will move from C towards L_2. The proportion of solid to liquid will be given by the ratio $L_1L_2:CL_1$. The proportions of the minerals X and Y in Figure 3.19(b) is given by YC:XC. In the case of the variation diagrams shown above, the predicted trends are straight. However, this not always the case and minerals showing solid solution may produce curved trends during fractionation. This is more difficult to quantify.

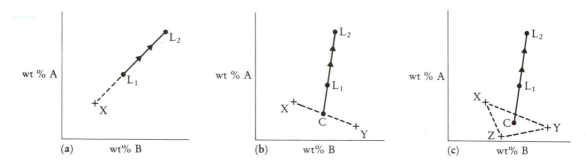

(a)   wt% B   (b)   wt% B   (c)   wt% B

Figure 3.19   Bivariate plot showing extract calculations for crystal fractionation. (a) Mineral X is removed from liquid L_1 and the liquid composition moves from L_1 to L_2. (b) Mineral extract C (made up of minerals X and Y) is removed from liquid L_1 and drives the liquid composition to L_2. (c) Mineral extract C (made up of the minerals X, Y and Z) is removed from liquid L_1 and drives the resultant liquid composition to L_2.

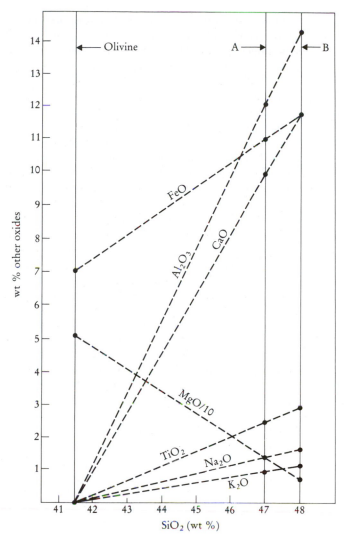

Figure 3.20   Addition–subtraction diagram for rocks A and B. The back-projection of trends for Al_2O_3, CaO, TiO_2, Na_2O and K_2O are reduced to zero and converge at SiO_2 41.5 %. MgO and FeO values at SiO_2 41.5 % indicate the composition of the olivine removed from rock A to produce composition B (after Cox *et al.*, 1979).

An extract calculation for a partial melting trend will not resolve into exact mineralogical constituents, whereas one based on a fractional crystallization trend should resolve exactly. Unfortunately these observations may not be as useful as they first appear, for Cox *et al.* (1979) point out that extract calculations are inexact when the minerals plotted show complex solid solution and that there are statistical uncertainties in fitting a straight line to a trend on a variation diagram. Thus in practice the differences between the effects of partial melting and fractional crystallization will be difficult to observe because of the imprecision of the method.

Addition–subtraction diagrams   An alternative approach to identifying the composition of the solid phase is to use an addition–subtraction diagram to calculate the composition of the phase(s). In this case, rather than just using two elements, the entire major element chemistry of two or more rocks is used. The method is illustrated in Figure 3.20, in which the oxide concentrations of two rocks are plotted on a Harker diagram showing seven superimposed oxides. Back-projection shows that five of the elements converge and reduce to zero at 41.5 % SiO_2, consistent with olivine control. The composition of the olivine can be estimated from the diagram and a simple calculation shows that composition A can be converted to B by the removal of 15 % olivine. A similar explanation can be given to the data from Kilauea Iki lava lake in Figure 3.14.

In some volcanic rocks, particularly members of the calc–alkaline series, there may be a very large number of phenocrysts (olivine–clinopyroxene–biotite–plagioclase–K–feldspar–sphene–apatite–magnetite). In this case graphical methods may not easily produce a solution and the calculation is better handled by a computer. It is important to note, however, that there may not be a unique solution to more complex extract calculations. In andesites, for example, there is the ambiguity that the mineral assemblage (plagioclase–olivine or orthopyroxene–augite–magnetite) is chemically equivalent to hornblende (Gill, 1981).

Extract calculations may also be limited in their use if (1) the liquid line of descent is actually a mix of several lines; (2) solid solution changes the composition of the crystallizing phases during fractionation; (3) the phenocrysts present in magma are not representative of the fractionating phases.

Trends showing
an inflection

Some variation diagrams show segmented trends. In this case the inflection is generally taken to indicate either the entry of a new phase during crystal fractionation or the loss of a phase during partial melting. Figure 3.21 shows a CaO–MgO variation diagram for Hawaiian lavas (Peterson and Moore, 1987). Below about 7 % MgO, CaO correlates positively with MgO, indicating the removal of CaO and MgO from the liquid in the coprecipitation of plagioclase and clinopyroxene. Above 7 % MgO, CaO and MgO correlate negatively because this part of the trend is controlled entirely by olivine. Inflections are not visible on all variation diagrams of a rock series and will only be apparent where the chemistry of the extract is reflected in the plotting parameters. However, when inflections are present they should be located at the same point in the rock series in each case. Inflections are most obvious where the number of fractionating minerals is small, such as in basaltic melts. In calc–alkaline volcanic rocks, where the number of fractionating minerals is large, the entry or exit of a single phase may not sufficiently affect the bulk chemistry of the melt to feature on a variation diagram.

Scattered trends

Variation diagrams sometimes show a cloud of data points rather than a neat linear trend. In the case of sedimentary rocks, this may be a function of the mixing processes leading to the formation of the sediment. In igneous rocks, however, where liquid–crystal equilibria are controlling compositions it is important to consider that some of the possible causes of scatter for this may throw further light on the processes. Similarly, in metamorphic rocks scattered trends may reflect the geochemical imprint of a metamorphic process on earlier igneous or sedimentary processes.

Some common reasons for scattered trends on varation diagrams for igneous rocks are:

(1) Not sampling liquid compositions. In highly porphyritic volcanic rocks much of the 'noise' in the data may be due to the accumulation of phenocrysts. In the case of many plutonic rocks it is very difficult to prove from field observations that the samples collected represent liquid compositions and in some cases it is highly improbable.

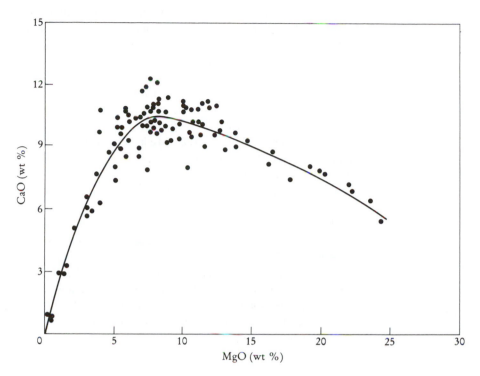

Figure 3.21   Bivariate plot of CaO vs MgO for basalts and related rocks from Hawaii. The change in slope at about MgO 7 % suggests a change in the fractionating phases at this point (after Peterson and Moore, 1987; courtesy of the US Geological Survey).

(2)   The samples are not from a single magma. This can be true even for lavas from a single volcano.

(3)   A changing fractionation assemblage during fractional crystallization.

(4)   Sampling procedures; parameters such as sample size relative to the grain size of the rock, sample heterogeneity and the number of samples collected are all relevant in different contexts. The problem of sampling bias is well illustrated by Neilsen (1988), who emphasizes the necessity for detailed sampling if the process to be resolved takes place on a fine scale.

(5)   Uncertainty in the analytical measurements. This may be due to the fact that analyses were made using different techniques or in different laboratories — both practices to be avoided if the data are to be plotted on the same diagram.

A computer-based approach to mixing calculations   There are a number of computer programs which can be used to interpret trends on variation diagrams and to solve mixing problems of the type described above. Fractional crystallization, for instance, may be expressed as:

rock A = rock B − (mineral X + mineral Y + mineral Z)

In this case a graphical solution will not yield a precise result whereas a computer-based, iterative mathematical procedure can estimate the proportions of the relative fractionating minerals. Similar formulations for mixing resulting from assimilation and partial melting may also be derived and these are equally amenable to computer solution.

Mixing programs of this type have been described by Stormer and Nicholls (1978) and Le Maitre (1981); these works include the computer source code in FORTRAN IV. Their aim is to minimize the difference between a measured rock composition and a composition calculated on the basis of a mixing hypothesis. The success of any model is estimated from the residuals of the calculation — the difference between the actual and calculated compositions. In Table 3.5 the results of a mixing calculation for the Columbia River basalts are presented (Wright, 1974). The hypothesis to be tested is whether the highly differentiated Umatilla 'basalt' is derived from the less evolved Lolo basalt. The differences between the calculated composition and the actual composition of the Lolo basalt, using the mixing program of Wright and Doherty (1970), suggest that the solution is acceptable. Examination of the residuals (the 'difference' in Table 3.5) confirms this, for they are small and the sum of the squares of the residuals is also very small.

Before accepting the results of such a calculation, however, they should be evaluated petrologically and the postulated fractionating phases (their relative proportions and their compositions) should be compared with the phenocrysts present in the lava suite. Where the rock has a complex history the mixing may be better formulated as a series of steps. Furthermore, it is important to stress that whilst a computer solution will produce a best-fit result, the result is not necessarily unique. Accordingly many workers seek to test mixing models initially proposed on the basis of major element chemistry with trace element data.

3.3.4   Modelling major element processes in igneous rocks

An alternative to the deductive use of variation diagrams in explaining petrological processes is an inverse approach in which the major element chemistry of an igneous suite is predicted from an initial starting composition. This type of major element modelling has been used chiefly to investigate fractional crystallization, although it also has other applications. The aim is to calculate, in a given silicate

Table 3.5   Differentiation of Columbia River basalt (Wright, 1974)

Lolo basalt = Umatilla basalt + olivine + augite + plagioclase + Fe–Ti oxides + apatite

	Proportion (%)	SiO_2	TiO_2	Al_2O_3	FeO	MnO	MgO	CaO	Na_2O	K_2O	P_2O_5
Umatilla basalt	44.76 %	54.94	2.64	13.87	12.58	0.22	2.70	6.27	3.24	2.60	1.02
Olivine	3.77 %	36.41	0.02	0.83	30.52	0.25	31.65	0.35	0.00	0.00	0.00
Augite	18.73 %	51.89	0.91	1.61	13.65	0.32	14.73	16.63	0.21	0.06	0.06
Plagioclase	25.16 %	54.57	0.00	28.95	0.00	0.00	0.00	11.16	5.14	0.21	0.00
Fe–Ti oxides	6.95 %	0.10	27.38	1.53	68.83	0.44	1.76	0.00	0.00	0.00	0.00
Apatite	0.67 %	0.00	0.00	0.00	0.00	0.00	0.00	56.00	0.00	0.00	44.00
Calculated composition		49.41	3.25	13.93	14.11	0.20	5.28	9.11	2.78	1.22	0.76
Lolo basalt		49.34	3.24	13.91	14.10	0.26	5.28	9.12	2.81	1.25	0.76
Difference (Lolo − calculated)		−0.07	−0.01	−0.02	−0.01	0.06	0.00	0.01	0.03	0.03	0.00
Sum of squares		0.011									

liquid, the nature of the first crystallizing phase, its composition and temperature of crystallization and the crystallization sequence of subsequent phases.

Three different approaches have been used. Firstly, the distribution of the major elements between mineral phases and a coexisting silicate melt may be calculated from experimental phase equilibrium data using regression techniques. Secondly, mineral–melt equilibria can be determined from mineral–melt distribution coefficients. A third, less empirical and more complex, approach is to use equilibrium thermodynamic models for magmatic systems. These require a thermodynamically valid mixing model for the liquid and an internally consistent set of solid–liquid thermochemical data.

The semi-empirical regression method was used by Nathan and Van Kirk (1978) to relate liquidus temperature to melt composition. From this relationship they were able to determine mineral compositions and the fractionating mineral assemblage at 1 atm pressure in both mafic and felsic liquids. Hostetler and Drake (1980) also used a regression technique but calculated solid–liquid distribution coefficients for eight major element oxides in the silicate melt. This permitted the calculation of phase equilibria for melts containing olivine, plagioclase and pyroxene from the melt composition but did not provide information on liquidus temperatures.

The alternative to the semi-empirical experimental approach to major element modelling is the thermodynamic modelling of silicate melts as described by Bottinga *et al.* (1981), Ghiorso (1985) and Ghiorso and Carmichael (1985). Ghiorso (1985) has developed an algorithm for chemical mass transfer in magmatic systems which predicts melt composition, mineral proportions and mineral compositions, and Ghiorso and Carmichael (1985) have demonstrated its usefulness when applied to fractional crystallization and assimilation in mafic melts at a range of pressures. Thermodynamic modelling, however, has an insatiable appetite for high-quality thermochemical data which do not exist for many minerals of interest in magmatic systems, thus severely limiting the applicability of this approach.

The chemical modelling of partial melting is even more difficult than the processes described above, for there is no general theory of melting which can cope with the multiphase, multicomponent nature of the Earth's crust and mantle at a range of pressures. Hanson and Langmuir (1978) and Langmuir and Hanson (1980) modelled basaltic systems from single-element and single-component mineral–melt distribution coefficients. These are combined with mass balance considerations and the stoichiometry of the mineral phases to calculate phase equilibria. Particularly interesting is their model for the partial melting of mantle pyrolite at 1 atmosphere pressure. Using the equations of Roeder and Emslie (1970) for the partitioning of magnesium and iron between olivine and melt, they calculated the abundances of MgO and FeO in the resultant melts and residual solids. These results are presented on an MgO–FeO cation % diagram which shows a field of melts and of residual solids, both contoured for percentage partial melting and temperature (Figure 3.22). Superimposed on the melt field are fractional crystallization trends for olivine in melts of differing composition. The diagram in Figure 3.22 cannot be used to define uniquely a partial melting trend from a given source, but it does delimit a field of permissible melts and for primary melt compositions can give information on the liquidus temperature and fraction of partial melting of the source. In addition, olivine fractional crystallization paths may be plotted for a given melt composition and the difference between equilibrium and fractional crystalliza-tion trends demonstrated (Langmuir and Hanson, 1980; Francis, 1985).

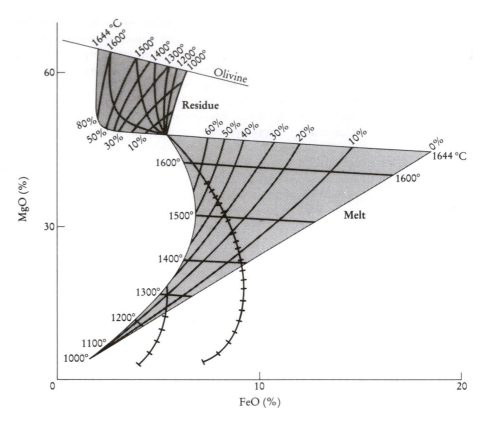

Figure 3.22   Fields of partial melt and residue, calibrated in liquidus temperature and percentage melting, for the partial melting of mantle pyrolite plotted on a FeO–MgO diagram expressed in cation mol % (calculated after the method of Hanson and Langmuir, 1978). The parent composition is where the two fields meet. The curved lines with small ticks show the trend of olivine fractional crystallization; the ticks are at 5 % intervals (from Langmuir and Hanson, 1980).

3.3.5   Discussion

Finally, it should be remembered that variation diagrams which utilize only the major elements have their limitations. Rarely can geological processes be uniquely identified from variation diagrams which use the major elements alone, and diagrams incorporating either trace elements or isotopes, as discussed in succeeding chapters, must also be employed.

3.4   Diagrams on which rock chemistry can be plotted together with experimentally determined phase boundaries

A number of igneous systems have been sufficiently well determined in the laboratory to allow the geochemist to interpret natural rock compositions in the light of experimentally determined phase boundaries. Phase diagrams of this type serve two useful functions. Firstly, they allow natural rock data to be projected onto them for interpretive purposes. Secondly, they allow the experimental results of different workers to be plotted in the same projection and thus compared. This becomes particularly pertinent when experiments in the same system have been carried using different experimental techniques and different starting materials. Unfortunately, in the maze of experimental data there is a wide variety of plotting and projection procedures for similar experimental systems. Thus similar diagrams showing the same data may appear different simply as a function of the projection procedure. The aim of this section therefore is to describe some of the main diagrams and projection schemes used in plotting experimental and natural rock data. A summary of the systems described is given in Box 3.3.

Box 3.3

Granite systems

Albite–orthoclase–silica
  ± Water
  + Anorthite

Nepheline syenite system

Nepheline–kalsilite–silica

Basaltic systems

CMAS
  Diagrams of O'Hara (1968)
Yoder–Tilley normative basalt tetrahedron
  Projections in Ol–Pl–Di–Q
  The normative Ne–Di–Ol–Hy–Q diagram
  The low-pressure tholeiitic phase diagram of Cox *et al.* (1979)

Calc–alkaline systems

The olivine–clinopyroxene–silica projection of Grove *et al.* (1982)
The olivine–diopside–quartz + orthoclase projection of Baker and Eggler (1983)

In many of the experiments on granitic and basaltic systems the aim of the experiment has been to determine the composition of minimum melts at varying pressures in systems of increasing compositional complexity. In this way the composition of first melts can be determined. Thus natural rock compositions presumed to represent primary liquid compositions may be compared with the composition of experimentally determined primary magmas. In many cases experimental petrologists have conducted their experiments on simplified rock compositions and so analyses of natural rocks have to be recalculated into a form that is appropriate to plot on the phase diagram. The success of this approach depends upon the extent to which the system under investigation matches the natural rock composition; to a first approximation this may be measured by the proportion of the rock composition which can be used in the projection. As is shown below, however, sometimes the presence of only a few per cent of an additional component may dramatically change the position of the phase boundaries.

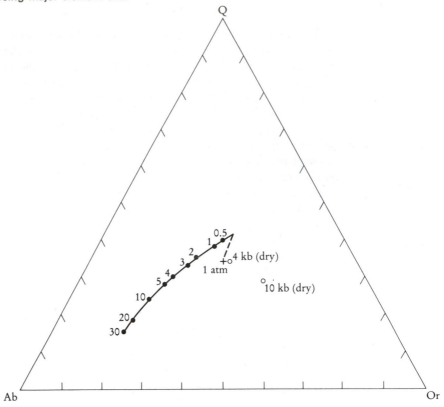

Figure 3.23   Minima and eutectics in the system albite–quartz–orthoclase–H_2O projected from H_2O onto the plane Ab–Q–Or for 1 atm (+) and 0.5 to 30 kb pressure (•). Also shown (○) are the minima for the anhydrous system at 4 and 10 kb pressure. The sources of data and the plotting positions are listed in Table 3.6.

3.4.1   The normative albite–orthoclase–quartz diagram — the 'granite system'

Tuttle and Bowen (1958) demonstrated a marked coincidence between the compositions of natural rhyolites and granites containing more than 80 wt % normative albite, orthoclase and quartz and the normative compositions of experimentally determined minima in the system albite–orthoclase–quartz–H_2O. Their observations provided a way for igneous petrologists to attempt to correlate experimental information with analytical data projected into the system Ab–Or–Q–H_2O. Tuttle and Bowen (1958) and subsequent workers have determined the compositions at which the phases quartz, orthoclase and albite coexist with a water-saturated melt at a variety of pressures (see Table 3.6) and plotted their results as a projection onto the anhydrous base of the tetrahedron Ab–Or–Q–H_2O (Figure 3.23). The plotting procedure requires three steps:

(1)   calculation of the CIPW norm from the chemical analysis;
(2)   the summation of the normative values of albite, orthoclase and quartz;
(3)   the recasting of these values as a percentage of their sum.

These values are plotted on a triangular diagram. The results of these experiments

Table 3.6   Plotting coordinates of minima and eutectics in the 'granite' system

Pressure (kb)	Temperature (°C)	Composition (wt %)			H₂O (wt %)		Reference
		Ab	Or	Q			
System: Albite–orthoclase–quartz–H₂O							
0.001	990	33	33	34	0.0	Minimum	Schairer and Bowen (1935)
0.490	770	30	30	40	3.0	Minimum	Tuttle and Bowen (1958)
0.981	720	33	29	38	4.4	Minimum	Tuttle and Bowen (1958)
1.961	685	39	26	35	6.5	Minimum	Tuttle and Bowen (1958)
2.942	665	42	25	33	8.3	Minimum	Tuttle and Bowen (1958)
3.923	655				9.7	Eutectic	Tuttle and Bowen (1958)
4.000	655	47	23	30	9.9	Eutectic	Steiner *et al.* (1975)
5.000	640	50	22	28	11.0	Eutectic	Luth *et al.* (1964)
10.000	620	56	21	23	17.0	Eutectic	Luth *et al.* (1964)
20.000	630	63	19	18	21.0	Eutectic	Huang and Wyllie (1975)
30.000	680	67	18	15	24.5	Eutectic	Huang and Wyllie (1975)
System: Albite–orthoclase–quartz (dry)							
4.000	1000	32	34	34	0.00	Minimum	Steiner *et al.* (1975)
10.000	1070	26	45	29	0.00	Minimum	Huang and Wyllie (1975)
System: Albite–orthoclase–quartz–anorthite–H₂O							
1.000	730	32	29	39 An₃ plane		nd	Piercing point	James and Hamilton (1969)
1.000	745	22	36	42 An₅ plane		nd	Piercing point	James and Hamilton (1969)
1.000	780	11	42	47 An₇.₅ plane		nd	Piercing point	James and Hamilton (1969)
System: Albite–orthoclase–quartz–H₂O–F							
1.000	690	45	26	29 1 % F		*ca* 4.0	Minimum	Manning (1981)
1.000	670	50	25	25 2 % F		*ca* 4.0	Minimum	Manning (1981)
1.000	630	58	27	15 4 % F		*ca* 4.0	Eutectic	Manning (1981)

nd, not determined

show that the quartz–alkali feldspar boundary moves away from the quartz apex with increasing pressure from 1 to 10 kb. A lesser expansion is observed between 10 and 30 kb. At approximately 3.5 kb, 660 °C the quartz–alkali feldspar field boundary intersects the crest of the alkali feldspar solvus and the liquid at this point coexists with quartz, orthoclase and albite (Merrill *et al.*, 1970). At 30 kb the assemblage is coesite – sanidine hydrate – jadeiite. Thus a direct comparison can be made between experimentally determined phase boundaries and natural rock compositions.

A number of authors have urged great caution with the above procedure when applied to granites (Luth, 1976; Steiner *et al.*, 1975), for there are several fundamental differences between experimental conditions and the natural plutonic environment. They emphasize that the bulk composition of a rock cannot be used alone to estimate uniquely its crystallization history *and* the pressure, temperature and water activity of the melt. The result obtained will be ambiguous and parameters such as pressure, temperature and water activity must be determined independently. The interpretive use of the Ab–Or–Q diagram must therefore be restricted to a generalized, qualitative description of processes with broad rather than specific applicability. The reasons for this are outlined below. Firstly, in most experimental investigations compositions are projected from H_2O onto the plane Ab–Or–Q and it is assumed that the melt is water-saturated. Secondly, experiments in the system Ab–Or–Q are not directly applicable to natural rocks because they do not take account of additional components such as anorthite, ferromagnesian minerals or phases representing an excess of alumina over alkalis. These have a considerable effect on the position of the determined eutectics. Thirdly, it is important to know whether the bulk compositions sampled represent igneous liquid compositions or whether they are in part crystal cumulates.

Since few natural rocks approximate to the system Ab–Or–Q–H_2O, in the sections that follow we discuss experimental results which attempt to approximate more closely to natural compositions. We look at the effects of reducing the water content of the melt and adding anorthite to the melt, thus extending the applicability of this system to granodiorites and tonalites.

Water-undersaturated equilibria

There are few data for water-undersaturated equilibria in the Ab–Or–Q–H_2O system, particularly for the ternary minima. The best data are those of Steiner *et al.* (1975), who investigated the system at 4 kb and presented results for the water-saturated and the dry systems (see Table 3.6). Luth (1969) has estimated the position of the 10 kb dry minimum and Huang and Wyllie (1975) have estimated the position of the 30 kb dry quartz–alkali feldspar field boundary. Figure 3.23 shows the positions of the minima in the dry system at 4 kb and 10 kb, which may be compared with positions of the eutectics in the hydrous system.

The presence of anorthite

The addition of anorthite to the 'granite' system shifts compositions into the granodiorite and tonalite fields. This was investigated by James and Hamilton (1969) at 1 kb who found that the position of the piercing point minimum shifts towards the Or–Q side of the projection with increasing anorthite, indicating an increase in the primary phase volume of plagioclase (Figure 3.24). Winkler (1976) also emphasized the importance of plagioclase, especially when considering partial melting in felsic rocks, although Johannes (1980, 1983, 1984) has been unable to duplicate Winkler's results and considers that they represent metastable data. There are, therefore, few reliable data from which sensible interpretations may be made of granodiorites and tonalites; any such data are for water-saturated equilibria.

3.4.2   The silica-undersaturated portion of the normative nepheline–kalsilite–silica diagram — the 'nepheline syenite' system

Experiments by Schairer (1950), Hamilton and MacKenzie (1965), Taylor and MacKenzie (1975) and Morse (1969) have determined the positions of the nepheline syenite minima at pressures of 1 atm, 1 kb, 2 kb and 5 kb, respectively, in the water-saturated system. These data are presented in Figure 3.25 as CIPW normative wt %, in a projection from H_2O onto the anhydrous base of the system, the plane nepheline–kalsilite–silica. Unlike the granite system, the position of the

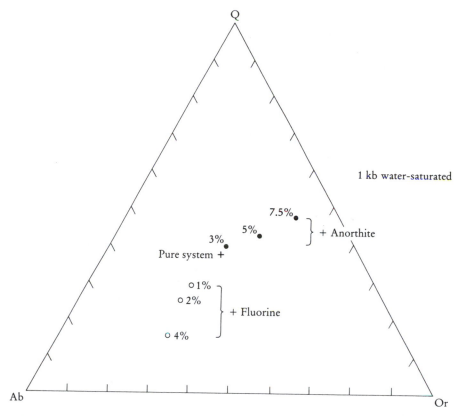

Figure 3.24 Projections onto the plane albite–quartz–orthoclase from H_2O at 1 kb. The pure system is represented by a cross (+). ● Piercing points for the planes 3 %, 5 % and 7.5 % anorthite; o, minima and eutectics for the fluorine-bearing system. The sources of data and the plotting positions are listed in Table 3.6.

minimum does not change greatly with increasing pressure and shifts only slightly towards the nepheline corner (Table 3.7, Figure 3.25) although the temperature of the minimum falls almost 400 °C over the pressure interval 0 to 5 kb. In addition there is a dramatic decrease in the size of the leucite field.

Hamilton and MacKenzie (1965) show that rocks with 80 % or more normative nepheline, albite and orthoclase plot close to the 1 kb nepheline–syenite minimum (albite and orthoclase are recalculated as nepheline + 2 silica, and kalsilite + 2 silica, respectively). However, at best, these experimental data lend themselves to the qualitative interpretation of natural alkaline rocks.

3.4.3 Basaltic experimental systems

There are two main projection schemes that are used for the major element compositions of basaltic rocks. One is the CMAS system, in which an analysis of a basalt is approximated by the four oxides CaO–MgO–Al_2O_3–SiO_2. The other is a normative scheme based upon the main minerals observed in basalts — nepheline–diopside–olivine–anorthite and quartz. This projection is based upon the classification scheme for basalts proposed by Yoder and Tilley (1962) in which the

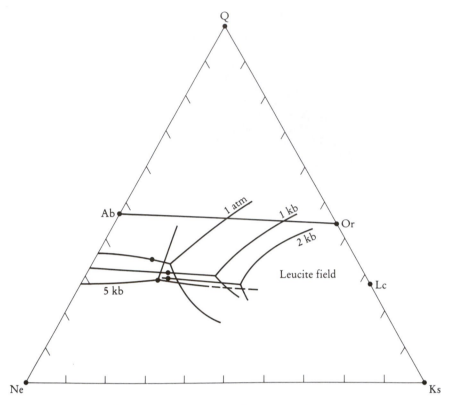

Figure 3.25   Projections onto the plane nepheline–quartz–kasilite from H_2O in the nepheline syenite system at 1 atm 1, 2 and 5 kb pressure, showing the contraction of the leucite field with increasing pressure and the position of the minima (•). The data sources and plotting positions are given in Table 3.7.

minerals nepheline–diopside–olivine and quartz are plotted at the apices of a tetrahedron (Figure 3.26a).

CMAS diagrams   The components of the CMAS system (CaO–MgO–Al_2O_3–SiO_2) comprise about 70–85 wt % of most basalts and more than 90 wt % of most mantle peridotites. For this reason the CMAS system is used by experimental petrologists as a simplified analogue of more complex basalt and mantle systems. The CMAS projection

Table 3.7   Minima in the nepheline syenite system

Pressure (kb)	Temperature (°C)	Composition (wt %)			Reference
		Ne	Ks	Q	
0.001	1020	51	15	34	Schairer (1950)
0.981	750	50	19	31	Hamilton and MacKenzie (1965)
2.000	710	51	20	29	Taylor and MacKenzie (1975)
5.000	635	53	19	28	Morse (1970)

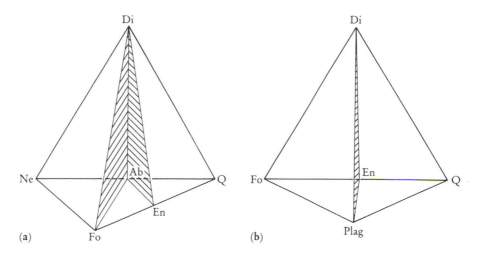

(a)   (b)

Figure 3.26   (a) The Yoder–Tilley basalt tetrahedron. The plane normative Di–Ab–En is the plane of silica saturation and the plane normative Di–Ab–Fo is the critical plane of silica undersaturation. (b) The silica-saturated part of the Yoder–Tilley basalt tetrahedron shown in (a) expanded into a tetrahedron showing the plane of silica saturation.

provides an excellent framework in which the possible melting behaviour of upper mantle materials may be discussed, and is a powerful tool in constructing petrological models. It may be used to compare the chemistry of particular rock suites with experimentally determined phase boundaries at low and high pressure. Partial melting trends may be identified from a linear array of rock compositions projecting through the plotted source composition, and fractional crystallization trends may be identified as linear arrays projecting through the composition of the fractionating phase(s).

(a) Projecting rock compositions into CMAS   The CMAS system was first used for mantle and basaltic compositions by O'Hara (1968), who proposed a polybaric phase diagram for dry natural basalts and ultramafic rocks up to pressures of 30 kb. He devised a scheme whereby natural rock compositions could be presented in such a way as to be directly comparable with experimental results in the system $CaO–MgO–Al_2O_3–SiO_2$ (CMAS). Weight % oxides are converted to molecular proportions and the plotting parameters are calculated as follows:

$$C = (\text{mol. prop. } CaO - 3.33\ P_2O_5 + 2Na_2O + 2K_2O) \times 56.08$$
$$M = (\text{mol. prop. } FeO + MnO + NiO + MgO - TiO_2) \times 40.31$$
$$A = (\text{mol. prop. } Al_2O_3 + Cr_2O_3 + Fe_2O_3 + Na_2O + K_2O + TiO_2) \times 101.96$$
$$S = (\text{mol. prop. } SiO_2 - 2Na_2O - 2K_2O) \times 60.09$$

A model calculation is given in Table 3.8. A simplification of the calculation scheme which ignores the minor elements Ti, Cr and Ni is given by Adam (1988).

Rock compositions are usually presented in one of three projections, chosen to include the important mineral phases and to minimize any distortion from the projection. The most used projections are:

(a) from olivine into the plane CS–MS–A;
(b) from enstatite into the plane $M_2S–A_2S_3–C_2S_3$;
(c) from diopside into either the plane $C_3A–M–S$ or CA–M–S.

Table 3.8   Calculation scheme for the CMAS projection

	Wt % oxide of rock	Molecular weight	Molecular proportions
SiO_2	46.95	60.09	0.7813
TiO_2	2.02	79.90	0.0253
Al_2O_3	13.10	101.96	0.1285
Fe_2O_3	1.02	159.69	0.0064
FeO	10.07	71.85	0.1402
MnO	0.15	70.94	0.0021
MgO	14.55	40.30	0.3610
CaO	10.16	56.08	0.1812
Na_2O	1.73	61.98	0.0279
K_2O	0.08	94.20	0.0008
P_2O_5	0.21	141.95	0.0015
Total	100.04		

CMAS plotting parameters:

C = 13.110
M = 19.269
A = 19.261
S = 43.493

Projection parameters for the olivine projection from olivine into CS–MS–A:
Using the equation of Cox et al. (1979)

$$C_cM_mA_aS_s + p\, M_{57.3}S_{42.7} = x C_{48.3}S_{51.7} + y M_{40.1}S_{59.9} + z A_{100}$$

where p is the amount of olivine required to bring the rock into the required plane, x,y,z when recast as percentages are the plotting parameters for CS, MS and A, and c, m, a and s are the calculated values for C, M, A and S for the rock.

Balancing C,	$48.3x = 13.11$
Balancing M,	$40.1y = 57.3p + 19.269$
Balancing A,	$100z = 19.261$
Balancing S,	$51.7x + 59.9y = 43.493 + 42.7p$

		(%)
x	0.271	27.952
y	0.507	52.212
z	0.193	19.835
Sum	0.971	

Equation for projection from orthopyroxene into M_2S–C_2S_3–A_2S_3:

$$C_cM_mA_aS_s + p\, M_{40.1}S_{59.9} = x\, M_{57.3}S_{42.7} + y\, C_{38.4}S_{61.6} + z\, A_{53.1}S_{46.9}$$

Equation for projection for diopside into C_3A–M–S:

$$C_cM_mA_aS_s + p\, C_{25.9}M_{18.6}S_{55.5} = x\, C_{62.3}A_{37.7} + y\, M_{100} + z\, S_{100}$$

(from Cox *et al.*, 1979).

The olivine projection is illustrated in Figure 3.27. This plane contains all the pyroxene solid solutions and the garnet solid solutions. The olivine–plagioclase piercing point is the point at which the olivine plagioclase join cuts the CS–MS–A

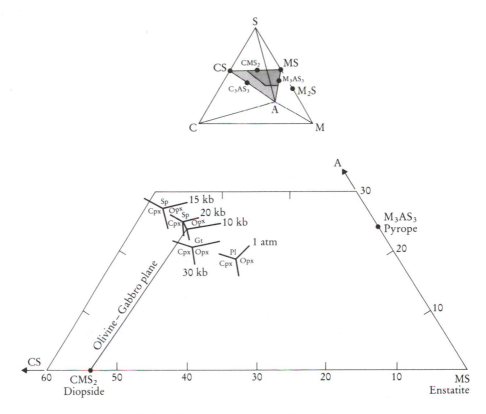

Figure 3.27   The CMAS projection for basaltic and mantle compositions. Projection from olivine (M_2S) onto the plane CS–MS–A in the CMAS system showing invariant peritectic points for 1 atm, and 10, 15, 20 and 30 kb. The olivine–gabbro plane divides the diagram into nepheline normative compositions (enstatite-poor) and tholeiitic compositions (enstatite-rich). The stability fields of the different phases are indicated (Pl, plagioclase; Opx, orthopyroxene; Cpx, clinopyroxene; Sp, spinel; Gt, garnet) and in addition all fields include olivine. The inset shows the relative positions of olivine and the plane of projection in the olivine projection of the CMAS system. M_2S is olivine, MS enstatite; all feldspar is projected as equivalent anorthite (CAS_2), all hercynite, ulvospinel, magnetite and chromite as spinel (MA), all Fe–Ni–Mg olivines as forsterite (M_2S), all garnet plots along the grossular (C_3AS_3)–pyrope (M_3AS_3) join, and all clinopyroxenes along the diopside (CMS_2)–Ca–Tschermak's molecule (CAS) join.

plane and the line which joins this piercing point to diopside (the olivine–gabbro plane) is the plane of silica saturation. This olivine–gabbro plane divides the diagram into Ne-normative compositions on the En-poor side and tholeiitic compositions on the En-rich side. An example of the calculation procedure for projecting a basaltic composition from olivine onto the CS–MS–A plane is given in Table 3.8.

(b) Interpreting CMAS diagrams   For a projection to be useful in interpreting crystal–liquid equilibria, it must be made from a phase which is present in the melt, for otherwise the observed trends are meaningless. Secondly, the projection should not be made from a phase at an oblique angle to the projection plane, for then trends which are simply a function of the oblique projection can be misinterpreted and given geological significance where there is none.

One of the problems with the CMAS projection is that, whilst it uses all the chemical constituents of a rock analysis and so is applicable to natural rocks, the effects of individual components cannot be easily identified. The converse problem derives from the small number of components used in the experimental system, for the effects of small amounts of additional components on the position of the phase boundaries is largely unknown. Na is likely to have the most important effect (Thompson, 1987), but Fe (Herzberg, 1992), H_2O (Adam, 1988) and possibly Ti are also likely to influence the position of the phase boundaries.

Diagrams based upon the Yoder– Tilley (1962) CIPW normative tetrahedron

The normative nepheline–diopside–olivine–quartz tetrahedron (Figure 3.26a) originally proposed by Yoder and Tilley (1962) for the classification of basalts has been extensively used for comparing the results of experimental studies on basalts with rock analyses. Various projection procedures are in use.

(a) Projections in the tholeiitic basalt tetrahedron Ol–Pl–Di–Q Three slightly different algorithms have been devised for plotting experimental data for tholeiites in the silica-saturated part of the Yoder–Tilley tetrahedron (Figure 3.26b). The results of these calculation schemes are presented in Table 3.9.

Presnall *et al.* (1979) recalculated rock compositions as CIPW norms, setting the $Fe^{2+}/(Fe^{2+} + Fe^{3+})$ ratio to 0.86. The mineral proportions are expressed as mol per cent and plotting parameters are defined as follows:

Pl = normative (An + Ab)
Ol = normative (Fo + Fa)
Q = normative SiO_2
Di = normative (Di + Hed)

Hypersthene is allocated to olivine and quartz. The main projections in the tetrahedron Di–Ol–Pl–Q (Figure 3.26b) are from diopside onto the Pl–Ol–Q face, and from plagioclase onto the Di–Ol–Q face. In the case of the plagioclase projection, the proportions of Di, Ol and Q are normalized to their sum and plotted on a molecular basis. The diopside projection is calculated in a similar manner. A model calculation is given in Table 3.9.

Walker *et al.* (1979) developed a different algorithm for plotting data in the same projections as Presnall *et al.* (1979), although Presnall and Hoover (1984) noted that the end result was similar. Weight % oxides are divided by their molecular weight to obtain molecular proportions and the plotting parameters are calculated from the molecular proportions as follows:

PLAG = Al_2O_3 + Na_2O + K_2O
DI   = CaO – Al_2O_3 + Na_2O + K_2O
OL   = (FeO + MgO + MnO + $2Fe_2O_3$ + Al_2O_3 – CaO — Na_2O — K_2O)/2
SIL   = SiO_2 – (Al_2O_3 + FeO + MgO + MnO + 3CaO + $11Na_2O$ + $11K_2O$ + $2Fe_2O_3$)/2

They use the notation DI–OL–SIL and OL–SIL–PLAG for projections from plagioclase onto the Di–Ol–Q face and from diopside onto the Pl–Ol–Q face. In the case of the plagioclase projection (Table 3.9), the proportions of DI, OL and SIL are then normalized to their sum and plotted on a molecular basis.

Elthon (1983) proposed a third algorithm arguing that chemical trends in

Table 3.9   Projection procedures for basaltic compositions within the Yoder–Tilley tetrahedron

	Wt % oxide of rock	Molecular weight	Molecular proportions		Wt % CIPW norm	Molecular weight	Norm/ mol. wt	Norm (mol %)
SiO_2	50.68	60.09	0.8434	Q	0.00	60	0.000	0.00
TiO_2	0.73	79.90	0.0091	Or	4.61	556	0.008	2.27
Al_2O_3	14.17	101.96	0.1390	Ab	21.37	524	0.041	11.14
Fe_2O_3	0.00	159.69	0.0000	Ne	0.00	284	0.000	0.00
FeO	12.29	71.85	0.1711	An	24.96	278	0.090	24.53
MnO	0.22	70.94	0.0031	Di	8.34	216	0.039	10.55
MgO	8.85	40.30	0.2196	Hed	6.28	248	0.025	6.92
CaO	8.77	56.08	0.1564	En	12.53	200	0.063	17.12
Na_2O	2.53	61.98	0.0408	Fs	9.44	264	0.036	9.77
K_2O	0.78	94.20	0.0083	Fo	3.79	140	0.027	7.40
P_2O_5	0.06	141.95	0.0004	Fa	3.15	204	0.015	4.22
				Mt	2.96	232	0.013	3.49
Total	99.08			Il	1.38	152	0.009	2.48
				Ap	0.12	310	0.000	0.11
				Sum	98.93		0.366	100.00

Plotting procedure of Presnall et al. (1979)
Allocate hypersthene to olivine and quartz (1 mole Hy = 1 mole Ol + 1 mole Q):

	(mol %)
Pl = normative (An + Ab)	35.67
Ol = normative (Fo + Fa) + Hy	38.50
Q = normative SiO_2 + Hy	26.89
Di = normative (Di + Hed)	17.47

Plagioclase projection:		(%)
Di	17.47	21.08
Ol	38.50	46.47
Q	26.89	32.45
Sum	82.86	

Plotting procedure of Walker et al. (1979)

PL = Al_2O_3 + Na_2O + K_2O
DI = CaO − Al_2O_3 + Na_2O + K_2O
OL = (FeO + MgO + MnO + $2Fe_2O_3$ + Al_2O_3 − CaO − Na_2O − K_2O)/2
SIL = SiO_2 − (Al_2O_3 + FeO + MgO + MnO + 3CaO + $11Na_2O$ + $11K_2O$ + $2Fe_2O_3$)/2

PL	0.188
DI	0.067
OL	0.164
SIL	0.072

(*Continued*)

Table 3.9 (Continued)

Plagioclase projection:		(%)
DI	0.067	22.0
OL	0.164	54.1
SIL	0.072	23.9
Sum	0.303	

Plotting procedure of Elthon (1983)

$PL = Al_2O_3 + Fe_2O_3$
$DI = CaO + Na_2O + K_2O - Fe_2O_3 - Al_2O_3$
$OL = (FeO + MgO + MnO - TiO_2 + Al_2O_3 + Fe_2O_3 - CaO - Na_2O - K_2O)/2$
$SIL = SiO_2 - (FeO + MgO + MnO - TiO_2 + Al_2O_3 + Fe_2O_3 + 3CaO + 3Na_2O + 3K_2O)/2$

PL	0.139
DI	0.067
OL	0.318
SIL	0.273

Plagioclase projection:		(%)
DI	0.067	10.11
OL	0.318	48.35
SIL	0.273	41.55
Sum	0.658	

oceanic basalt suites projected in the olivine–clinopyroxene–silica plane are greatly improved if the plagioclase feldspars are separated along the anorthite–albite join and do not plot at a single point. In this projection, therefore, plagioclase compositions are spread along the silica–anorthite edge of the Di–Ol–An–Q tetrahedron. The normative mineralogy is projected onto the planes CPX–OLIVINE–SILICA and OLIVINE–SILICA–PLAG and the plotting parameters are calculated from the molecular proportions, as in the case of Walker *et al.* (1979) as follows:

$$PLAG = Al_2O_3 + Fe_2O_3$$
$$DI = CaO + Na_2O + K_2O - Al_2O_3 - Fe_2O_3$$
$$OL = [(FeO + MgO + MnO - TiO_2) - (CaO + Na_2O + K_2O) + (Fe_2O_3 + Al_2O_3)]/2$$
$$SIL = SiO_2 - [(FeO + MgO + MnO - TiO_2) + (Al_2O_3 + Fe_2O_3) + 3(CaO + Na_2O + K_2O)]/2$$

Fe^{3+}/Fe^{2+} is assumed to be 0.10. This method of projection does produce different plotting positions from the algorithms of Presnall *et al.* (1979) and Walker *et al.* (1979) and a model calculation is given in Table 3.9.

(b) The normative Ne–Di–Ol–Hy–Q diagram   This diagram represents the left-hand face of the Yoder–Tilley (1962) tetrahedron together with the front face

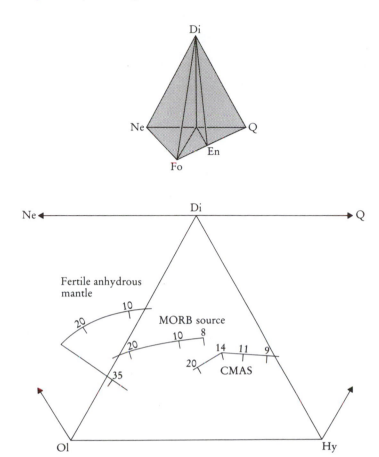

Figure 3.28   Normative nepheline–olivine–diopside–hypersthene–quartz projection after Thompson (1984), showing the compositions of initial melts in the CMAS system, MORB source and fertile anhydrous mantle at a range of pressures (in kb). The inset shows the relationship of the diagram to the surface of the Yoder–Tilley basalt tetrahedron.

expanded into two equilateral triangles. It was presented in Section 3.2.2 above as a means of classifying basalts using their CIPW normative compositions (Thompson, 1984). However, it was also used by Thompson to display experimental data for both saturated and undersaturated basalts (Thompson, 1984, 1987) and to show the changing composition of initial melts of different mantle compositions, produced at different pressures (Figure 3.28). CIPW normative compositions calculated on a wt % basis are plotted on one or other of the three triangles. Fe_2O_3 is calculated as 15 % of the total iron content.

(c) The low-pressure tholeiitic basalt phase diagram (Cox et al., 1979).   Cox et al. (1979) proposed a low-pressure phase diagram based upon the silica-poor part of normative basalt system Ol–Cpx–Pl–Q. The diagram is based on the CIPW normative composition of a tholeiitic (i.e. hypersthene normative) basalt which is projected from SiO_2 onto the Fo–Ab–Di plane, i.e. the plane of silica saturation, of the Yoder–Tilley (1962) tetrahedron (Figure 3.29). The phase diagram is useful for estimating the phases present in the initial stages of low-pressure crystallization of a given tholeiite. In constructing the diagram, the Fo–Ab–Di–Q tetrahedron

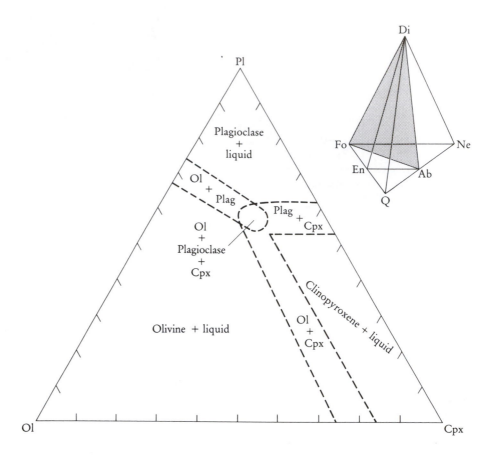

Figure 3.29   The low-pressure tholeiitic basalt phase diagram after Cox *et al.* (1979). The inset shows that the projection is from normative quartz onto the critical plane of silica undersaturation modified to olivine–plagioclase–clinopyroxene in the Yoder–Tilley basalt tetrahedron.

becomes the Ol–Pl–Cpx–Q tetrahedron. The hypersthene content of the norm is recalculated as an equivalent amount of olivine and quartz, and the plotting parameters are then calculated from the norm as follows:

plagioclase = normative anorthite + albite
clinopyroxene = normative diopside
olivine = normative olivine + that recalculated from hypersthene
quartz = normative quartz + that recalculated from hypersthene

The four parameters are calculated as percentages of the total. At this point inappropriate analyses should be screened out. This is when

normative plagioclase is < An_{50}
ratio $(FeO + Fe_2O_3)/(MgO + FeO + Fe_2O_3) > 0.7\,\%$
$K_2O > 1.0\,\%$
% quartz (as calculated above) > 10 %

If the rock is appropriate the plotting parameters plagioclase, olivine and clinopyroxene are recalculated to 100 % and plotted onto the triangular phase diagram.

(d) Problems with CIPW normative projections   A word of caution is necessary when using CIPW normative projections; Presnall *et al.* (1979) noted that small uncertainties in the chemical analysis may translate into large shifts in the projected normative composition. This is particularly acute for uncertainties in Na_2O and SiO_2, where the direction of shift is parallel to an identified fractional crystallization trend. This is also true for the algorithm of Elthon (1983). Thus Presnall *et al.* (1979) make the point that some identified fractional crystallization trends observed in ocean-floor tholeiitic glasses could be an artefact resulting from analytical uncertainty. Presnall and Hoover (1984) elaborate on this and show that the variation in normative quartz in CIPW normative projections of basaltic glass compositions may be due to analytical uncertainty, although this feature is not apparent in the projection scheme of Elthon (1983).

3.4.4   Experimental systems for calc–alkaline rocks

Phase relations and rock compositions of andesitic and related magmas of the calc–alkaline suite can also be represented by the diopside–olivine–albite–quartz tetrahedron — the silica-saturated part of the Yoder–Tilley (1962) basalt tetrahedron.

The olivine–clinopyroxene–silica projection of Grove et al. (1982)   Grove *et al.* (1982, 1983) developed a projection scheme to present low-pressure experimental data for andesites in which the phases plagioclase, olivine, augite, pigeonite and orthopyroxene are present. The diagram is useful for calc–alkali series rocks but may not be applicable for low-alkali–silica-rich suites. The projection scheme is similar to that of Walker *et al.* (1979) for tholeiitic basalts. It is based upon a modification of the CMAS projection in which wt % oxide values are converted to molecular proportions, alkalis and alumina are converted to $NaO_{[0.5]}$, $KO_{[0.5]}$ and $AlO_{[1.5]}$, and the mineral components are calculated as molecular proportions as follows:

$$\text{sum} = SiO_2 - CaO - 2(KO_{[0.5]} + NaO_{[0.5]} + Cr_2O_3 + TiO_2)$$
$$Q = (SiO_2 - 0.5(FeO + MgO) - 1.5CaO - 0.25\,AlO_{[1.5]} - 2.75$$
$$(NaO_{[0.5]} + KO_{[0.5]} + Cr_2O_3 + 0.5TiO_2)/\text{sum}$$
$$Pl = 0.5(AlO_{[1.5]} + NaO_{[0.5]} + KO_{[0.5]})/\text{sum}$$
$$Ol = 0.5(FeO + MgO) + 0.5(AlO_{[1.5]} - KO_{[0.5]} - NaO_{[0.5]}$$
$$- CaO - 2TiO_2 - Cr_2O_3)/\text{sum}$$
$$Cpx = CaO - 0.5AlO_{[1.5]} + 0.5(KO_{[0.5]} + NaO_{[0.5]})/\text{sum}$$
$$Or = KO_{[0.5]}/\text{sum}$$
$$Sp = (Cr_2O_3 + TiO_2)/\text{sum}$$

The values are normalized and projected from plagioclase onto the plane quartz–olivine–clinopyroxene.

The projections of Baker and Eggler (1983, 1987)   Baker and Eggler (1983) use a modification of the projection procedure of Walker *et al.* (1979) in which the projection is made either from magnetite + plagioclase onto olivine–diopside–(quartz + orthoclase) or from magnetite + olivine onto diopside–plagioclase–(quartz + orthoclase) (Figure 3.30). Quartz and orthoclase are combined in these diagrams to improve upon the projection schemes of Walker *et al.* (1979) and Grove *et al.* (1983), which both wrongly imply that the residual melts are saturated with orthoclase. The procedure is as follows:

The ferric iron content of the rock is calculated from Sack *et al.* (1980) for $T =$ 1150 °C and fO_2 defined by the Ni–NiO buffer. Under the specific conditions given here, using the temperature — oxygen fugacity relationship given by Eugster and Wones (1962) for the Ni–NiO buffer, the expression reduces to:

$$\ln (X_{Fe_2O_3}^{liq}/X_{FeO}^{liq}) = 2.9286 - 2.15036X_{SiO_2} - 8.35163X_{Al_2O_3} - 4.4951X_{FeO_{(total)}} - 5.4364X_{MgO} + 0.0731X_{CaO} + 3.5415X_{Na_2O} + 4.1869X_{K_2O}$$

(where X_{SiO_2} etc = the mole fraction of SiO_2, and so on). An alternative version of this equation is given by Kilinc *et al.* (1983). Oxide wt % values are converted to molar proportions:

magnetite = Fe_2O_3
plagioclase = $Al_2O_3 + Na_2O + K_2O$
diopside = $CaO + Al_2O_3 + Na_2O + K_2O$
olivine = $(FeO + Fe_2O_3 + MgO + MnO + Al_2O_3 - CaO - Na_2O - K_2O)/2$
quartz + orthoclase = $SiO_2 - 2K_2O - (Al_2O_3 + FeO - Fe_2O_3 + MnO + MgO + 3CaO + 11Na_2O + 3K_2O)/2$

Baker and Eggler (1983) note that this projection is very sensitive to small differences in Na_2O concentrations.

3.4.5   Discussion

The results of experimental petrology play a major role in our understanding of the origin of igneous rocks. Nevertheless, on their own they can rarely give definitive

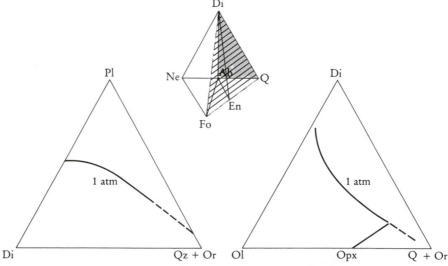

Figure 3.30   Projections from normative magnetite + olivine onto diopside–plagioclase–(quartz + orthoclase) (stippled surface on inset diagram) and from normative magnetite and plagioclase onto olivine–diopside–(quartz + orthoclase) (hatched surface on inset diagram) after Baker and Eggler (1983). The 1 atm liquidus boundary curve is also shown.

answers. Rather they point the way to less probable and more probable options. Thus the diagrams presented in this section have their place in elucidating the origin of igneous rocks but their interpretation is subject to a large number of uncertainties. This means that the results of experimental studies should always be taken together with the constraints of other geochemical investigations.

Chapter 4

Using trace element data

4.1 Introduction

A trace element may be defined as an element which is present in a rock in concentrations of less than 0.1 wt %, that is less than 1000 parts per million (ppm). Sometimes trace elements will form mineral species in their own right but most commonly they substitute for major elements in the rock-forming minerals.

Trace element studies have become a vital part of modern petrology and are more capable of discriminating between petrological processes than are the major elements. Particularly important is the fact that there are mathematical models to describe trace element distributions which allow the quantitative testing of petrological hypotheses. These are most applicable to processes controlled by crystal–melt or crystal–fluid equilibria.

In this chapter we first develop some of the theory behind the distribution of trace elements and explain the physical laws used in trace element modelling. Then various methods of displaying trace element data are examined as a prelude to showing how trace elements might be used in identifying geological processes and in testing hypotheses.

4.1.1 Classification of trace elements according to their geochemical behaviour

Trace elements are often studied in groups, and deviations from group behaviour or systematic changes in behaviour within the group are used as an indicator of petrological processes. The association of like trace elements also helps to simplify what can otherwise be a very unwieldy data-set. Trace elements are normally classified either on the basis of their position in the periodic table or according to their behaviour in magmatic systems.

Trace element groupings in the periodic table
Several groups of elements in the periodic table are of particular geochemical interest (Figure 4.1). The most obvious in this respect are the elements with atomic numbers 57 to 71, the lanthanides or **rare earth elements (REE)** as they are usually called in geochemistry. Other groups are the **platinum group elements (PGE)** (atomic numbers 44 to 46 and 76 to 79) also known as the **noble metals** if they include Au, and the **transition metals** (atomic numbers 21 to 30). In geochemistry, this latter term is usually restricted to the first transition series and includes two major elements, Fe and Mn.

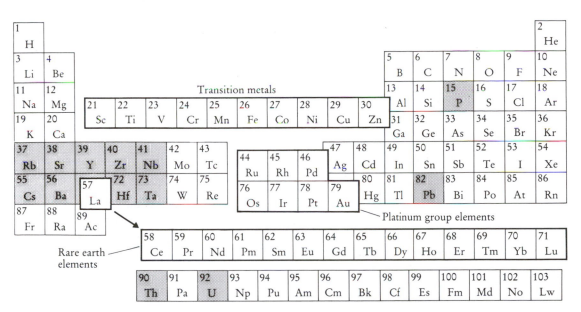

Figure 4.1   The periodic table of the elements, showing three main groups of trace elements, which are often treated together in geochemistry because of their relative positions in the table. These are the elements of the first transition series, the platinum group elements and the rare earth elements. The elements which are shaded are also important trace elements in geochemistry.

The elements in each of these respective groups have similar chemical properties and for this reason are expected to show similar geochemical behaviour. This is not always the case, however, because geological processes can take advantage of subtle chemical differences and fractionate elements of a group one from the other. Thus one of the tasks of trace element geochemistry is to discover which geological processes may have this effect and to quantify the extent of a particular process.

Trace element behaviour in magmatic systems   When the Earth's mantle is melted, trace elements display a preference either for the melt phase or the solid (mineral) phase. Trace elements whose preference is the mineral phase are described as **compatible**, whereas elements whose preference is the melt are described as **incompatible** — i.e. they are incompatible in the mineral structure and will leave at the first available opportunity. Incompatible elements have also been called **hygromagmatophile**, a term first introduced by Treuil and Varet (1973).

In detail there are degrees of compatibilty and incompatibility and trace elements will vary in their behaviour in melts of a different composition. For example, P is incompatible in a mantle mineralogy and during partial melting will be quickly concentrated in the melt. In granites, however, even though P is present as a trace element, it is compatible because it is accommodated in the structure of the minor phase apatite.

It is sometimes helpful to subdivide the incompatible elements on the basis of their charge/size ratio. This property is often described as field strength and may be thought of as the electrostatic charge per unit surface area of the cation. It is also described as the ionic potential of an element and is quantified as the ratio of the valence to the ionic radius. Figure 4.2 shows a plot of ionic radius vs charge for most of the trace elements studied in geochemistry. Small highly charged cations are known as **high field strength (HFS)** cations (ionic potential > 2.0) and large cations of small charge are known as **low field strength** cations (ionic potential < 2.0). Low field strength cations are also known as **large ion lithophile elements (LILE)**. Elements with small ionic radius and a relatively low charge tend to be compatible. These include a number of the major elements and the transition metals. Figure 4.2 shows the main groupings of trace elements and highlights the similarity in ionic size and charge between some element groups. Elements with the same ionic charge and size are expected to show very similar geochemical behaviour.

High field strength cations include the lanthanides Sc and Y, and Th, U, Pb, Zr, Hf, Ti, Nb and Ta. The element pairs Hf and Zr, and Nb and Ta, are very similar in size and charge and show very similar geochemical behaviour. Low field strength, large ion lithophile cations include Cs, Rb, K and Ba. To these may be added Sr, divalent Eu and divalent Pb — three elements with almost identical ionic radii and charge.

4.2   Controls on trace element distribution

Most modern quantitative trace element geochemistry assumes that trace elements are present in a mineral in solid solution through substitution and that their concentrations can be described in terms of equilibrium thermodynamics. Trace elements may mix in either an ideal or a non-ideal way in their host mineral. Their very low concentrations, however, lead to relatively simple relationships between composition and activity. When mixing is ideal the relationship between activity and composition is given by **Raoult's Law**, i.e.

$$a_i = X_i \tag{4.1}$$

where a_i is the activity of the trace element in the host mineral and X_i is its composition.

If the trace element interacts with the major components of the host mineral, the activity will depart from the ideal mixing relationship and at low concentrations the activity composition relations obey **Henry's Law**. This states that at equilibrium the activity of a trace element is directly proportional to its composition:

$$a_i^j = k_i^j X_i^j \tag{4.2}$$

where k_i^j is the Henry's Law constant — a proportionality constant (or activity coefficient) for trace element i in mineral j. Henry's Law seems to apply to a wide range of trace element concentrations (Drake and Holloway, 1981) although at very low concentrations (< 10 ppm) there are deviations from Henry's Law behaviour (Harrison and Wood, 1980). Henry's Law also ceases to apply at very high concentrations, although the point at which this takes place cannot be easily predicted and must be determined for each individual system. In the case where trace elements form the essential structural constituent of a minor phase, such as Zr in zircon, Henry's Law behaviour does not strictly apply.

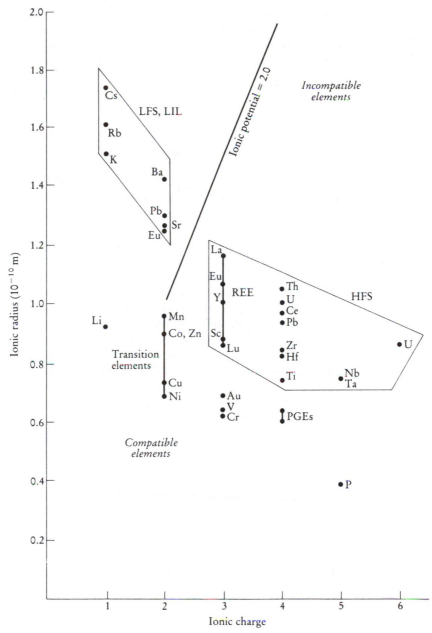

Figure 4.2   Plot of ionic radiius vs ionic charge for trace elements of geological interest. An ionic potential (charge/size ratio) of 2.0 subdivides the incompatible elements into low field strength (LFS) elements, also known as large ion lithophile elements (LIL) and high field strength elements (HFS). Compatible elements are placed towards the bottom, left-hand corner of the diagram. The ionic radii are from Shannon (1976) and are quoted for eight-fold coordination to allow a comparison between elements. Some of the first transition series metals (transition elements) and the PGE elements, are quoted for six-fold coordination.

The relatively simple mixing relationships between trace elements and major elements in their host minerals mean that the distribution of trace elements between minerals and melt can be quantified in a simple way, as outlined below.

4.2.1   Partition coefficients

The distribution of trace elements between phases may be described by a **distribution coefficient** or **partition coefficient** (McIntire, 1963). The Nernst distribution coefficient is used extensively in trace element geochemistry and describes the equilibrium distribution of a trace element between a mineral and a melt. The Nernst distribution coefficient is defined by:

$$Kd = C_{\text{element}i}^{\text{mineral}} / C_{\text{element}i}^{\text{melt}} \qquad\qquad [4.3]$$

where Kd is the Nernst distribution coefficient, and C is the concentration of the trace element i in ppm or wt %. An example would be 500 ppm Sr in a plagioclase phenocryst and 125 ppm Sr in the glassy matrix of the lava giving a plagioclase/silicate melt Kd of 4 for Sr. The Nernst distribution coefficient as defined above includes the Henry's Law constants for trace element i in the mineral and in the melt and is a function of temperature, pressure and composition of the melt, but is controlled neither by the concentration of the trace element of interest nor by the concentration of other trace elements. Similar partition coefficients may be written for mineral–fluid or mineral–mineral distributions. A mineral/melt partition coefficient of 1.0 indicates that the element is equally distributed between the mineral and the melt. A value of greater than 1.0 implies that the trace element has a 'preference' for the mineral phase and in the mineral–melt system under investigation is a compatible element. A value of less than 1.0 implies that the trace element has a 'preference' for the melt and is an incompatible element.

An alternative mode of formulation of the partition coefficient (although not commonly used) is the **two-component partition coefficient**. This may be used when the trace element is replacing an identified major element in the host mineral. A good example would be Ni substitution for Mg in olivine. In this case the partition coefficient (K_{D}) is defined by the expression

$$K_{\text{D}}\,(i/j) = [X_i/X_j]^{\text{Solid}}/[X_i/X_j]^{\text{Liquid}} \qquad\qquad [4.4]$$

where i is the trace element and j is the element in the host mineral which is replaced by i, and X is concentration either in wt % or mol %.

Two-element partition coefficients have the advantage that they do not vary as extensively as single-element partition coefficients with changes in melt composition.

A **bulk partition coefficient** is a partition coefficient calculated for a rock for a specific element from the Nernst partition coefficients of the constituent minerals and weighted according to their proportions. It is defined by the expression

$$D_i = x_1 Kd_1 + x_2 Kd_2 + x_3 Kd_3 \ldots \qquad\qquad [4.5]$$

where D_i is the bulk partition coefficient for element i, and x_1 and Kd_1 etc. are the percentage proportion of mineral 1 in the rock and the Nernst partition coefficient for element i in mineral 1 respectively.

In a rock containing 50 % olivine, 30 % orthopyroxene and 20 % clinopyroxene, the bulk partition coefficient (D) for the trace element i would be

$$D_i = 0.5\,Kd_i^{\text{ol}} + 0.3\,Kd_i^{\text{opx}} + 0.2\,Kd_i^{\text{cpx}}$$

Measuring partition coefficients   Partition coefficients can be determined in natural systems from the analysis of minerals and their glassy matrix in rapidly cooled volcanic rocks. Provided sufficient care is given to obtaining a clean mineral separate from unzoned minerals and a sufficiently sensitive analytical technique is used, mineral/matrix or phenocryst/

matrix partition coefficients can be very reliable and are frequently used. Many of the early mineral/melt partition coefficient measurements are of this type (e.g. Philpotts and Schnetzler, 1970).

An alternative to using natural systems is to use experimental data in which synthetic or natural starting materials are doped with the element of interest. This approach has the advantage that variations in temperature and pressure can be more carefully monitored than in natural systems. However, it is important to attempt to establish Henry's Law behaviour when determining trace element partition coefficients, for this then allows the result to be extrapolated to other compositions and use to be made of the result in petrogenetic modelling (see e.g. Dunn, 1987). Irving (1978) gives an excellent review of experimental determinations of partition coefficients up until 1978.

Physical controls on the value of partition coefficients in mineral / melt systems

Many geochemistry texts contain compilations of mineral/melt partition coefficients for use in trace element modelling. However, great care must be taken in applying these data, for experimental studies have shown that the Nernst partition coefficient can vary extensively according to the temperature, pressure, composition and oxygen activity of the melt. Disentangling these separate effects in experimental studies and then taking full account of them in petrogenetic modelling can be a serious problem. In an elegant study based on a very large number of experiments, Green and Pearson (1986) showed how the partition coefficients for the REE in sphene vary according to temperature, pressure and rock composition (Figure 4.3). Their work illustrates how meaningless a single mean value for a partition coefficient can be, even when the melt composition has been specified. However, we are not always in the fortunate position of having as much information available as this and it is often necessary to 'make do' with the available data. Below we discuss the extent to which different variables may affect partition coefficients.

(a) Composition Without doubt, melt composition is the most important single factor controlling mineral/melt partition coefficients. This was demonstrated in studies by Watson (1976) and Ryerson and Hess (1978), who showed that elements partitioned between immiscible acid and basic melts show distinct preferences for one or other type of melt. It is for this reason that the partition coefficients listed in Tables 4.1 to 4.3 are grouped according to rock type and the silica content of the melt. The composition control of mineral/melt partition coefficients between the REE and hornblende is illustrated in Figure 4.4.

(b) Temperature A number of experimental studies show that partition coefficients are a function of temperature (Figure 4.3). For example, Dunn (1987) found that the partition coefficients for Lu between olivine and basalt, and Lu and Hf between clinopyroxene and basalt, all decrease with increasing temperature.

Sometimes unravelling the separate effects of temperature and composition can be difficult, especially where the liquidus temperature of a melt is a function of composition. Such is the problem with Ni partitioning between olivine and a basaltic melt. Two experimental studies, published at the same time, seem to show conflicting results. Leeman and Lindstrom (1978) showed that the prime control on the olivine partition coefficient for Ni in a natural basaltic melt was temperature whilst Hart and Davis (1978) showed that there is clear inverse correlation between the melt composition and partition coefficient. To resolve the apparent conflict

Table 4.1 Mineral/melt partition coefficients for basaltic and basaltic andesite liquids

	Olivine	Ortho-pyroxene	Clinopyroxene	Hornblende	Phlog-opite	Plagioclase	Garnet	Mag-netite	Sphene
Rb	0.0098	0.022	0.031	0.29	3.06	0.071	0.042		
Sr	0.0140	0.040	0.060	0.46	0.081	1.830	0.012		
Ba	0.0099	0.013	0.026	0.42	1.090	0.230	0.023		
K	0.0068	0.014	0.038	0.96		0.170	0.015		

	Olivine	Ortho-pyroxene	Clinopyroxene	Hornblende	(6)	Phlog-opite	Plagioclase	(6)	Garnet	(7)	(8)	Mag-netite	Sphene
(Ref)													
Y	0.010	0.18	0.900	1.00		0.03	0.030		9.00			0.20	
Ti	0.020	0.10	0.400	1.50		0.90	0.040		0.30			7.50	
Zr	0.012	0.18	0.100	0.50	1.5640	0.60	0.048	0.0121	0.65			0.10	
Hf	0.013		0.263	0.50	1.5335		0.051	0.0115	0.45	0.140	0.250	2.0–4.0	
Nb	0.010	0.15	0.005	0.80		1.00	0.010		0.02			0.40	4.65
Ta			0.013						0.06			1.0–10	13
Th			0.030	0.50			0.010						
U	0.002		0.040	0.10			0.010						

	Olivine (1)	(2)	Opx (1)	Cpx (1)	(2)	Hbl (3)	(4)	(6)	Phlog (1)	Plag (1)	(2)	(6)	Gt (1)	(7)	(8)	Mag (5)
(Ref)																
La		0.0067			0.056		0.25	0.5442			0.190	0.1477		0.001	0.026	1.5–3.0
Ce	0.0069	0.0060	0.02	0.15	0.092	0.20	0.32	0.8430	0.034	0.120	0.111	0.0815	0.03	0.007	0.051	1.3–3.0
Pr																
Nd	0.0066	0.0059	0.03	0.31	0.230	0.33		1.3395	0.032	0.081	0.090	0.0551	0.07	0.026		1.0–3.0
Sm	0.0066	0.0070	0.05	0.50	0.445	0.52	1.40	1.8035	0.031	0.067	0.072	0.0394	0.29	0.102	0.600	1.1–2.2
Eu	0.0068	0.0074	0.05	0.51	0.474	0.40	1.20	1.5565	0.030	0.340	0.443	1.1255	0.49	0.243	1.000	0.6–1.5
Gd	0.0077	0.0100	0.09	0.61	0.556	0.63		2.0165	0.03	0.063	0.071	0.0310	0.97	0.680	2.100	
Tb					0.570		1.30							0.705		1.0–2.0
Dy	0.0096	0.0130	0.15	0.68	0.582	0.64		2.0235	0.030	0.055	0.063	0.0228	3.17	1.940	4.100	
Ho														1.675	13.200	
Er	0.0110	0.0256	0.23	0.65	0.583	0.55		1.7400	0.034	0.063	0.057	0.0202	6.56	4.700		
Tm																1.0–2.0
Yb	0.0140	0.0491	0.34	0.62	0.542	0.49	1.20	1.6420	0.042	0.067	0.056	0.0232	11.50	6.167	35.60	0.9–1.8
Lu	0.0160	0.0454	0.42	0.56	0.506	0.43	1.10	1.5625	0.046	0.06	0.053	0.0187	11.90	6.950	41.00	

	Olivine	Ortho-pyroxene	Clinopyroxene	Hornblende	Garnet			Mag-netite
Ni	5.9–29	5	1.5–14	6.8				29.0
Co	6.60	2–4	0.5–2.0	2.00	0.7–1.8	0.955	0.660	7.4
V	0.06	0.6	1.35	3.40				26.0
Cr	0.70	10	34	12.5	0.6–2.9	1.345	0.060	153.0
Sc	0.17	1.2	1.7–3.2	2.2–4.2		8.500	2.600	
Mn	1.45	1.4	0.3–1.2					

Data from compilation of Arth (1976); compilation of Pearce and Norry (1979); Green *et al.* (1989); Schock (1979); Fujimaki *et al.* (1984); Dostal *et al.* (1983); compilation of Henderson (1982); Leeman and Lindstrom (1978); Lindstrom and Weill (1978); Green and Pearson (1987). REE data: (1) compilation of Arth (1976); (2) Fujimaki *et al.* (1984); (3) basaltic compositions (Arth, 1976), Eu from Green and Pearson (1985a); (4) basaltic andesite (Dostal *et al.*, 1983); (5) Schock (1979); (6) mean of two basaltic andesites $SiO_2 = 55\%$ and 57 % from Fujimaki *et al.* (1984); (7) mean of 2 basalts + 1 basanite + 1 alkali olivine basalt (Irving and Frey, 1978); (8) hawaiite ($SiO_2 = 48.7\%$) (Irving and Frey, 1978).

Table 4.2 Mineral/melt partition coefficients for andesitic liquids

	Olivine	Orthopyroxene			Clinopyroxene			Hornblende	Plagioclase				Garnet	Magnetite	Sphene
(Ref.)				(8)	(7)		(8)	(3,8)		(5)	(8)	(7)	(7)	(7)	(4)
Rb				0.022	0.020		0.013	0.040			0.053	0.070	0.010	0.010	
Sr				0.032	0.080		0.033	0.2–0.4		2.82	1.600	1.800		0.010	0.060
Ba				0.013	0.020		0.040	0.100		0.503	0.155	0.160		0.010	
K				0.014	0.020		0.011	0.33			0.117	0.110	0.010	0.010	

	Olivine	Orthopyroxene		Clinopyroxene		Hornblende	Plagioclase			Garnet	Magnetite	Sphene
(Ref.)	(6,7)	(6)	(7)	(6)	(7)	(6,7)	(6)	(5)	(7)	(2,6)	(6,7)	(4)
Y	0.010	0.450		1.500		2.500	0.060	0.013		11.00	0.500	
Ti	0.03	0.250		0.400		3.00	0.050			0.500	9.000	
Zr	0.010	0.046	0.100	0.162	0.270	1.400	0.013		0.010	0.500	0.200	
Hf		0.051	0.100	0.173	0.250		0.015			0.570		
Nb	0.010	0.350		0.300		1.300	0.025				1.000	6.100
Ta												17.000
Th	0.010		0.050		0.010	0.150			0.010		0.100	

	Orthopyroxene			Clinopyroxene			Hornblende	Plagioclase				Garnet	Magnetite	Sphene
(Ref.)	(1)	(7)	(8)	(1)	(7)	(8)	(3)	(1)	(5)	(8)	(7)	(2)	(7)	(4)
La	0.031			0.047			0.500	0.302	0.228			0.076		2.00
Ce	0.028	0.050	0.030	0.084	0.250	0.508		0.221	0.136	0.186	0.200		0.200	
Pr														
Nd	0.028		0.047	0.183		0.645		0.149	0.115	0.143				
Sm	0.028	0.100	0.082	0.377	0.750	0.954	1.2–3.0	0.102	0.077	0.117	0.110	1.250	0.300	10.000
Eu	0.028	0.120	0.069		0.800	0.681		1.214	0.079	0.376	0.310	1.520	0.250	
Gd	0.039		0.132	0.583		1.350		0.067	0.056	0.050		5.200		
Tb												7.100		
Dy	0.076		0.212	0.774		1.460		0.050	0.045	0.126				
Ho							1.5–3.0					23.800		10.000
Er	0.153		0.314	0.708		1.330		0.045	0.040	0.034				
Tm														
Yb	0.254	0.460	0.438	0.633	0.900	1.300	1.2–2.1	0.041		0.029	0.050	53.000	0.250	
Lu	0.323		0.646	0.665				0.039	0.046	0.031		57.000		6.000

	Olivine	Orthopyroxene	Clinopyroxene	Hornblende	Plagioclase	Garnet	Magnetite
(Ref.)	(7)	(7)	(7)	(7)	(7)	(2,7)	(7)
Ni	58.000	8.000	6.000	10.000	0.010	0.600	10.000
Co		6.000	3.000	13.000	0.010	1.800	8.00
V	0.080	1.100	1.100	32.000	0.010	8.000	30.000
Cr	34.000	13.000	30.000	30.000	0.010	22.000	32.000
Sc	0.300	3.000	3.000	10.000	0.010	3.900	2.000

(1) Fujikami *et al.* (1984): sample No. 8.

(2) Irving and Frey (1978): andesite, SiO_2= 60.79 wt %

(3) Green and Pearson (1985a): interpolated from Figure 3.

(4) REE: Green and Pearson (1983): at 7.5 kb; values increase with pressure. Nb,Ta: Green *et al.* (1989).

(5) Drake and Weill (1975): Eu value for Eu^{3+}.

(6) Pearce and Norry (1979): intermediate compositions.

(7) Compilation of Gill (1981): Zr in Cpx: Watson and Ryerson (1986).

(8) Philpotts and Schnetzler (1970): and Schnetzler and Philpotts (1970).

Table 4.3 Mineral/melt partition coefficients for dacitic and rhyolitic melts

(Ref.)	Orthopyroxene				Clinopyroxene				Hornblende		Biotite				Garnet	
	(2)	(3)	(4)	(7)	(2)	(3)	(4)	(7)	(1)	(2)	(1)	(2)	(3)	(4)	(1)	(5)
Rb	0.003				0.032					0.014	3.260	2.240	3.200	4.200	0.009	
Sr	0.009				0.516					0.022	0.120		0.447		0.015	
Ba	0.003		(1.10)		0.131		(1.40)			0.044	6.360	9.700	23.533	5.367	0.017	
K	0.002	0.605			0.037					0.081					0.200	
Cs													3.000	2.300		
Pb													0.767			
Y	1.000		<1.1		4.000		3.100			6.000		0.030	1.233		35.000	
Ti	0.400				0.700					7.000					1.200	
Zr	0.200			0.033	0.600			0.184	0.310	4.000			1.197		1.200	
Hf		0.200	(0.00)	0.031		0.633	(0.00)	0.247					0.703	0.600		3.300
Nb	0.800				0.800					4.000			6.367			
Ta		0.165	(1.14)			0.263	(0.75)						1.567	1.340		
Th		0.130	(6.53)			0.150	(5.99)						0.997	1.227		
U		0.145	(0.28)				(0.21)						0.773	0.167		
La		0.780	<0.4	0.015		1.110	0.600	0.015					5.713	3.180		0.390
Ce	0.150	0.930	<0.4	0.016	0.500	1.833	1.000	0.044	0.899	1.520	0.037	0.320	4.357	2.803	0.350	0.690
Pr																
Nd	0.220	1.250	<0.9	0.016	1.110	3.300	2.100	0.166	2.890	4.260	0.044	0.290	2.560	2.233	0.530	0.603
Sm	0.270	1.600	(7.87)	0.017	1.670	5.233	(10.65)	0.457	3.990	7.770	0.058	0.260	2.117	1.550	2.660	2.035
Eu	0.170	0.825	(2.85)		1.560	4.100	(5.00)	0.411	3.440	5.140	0.145	0.240	2.020	0.867	1.500	0.515
Gd	0.340			0.027	1.850			0.703	5.480	10.000	0.082	0.280			10.500	6.975
Tb		1.850	(5.50)			7.533	(9.25)						1.957	1.053		11.900
Dy	0.460	1.800	(3.85)	0.041	1.930	7.300	(8.90)	0.776	6.200	13.000	0.097	0.290	1.720	0.823	28.600	
Ho																28.050
Er	0.650			0.072	1.800			0.699	5.940	12.000	0.162	0.350			42.800	
Tm																
Yb	0.860	2.200	(2.35)	0.115	1.580	6.367	(4.55)	0.640	4.890	8.380	0.179	0.440	1.473	0.537	39.900	43.475
Lu	0.960	2.250	(2.70)	0.154	1.540	5.933	(4.30)	0.683	4.530	5.500	0.185	0.330	1.617	0.613	29.600	39.775
Ni																
Co			(140)				(72)							88.667		2.625
V																
Cr													19.650	5.233		3.700
Sc		18.000	(22.0)			53.000	(89.50)						13.633	15.567		15.950
Mn		45.500	(57.0)			32.667	(28.35)						124.530	10.367		

(1) Rb–K and REE Arth (1976), dacites; Hbl Zr value from Watson and Harrison (1983).

(2) Rb–K and REE: Arth (1976), rhyolites. Y–Nb Pearce and Norry (1979), acid magmas.

(3) Nash and Crecraft (1985), rhyolites — SiO$_2$ 71.9–76.2 wt %.

(4) Mahood and Hildreth (1983), high-silica rhyolites — SiO$_2$ 75–77.5 wt %. For pyroxenes, values in parentheses (Mahood & Hildreth, 1983); other values: Michael (1988).

(5) Irving and Frey (1978), dacites and rhyolites — SiO$_2$ 62.89–70.15 wt %.

(6) Fujimaki (1986), dacites — SiO$_2$ 63.21–64.86 wt %.

(7) Fujimaki *et al.* (1984), dacite No. 7 — SiO$_2$ 70.81 wt %.

(8) Nb,Ta: Green and Pearson (1987), trachyte. REE interpolated from Figure 3 of Green *et al.* (1989).

(9) Brooks *et al.* (1981).

Magnetite (2)	Ilmenite (3)	Ilmenite (3)	Quartz (3)	Plagioclase (1)	Plagioclase (2)	Plagioclase (3)	K-feldspar (2)	K-feldspar (3)	K-feldspar (4)	Apatite (2)	Apatite (6)	Zircon (4)	Zircon (6)	Sphene (8)	Allanite (4)	Allanite (9)	(Ref.)
			0.041	0.048	0.041	0.105	0.340	1.750	0.487								Rb
				2.840	4.400	15.633	3.870	5.400	3.760								Sr
			0.022	0.360	0.308	1.515	6.120	11.450	4.300								Ba
			0.013	0.263	0.100												K
			0.029			0.105		0.195	0.032			3.15					Cs
						0.972		2.473									Pb
2.000					0.100	0.130				40							Y
12.500			0.038	0.050						0.1							Ti
0.800					0.100	0.135		0.030		0.1	0.64						Zr
	1.883	3.100	0.030			0.148		0.033	0.017		0.73	3193.5	977.50		18.9		Hf
2.500					0.060					0.1				6.3			Nb
	3.167	106.000	0.008			0.035		0.010	0.019			47.50		16.5	3.1		Ta
	0.463	7.500	0.009			0.048		0.023	0.018			76.80			484.0	168.0	Th
	0.517	3.200	0.025			0.093		0.048	0.021			340.50			15.5	<6.7	U
	1.223	7.100	0.015			0.380		0.080	0.072		14.50	16.90	4.18	4.0	2594.5	820.0	La
	1.640	7.800	0.014	0.240	0.270	0.267	0.044	0.037	0.046	34.7	21.10	16.75	4.31		2278.5	635.0	Ce
																	Pr
	2.267	7.600	0.016	0.170	0.210	0.203	0.025	0.035	0.038	57.1	32.80	13.30	4.29		1620.0	463.0	Nd
	2.833	6.900	0.014	0.130	0.013	0.165	0.018	0.025	0.025	62.8	46.00	14.40	4.94	21.0	866.5	205.0	Sm
	1.013	2.500	0.056	2.110	2.150	5.417	1.130	4.450	2.600	30.4	25.50	16.00	3.31		111.0	81.0	Eu
				0.900	0.097	0.125	0.011			56.3	43.90	12.00	6.59			130.0	Gd
	3.267	6.500	0.017					0.025	0.033			37.00			273.0	71.0	Tb
	2.633	4.900	0.015	0.086	0.064	0.112	0.006	0.055	0.052	50.7	34.80	101.50	47.40		136.5		Dy
														19.0			Ho
				0.084	0.055		0.006			37.2	22.70	135.00	99.80				Er
																	Tm
	1.467	4.100	0.017	0.077	0.049	0.090	0.012	0.030	0.015	23.9	15.40	527.00	191.0		30.8	8.9	Yb
	1.203	3.600	0.014	0.062	0.046	0.092	0.006	0.033	0.031	20.2	13.80	641.50	264.5	10.0	33.0	7.7	Lu
																	Ni
									0.240			16.00			42.5		Co
																	V
	109.00	3.000										189.50			380.0		Cr
	10.633	5.900	0.012			0.053		0.023	0.040			68.65			55.9		Sc
	32.000	115.000	0.039			0.365			0.022			1.52			18.1		Mn

Leeman and Lindstrom (1978) formulated a complex partition coefficient that included the composition effect. They showed that Ni partitioning between olivine and basalt is temperature-dependent and concluded that, in this case, composition is less important than temperature in determining the partition coefficient.

(c) Pressure   One of the most convincing demonstrations of the effect of pressure on partition coefficients is the work of Green and Pearson (1983, 1986) on the

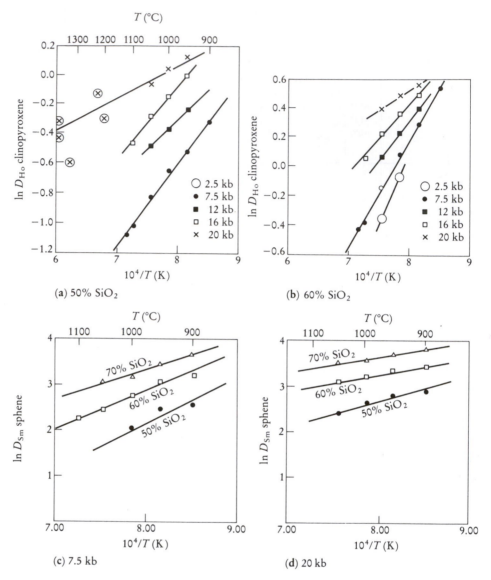

Figure 4.3 The combined effects of pressure, temperature and rock composition on a partition coefficient. (a) and (b) The partition coefficient for the REE Ho in clinopyroxene as a function of temperature, for pressures of 2.5, 7.5, 12, 16 and 20 kb in liquids with 50 wt % SiO_2 and 60 wt % SiO_2 (after Green and Pearson, 1985b). (c) and (d) The partition coefficient for Sm in sphene as a function of temperature for liquids with 50, 60 and 70 wt % SiO_2 at 7.5 kb and 20 kb pressure (after Green and Pearson, 1986). Both these diagrams may be used to interpolate a value for the partition coefficient for any pressure, temperature and composition within the experimental range. These values may then be extrapolated to other members of the REE series.

partitioning of REE between sphene and an intermediate silicic liquid. Within a small compositional range (56–61 wt % SiO_2) at 1000°C Green and Pearson showed that there is a measurable increase in partition coefficient with increasing pressure from 7.5 to 30 kb (Figure 4.5). One important aspect of this pressure effect is that

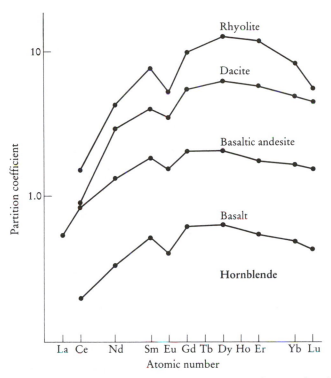

Figure 4.4   A plot of the partition coefficients for the rare earth elements between hornblende and melt (log scale) vs atomic number (normal scale) in basalt, basaltic andesite, dacite and rhyolite. There is a clear increase in partition coefficient with increasing silica content of the melt, amounting to an order of magnitude difference between basaltic and rhyolitic melts. Data from Tables 4.1 to 4.3.

high-level phenocryst/matrix pair partition coefficients may not be suitable for geochemical modelling of deep crustal and mantle processes. However, the effect of increased pressure and increased temperature are generally in an opposite sense and may to some extent cancel each other out.

(d) Oxygen activity   The most widely quoted example of the control of oxygen activity on a partition coefficient is that of the partitioning of Eu^{2+} between plagioclase and a basaltic melt (Drake and Weill, 1975). There is an order of magnitude difference in the partition coefficient for Eu between atmospheric conditions and the relatively reducing conditions found in natural basalts (Figure 4.6). This is because europium forms Eu^{2+} at low oxygen activities and Eu^{3+} at high oxygen activities. Eu^{2+} and Eu^{3+} behave very differently in their partitioning between plagioclase and a basaltic melt, for Eu^{2+} is much more compatible than Eu^{3+} in plagioclase. Thus at low oxygen activities partition coefficients for Eu between plagioclase and basaltic melts are high (generally > 1.0) and anomalous relative to the other REE (Figure 4.6), whereas at high oxygen activities partition coefficients for Eu are low and Eu behaves in a similar way to the other REE.

(e) Crystal chemistry   Onuma *et al.* (1968), Matsui *et al.* (1977) and Philpotts (1978) have shown that crystal structure exerts a major influence on trace element partitioning. Using a plot of partition coefficient (expressed as log to the base 10,

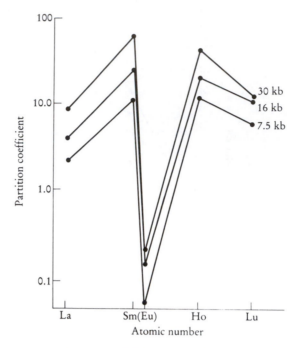

Figure 4.5   The partition coefficents for selected REE between sphene and a silicic melt of intermediate composition, plotted as a function of pressure. There is an increase in partition coefficient with increasing pressure in the range 7.5 to 30 kb (after Green and Pearson, 1983).

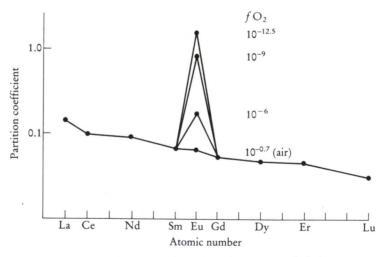

Figure 4.6   The partition coefficient for Eu between plagioclase and a basaltic melt plotted as function of oxygen activity (fO_2) compared with other REE (after Drake and Weill, 1975)

i.e. $\log_{10}$) vs ionic radius (in Ångstroms; $1 \text{ Å} = 10^{-10}$ m), they showed that the partition coefficients of elements carrying the same ionic charge in the same mineral/melt system exhibit a smooth curve (Figure 4.7). Curves for different ionic charge tend to be parallel to each other for the same mineral/melt system. Such

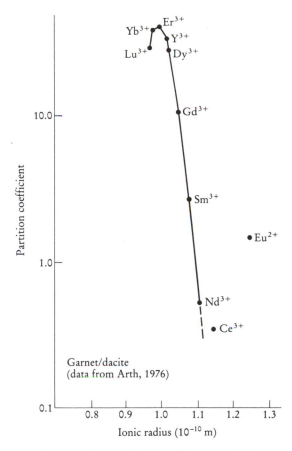

Figure 4.7   A plot of partition coefficient vs ionic radius (from Shannon, 1976) on an Onuma diagram, for the REE between garnet and dacite (data from Arth, 1976). The trivalent REE and Y (with the exception of Ce) define a smooth curve.

diagrams have become known as **Onuma diagrams**. Deviations from anticipated patterns may reveal controls on trace element partitioning other than those of the size and charge of the cation. Onuma diagrams can also be used to estimate the size of a distribution coefficient when measurements have been made for a similar element.

(f) Water content of the melt   Few studies have been carried out to examine explicitly the effects of the water content of a melt on trace element partitioning behaviour. However, Green and Pearson (1986) showed that in the case of the partitioning of the REE between sphene and silicate liquids the water content of the melt (0.9–29 mol % water) has no significant effect on measured partition coefficients.

(g) Selecting a partition coefficient   Clearly, the most important parameter controlling the partitioning of a trace element between a mineral and melt is the composition of the melt itself. Once this is established, a partition coefficient should be used whose pressure and temperature conditions of determination most closely match those of the system being investigated.

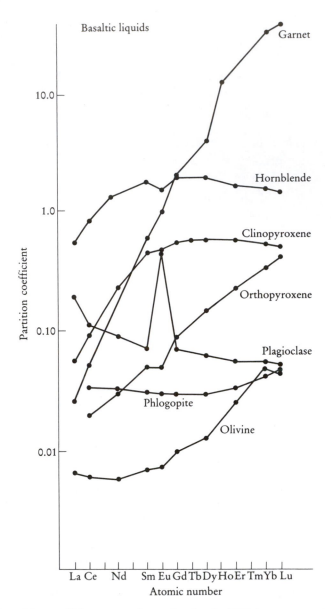

Figure 4.8   A plot of partition coefficient vs atomic number for the REE in common minerals in basaltic melts. Data from Table 4.1. (The hornblende data are for a basaltic andesite.)

Partition coefficients in basalts and basaltic andesites   In Table 4.1 partition coefficients are listed for trace elements in minerals in equilibrium with basaltic and basaltic andesite liquids. Following the TAS classification these are liquids with silica contents in the range 45–57 wt %. A summary of the REE partition coefficients is presented in Figure 4.8 as a plot of partition coefficient vs atomic number. The compilation in Table 4.1 is based upon a variety of published sources from experimental and phenocryst/matrix data. Averages are arithmetic means.

The several sets of REE data shown are generally in good agreement with each other. There is however a strong compositional effect on the *Kd* for the REE in hornblende (Green and Pearson, 1985a) and so the differences in published values for basalts (Arth, 1976) and basaltic andesites (Dostal *et al.*, 1983; Fujimaki *et al.*, 1984) may reflect a real difference in the partition coeficients. There may be a similar explanation for the differences in partition coefficient in garnet.

Colson *et al.* (1988) in a detailed study of trace element partitioning between olivine and silicic melt, and orthopyroxene and silicic melt, have shown that many partition coefficients are strongly dependent upon temperature and melt composition. They show that these partition coefficients vary according to ionic size and they have modelled temperature and composition dependence as a function of these variables. On the basis of their equations it is possible to predict partition coefficients between olivine and melt, and orthopyroxene and melt, for a wide range of tri- and di-valent cations under a variety of magmatic conditions.

Partition coefficients in andesites

Table 4.2 lists partition coefficients for trace elements between a range of rock-forming minerals and andesitic liquids (57–63 wt % SiO_2 in the TAS classification). A summary of the REE values is given in Figure 4.9. Many published sources of partition coefficients treat andesites and basalts together and a comparison of Figures 4.1 and 4.2 shows that the partition coefficients for the REE are similar in basaltic and andesitic liquids. Values for garnet, the pyroxenes and the value for Eu in plagioclase, however, are higher in andesites than in basalts. REE partition coefficients for hornblende are higher in andesites than basalts, but are comparable between andesites and basaltic andesites. Values for the REE in the pyroxenes vary between published sources; this may in part be due to compositional effects.

Partition coefficients in dacites and rhyolites

The compilation in Table 4.3 compares partition coefficients from a number of published sources for dacites, rhyodacites, rhyolites and high-silica rhyolites. These are rocks which have > 63 wt % SiO_2 in the TAS classification. A summary of the REE values is presented in Figure 4.10.

A comparison between the REE in rhyolites and in basaltic and andesitic liquids (Figures 4.8 to 4.10) shows that values for pyroxenes and hornblende are an order of magnitude higher in the rhyolites and that these minerals now show a small but measurable negative europium anomaly. The values for the light REE in garnet are also higher and the Eu anomaly in plagioclase is much increased.

Partition coefficients for the REE in any one of the ferromagnesian minerals are variable. This may in part be a function of changing melt composition although there are a number of other possible explanations. Firstly, the presence of mineral inclusions gives rise to very high partition coefficients in rhyolites with very high silica contents ($SiO_2 > 75$ %). This may also account for some of the differences in the sets of values for biotite. Secondly, it is possible that the changing iron/magnesium ratio of ferromagnesian minerals may also explain differences in partition coefficients, although there are currently insufficient data with which to evaluate this possibility. In the case of pyroxenes the light REE values of Mahood and Hildreth (1983) are probably in error and the alternative values of Michael (1988) are used here. Values for the feldspars are in reasonable agreement between the different published results although the LIL elements Sr, Ba, Rb, Eu and Pb^{2+} show very erratic results in the original data-sets and this is still apparent in the mean values in this compilation. This may in part be a temperature effect (Irving, 1978; Long, 1978) but may also be compositionally controlled, for Long (1978)

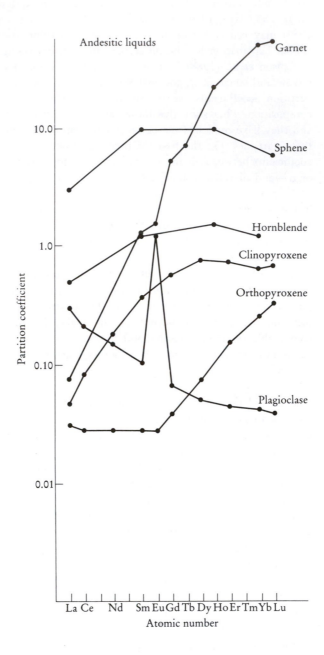

Figure 4.9   A plot of partition coefficient vs atomic number for the REE in common minerals in andesitic melts. Data from Table 4.2.

noted that the partitioning of Sr between alkali feldspar and a granitic melt is sensitive to the concentration of Ba in the melt. Partition coefficients for Eu in plagioclase increase with decreasing oxygen activity (Figure 4.6).

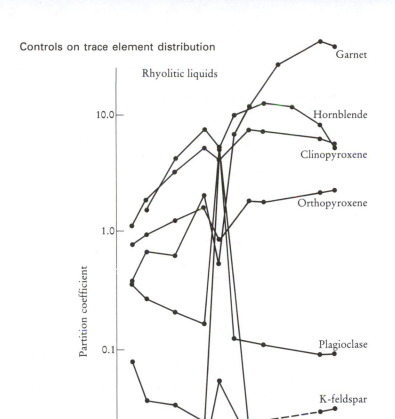

Figure 4.10 A plot of partition coefficient vs atomic number for the REE in common minerals in rhyolitic melts. Data from Table 4.3.

4.2.2 Geological controls on the distribution of trace elements

Clearly, a geochemical study of trace elements is only of use if we have some understanding of the way in which geological processes control their distribution. At present our knowledge is patchy, for some processes are well understood and there are mathematical models available to describe them. In these areas, principally those involving mineral/melt equilibria, trace element geochemistry has had a profound impact on geological thinking. Other areas, equally important but less amenable to the quantitative approach, are understood qualitatively. In this section we review a wide range of geological processes and discuss the way they control trace element distributions. Where there are appropriate mathematical models the equations are given and the terms used are defined in Box 4.1.

Box 4.1

Definition of terms used in equations for trace element partitioning between solid and melt (Section 4.2.2)

C_L	Weight concentration of a trace element in the liquid
$\bar{C}_L$	Average weight concentration of a trace element in a mixed melt
C_0	In partial melting, the weight concentration of a trace element in the original unmelted solid; in fractional crystallization, the weight concentration in the parental liquid
C_R	Weight concentration of a trace element in the residual solid during crystal fractionation
C_S	Weight concentration of a trace element in the residual solid after melt extraction
$_{ss}S_B$	Weight concentration of a trace element in a steady-state liquid after a large number of RTF cycles
C_A	Concentration of a trace element in the wallrock being assimilated during AFC processes
D_{RS}	Bulk distribution coefficient of the residual solids (see Eqn [4.5])
D_0	Bulk distribution coefficient of the original solids (see Eqn [4.5])
D	Bulk distribution coefficient of the fractionating assemblage during crystal fractionation
F	Weight fraction of melt produced in partial melting; in fractional crystallization, the fraction of melt remaining
f	Fraction of melt allocated to the solidification zone in *in situ* crystallization which is returned to the magma chamber
f'	A function of F, the fraction of melt remaining in AFC processes
Kd	Mineral/melt partition coefficient
M_L	Mass of the liquid in *in situ* crystallization
M_0	Total mass of the magma chamber *in situ* crystallization
n	Number of rock volumes processed during zone refining
P	Bulk distribution coefficient of minerals which make up a melt (see Eqn [4.9])
r	Ratio of the assimilation rate to the fractionation rate in AFC processes
x	Mass fraction of the liquid crystallized in each RTF cycle
y	Mass fraction of the liquid escaping in each RTF cycle

Element mobility   Any suite of rocks which has been subjected to hydrothermal alteration or metamorphism is likely to suffer element mobility. It is essential, therefore, in any trace element study to demonstrate first that element concentrations are undisturbed and original before inferences can be made about the petrogenesis of the rock group.

Trace element mobility is controlled by the mineralogical changes which take place during alteration and the nature of the fluid phase. As a generalization, incompatible elements which belong to the LFS group (Cs, Sr, K, Rb, Ba — Figure 4.2) are mobile, whereas the HFS elements are immobile. This latter group includes the REE, Sc, Y, Th, Zr, Hf, Ti, Nb, Ta and P (Pearce, 1983). In addition the transition metals Mn, Zn and Cu tend to be mobile, particularly at high temperatures (see Seewald and Seyfried, 1990), whilst Co, Ni, V and Cr are immobile. Such generalizations are normally valid, although many exceptions are documented. This may be illustrated with reference to the traditionally immobile REE. Humphries (1984) shows that there is no simple relationship between the degree of mobility of the REE and rock type or metamorphic grade, and emphasizes the mineralogical and fluid controls. For example, the REE may be more easily released from a glassy basalt during alteration than from a rock with the same composition but which is crystalline. Again the REE may be mobilized by halogen-rich or carbonate-rich mineralizing fluids in a rock in which they would otherwise be stable with respect to the movement of an aqueous fluid.

A special case of trace element mobilization is in the dehydration of subducted ocean floor, a process thought to be pertinent to the generation of calc-alkali magmas. Pearce (1983) has suggested that the elements Sr, K, Rb, Ba, Th, Ce, P and Sm may be mobile in such circumstances.

Partial melting Two types of partial melting process are commonly described in the geological literature and represent end-member models of natural processes. **Batch melting**, also known as equilibrium fusion and equilibrium partial melting, describes the formation of a partial melt in which the melt is continually reacting and re-equilibrating with the solid residue at the site of melting until mechanical conditions allow it to escape as a single 'batch' of magma. In **fractional melting**, also known as Rayleigh melting, only a small amount of liquid is produced and instantly isolated from the source. Equilibrium is therefore only achieved between the melt and the surfaces of mineral grains in the source region.

Which partial melting process is appropriate in a particular situation depends upon the ability of a magma to segregate from its source region, which in turn depends upon the permeability threshold of the source. The problem is discussed in some detail by Wilson (1989). Fractional melting may be an appropriate model for some basaltic melts, for recent physical models of melt extraction from the mantle indicate that very small melt fractions can be removed from their source region (McKenzie, 1985; O'Nions and McKenzie, 1988). More viscous, felsic melts have a higher permeability threshold and probably behave according to the batch melting equation. It is worth noting in passing that physical models of melt extraction describe melt fractions in terms of their *volume* whereas chemical models describe melt fractions in terms of their *mass*.

(a) Batch melting The concentration of a trace element in the melt C_L is related to its concentration in the unmelted source C_0 by the expression

$$C_L/C_0 = 1/[D_{RS} + F(1 - D_{RS})] \qquad [4.6]$$

and the concentration of a trace element in the unmelted residue C_S relative to the unmelted source C_0 is

$$C_S/C_0 = D_{RS}/[D_{RS} + F(1 - D_{RS})] \qquad [4.7]$$

where D_{RS} is the bulk partition coefficient (see Eqn [4.5]) of the **residual solid** and F is the weight fraction of melt produced. It should be noted that the bulk partition coefficient is calculated for the residual solids present at the instant the liquid is removed, so that solid phases that were present but are now melted out do not influence the trace element concentration in the liquid (Hanson, 1978). This formulation of the batch melting equation is very straightforward to use. If, however, a more complex formulation is sought, then the batch melting equation is expressed in terms of the original mineralogy of the source and the relative contributions each phase makes to the melt:

$$C_L/C_0 = 1/[D_0 + F(1 - P)] \qquad [4.8]$$

where D_0 is the bulk distribution coefficient at the onset of melting and P is the bulk distribution coefficient of the minerals which make up the melt. P is calculated from

$$P = p_1 Kd_1 + p_2 Kd_2 + p_3 Kd_3 + \dots \qquad [4.9]$$

where p_1 etc. is the normative weight fraction of mineral 1 in the melt and Kd_1 is the mineral–melt distribution coefficient for a given trace element for mineral 1.

In the case of modal melting (i.e. where the minerals contribute to the melt in proportion to their concentration in the rock), Eqn [4.8], simplifies to

$$C_L/C_0 = 1/[D_0 + F(1 - D_0)] \dots \qquad [4.10]$$

Even more complex formulations, which allow a phase to be consumed during melting, melt proportions to vary during partial melting and variations in partition coefficients, are given by Hertogen and Gijbels (1976) and Apted and Roy (1981).

Taking the simple case where D is calculated for the unmelted residue Eqn [4.6], the degree of enrichment or depletion relative to the original liquid (C_L/C_0) for different values of F is illustrated in Figure 4.11(a) for seven different values of D, the bulk distribution coefficient. When D is small, expression [4.6] reduces to $1/F$ and marks the limit to trace element enrichment for any given degree of batch melting (see shaded area on Figure 4.11a). When F is small, Eqn [4.6] reduces to $1/D$ and marks the maximum possible enrichment of an incompatible element and the maximum depletion of a compatible element relative to the original source. Small degrees of melting can cause significant changes in the ratio of two incompatible elements where one has a bulk partition coefficient of, say, 0.1 and the other 0.01, but at smaller values of D (0.01–0.0001) the discrimination is not possible.

Enrichment and depletion in the solid residue in equilibrium with the melt (Eqn [4.7]) is shown in Figure 4.11(b) for different values of F and D. Even small degrees of melting will deplete the residue significantly in incompatible elements. Compatible elements, however, at small degrees of melting remain very close to their initial concentrations. Cox *et al.* (1979) used this relationship to estimate the average content of compatible elements such as Ni and Cr in the upper mantle from the composition of ultramafic nodules from which a melt may have already been extracted.

(b) Fractional melting   There are two versions of the fractional melting equation. One considers the formation of only a single melt increment whilst the other considers the aggregated liquid formed by the collection of a large number of small melt increments. If it is assumed that during fractional melting the mineral phases enter the melt in the proportions in which they are present in the source, then the concentration of a trace element in the liquid relative to the parent rock *for a given melt increment* is given by the expression

$$C_L/C_0 = \frac{1}{D_0}(1 - F)^{(1/D_0-1)} \qquad [4.11]$$

where F is the fraction of melt already removed from the source and D_0 is the bulk

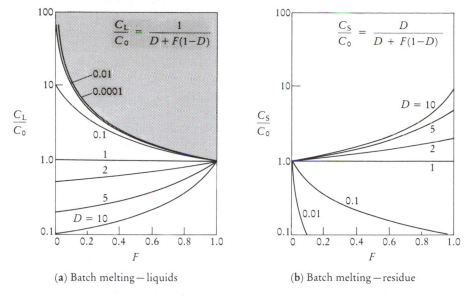

(a) Batch melting — liquids   (b) Batch melting — residue

Figure 4.11   (a) The enrichment of a trace element in a partial melt relative to its concentration in the source (C_L/C_0) during batch partial melting with changing degrees of melting (F). The numbered curves are for different values of the bulk partition coefficient D. At small degrees of melting, compatible elements are greatly depleted relative to the source whereas incompatible elements are greatly enriched to a maximum of $1/F$. The shaded region is the area in which enrichment is impossible. (b) Enrichment and depletion of a trace element in the residue relative to the original source (C_s/C_0) with changing degrees of melting (F) for different values of bulk partition coefficient (D).

partition coefficient for the original solid phases prior to the onset of melting. The equation for the residual solid is

$$C_S/C_0 = (1 - F)^{(1/D_0 - 1)} \tag{4.12}$$

The general expressions for the more probable case where minerals do not enter the melt in their modal proportions are given by

$$C_L/C_0 = \frac{1}{D_0}(1 - PF/D_0)^{(1/P - 1)} \tag{4.13}$$

and

$$C_S/C_0 = \frac{1}{(1 - F)}(1 - PF/D_0)^{1/P} \tag{4.14}$$

where P is the bulk distribution coefficient of the minerals which make up the melt and is calculated from Eqn [4.9].

The variation in trace element concentrations relative to the original liquid (C_L/C_0) during fractional melting for a single melt increment at different degrees of melting and for different values of D is shown for modal melting (Eqn [4.11]) in Figure 4.12(a). An enlargement of the region of interest ($F = 0$–0.1) is given in Figure 4.12(c). In the range 0–10 % melting, the changes in element concentrations relative to the original source are more extreme than in batch melting, although the limiting value of $1/D$ is the same.

Trace element concentrations in the original solid, momentarily in equilibrium with the liquid (Eqn [4.12]) are shown for small melt fractions in Figure 4.12(d). Incompatible elements are even more strongly depleted than in batch melting although compatible element concentrations are unchanged relative to the source. Where *several melt increments have collected together* the general expression is

$$\bar{C}_L/C_0 = \frac{1}{F} [1 - (1 - PF/D_0)^{1/P}] \qquad [4.15]$$

where $\bar{C}_L$ is the averaged concentration of a trace element in a mixed melt. In the case of modal melting Eqn [4.15] simplifies to

$$\bar{C}_L/C_0 = \frac{1}{F} [1 - (1 - F)^{1/D_0}] \qquad [4.16]$$

The numerical consequences of fractional melting where the melt increments are collected together in a common reservoir are illustrated in Figure 4.12(b). This type of fractional melting is indistinguishable from batch melting except for compatible elements at very large degrees of melting (cf. Figure 4.11a).

Crystal fractionation   Three types of fractional crystallization are considered here — equilibrium crystallization, Rayleigh fractionation and *in situ* crystallization.

(a) Equilibrium crystallization   The process of equilibrium crystallization describes complete equilibrium between all solid phases and the melt during crystallization. This is not thought to be a common process although the presence of unzoned crystals in some mafic rocks suggests that it may be applicable on a local scale in some mafic magmas. The distribution of trace elements during equilibrium crystallization is the reverse of equilibrium melting (page 121), and the equation therefore is

$$C_L/C_0 = 1/[D + F(1 - D)] \qquad [4.17]$$

In this case C_0 is redefined as the initial concentration of a trace element in the primary magma, F is the fraction of melt remaining and D is the bulk partition coefficient of the fractionating assemblage. The enrichment and depletion of trace elements relative to the orginal liquid may be deduced from Figure 4.11(a), the batch melting diagram, but in this case the diagram should be read from right to left.

(b) Fractional crystallization/Rayleigh fractionation   More commonly crystals are thought to be removed from the site of formation after crystallization and the distribution of trace elements is not an equilibrium process. At best, surface equilibrium may be attained. Thus, fractional crystallization is better described by the Rayleigh Law. Rayleigh fractionation describes the extreme case where crystals are effectively removed from the melt the instant they have formed. The equation for Rayleigh fractionation is

$$C_L/C_0 = F^{(D-1)} \qquad [4.18]$$

and the equation for the enrichment of a trace element relative to the original liquid in the crystals as they crystallize (the instantaneous solid) C_R is given by

$$C_R/C_0 = DF^{(D-1)} \qquad [4.19]$$

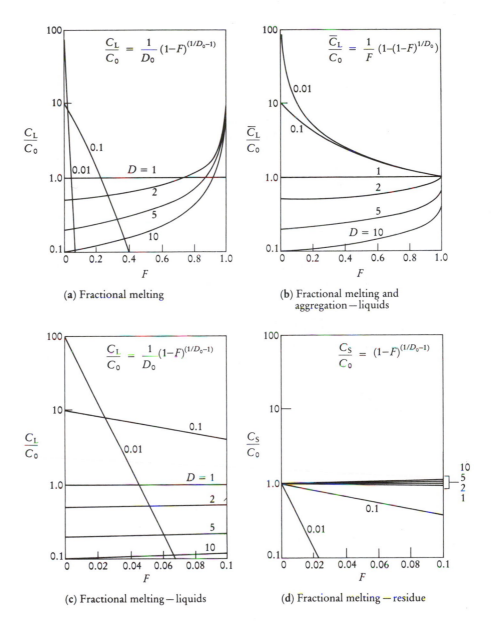

(a) Fractional melting

(b) Fractional melting and aggregation — liquids

(c) Fractional melting — liquids

(d) Fractional melting — residue

Figure 4.12   (a) The enrichment of a trace element in a melt relative to its source (C_L/C_0) as a function of the fraction of melting (F), during fractional melting for different values of bulk partition coefficient (D). During fractional melting only a very small melt fraction is produced and it is instantly removed from the source. (b) The enrichment of a trace element in a melt relative to its source ($\bar{C}_L/C_0$) as a function of fraction of melting (F) and bulk partition coefficient (D) during fractional melting. In this case small melt fractions are removed instantly from the source but aggregate together. This process produces very similar results to that of batch melting. (c) An enlargement of (a) between values of F of 0 and 0.1. (d) The enrichment of a trace element in the residual solid relative to the concentration in the original source (C_s/C_0) as a function of F between values of 0 and 0.1.

The equation for the mean enrichment of a trace element in the cumulate relative to the original liquid, i.e. the total residual solid C_R, is

$$C_R/C_0 = \frac{1 - F^D}{1 - F}$$

[4.20]

Rayleigh fractionation is illustrated in Figure 4.13(a), which shows the concentration of a trace element relative to its initial concentration in the liquid at differing values of F — the proportion of liquid remaining — for different values of D. For incompatible elements there is little difference between Rayleigh fractionation and equilibrium crystallization until more than about 75 % of the magma has crystallized, at which point the efficient separation of crystals and liquid becomes physically difficult. The limiting case for incompatible elements is where $D = 0$, in which case $C_L/C_O = 1/F$, the same as for equilibrium crystallization. It is therefore impossible to enrich a liquid beyond this point by fractional crystallization. Rayleigh fractionation is less effective than batch melting in changing the ratio of two incompatible elements, for the curves for 0.1 and 0.01 are very close together (Figure 4.13a). Compatible elements are removed from the melt more rapidly than in the case of equilibrium crystallization.

The concentration of trace elements in the instantaneous solid residue of Rayleigh fractionation is illustrated in Figure 4.13(b). More relevant, however, is

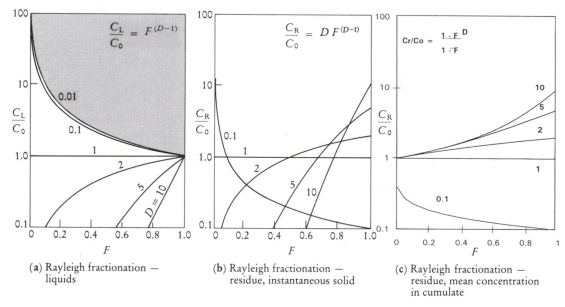

(a) Rayleigh fractionation — liquids

(b) Rayleigh fractionation — residue, instantaneous solid

(c) Rayleigh fractionation — residue, mean concentration in cumulate

Figure 4.13   (a) The enrichment of a trace element in a melt relative to its concentration in the parental melt (C_L/C_0) as a function of the fraction of remaining liquid (F) during Rayleigh fractionation for different values of bulk partition coefficient (D). The limiting value of C_L/C_0 is $1/F$ and the shaded area is the region of impossible values. (b) The enrichment of a trace element in the instantaneous solid, i.e. the solid which is produced and then immediately removed during crystal fractionation, relative to the parental liquid (C_R/C_0) as a function of the fraction of remaining liquid and bulk partition coefficient. (c) The enrichment of a trace element in the residue, i.e. the mean concentration in the cumulate, relative to the parental liquid (C_R/C_0) as a function of the fraction of remaining liquid and bulk partition coefficient (D). Diagrams (a) to (c) are read from right to left, as the value of F decreases with fractional crystallization.

the mean concentration in the cumulate given by Eqn [4.20] and shown in Figure 4.13(c). Relative to concentrations in the original liquid, incompatible element concentrations decrease and compatible element concentrations increase with progressive fractional crystallization.

In reality, of course, crystallization processes probably operate between the two extremes of equilibrium crystallization and fractional crystallization, which should be seen as limiting cases.

(c) In situ crystallization Over the past decade new ideas have emerged on how crystallization takes place in magma chambers and on how crystals and liquid are separated. The view that crystals will separate from a melt through gravitational settling has been strongly criticized and is probably strictly applicable only to ultramafic melts. Instead it is thought that crystallization takes place on the side walls of magma chambers and fractional crystallization is considered an *in situ* process in which residual liquid is separated from a crystal mush — rather than vice versa — in a solidification zone at the magma chamber margin. The solidification zone progressively moves through the magma chamber until crystallization is complete.

Langmuir (1989) has derived an equation which describes trace element distributions during *in situ* crystallization. In order to be able to compare the results with the effects of fractional and equilibrium crystallization, the equation presented below is for the special case where there is no trapped liquid in the solidification zone.

$$C_L/C_0 = (M_L/M_0)^{(f(D-1)/[D(1-f)+f])} \qquad [4.21]$$

M_O is the initial mass of the magma chamber and M_L is the mass of liquid so that the ratio M_L/M_O is equivalent to the term F — the fraction of melt remaining and used in equilibrium and fractional crystallization models; f is the fraction of magma allocated to the solidification zone which is returned to the magma chamber. Equation [4.21] is similar in form to the Rayleigh fractionation equation but with a more complex exponent.

The effects of *in situ* crystallization on the ratio of a trace element relative to the parental magma (C_L/C_0) for different melt fractions returned to the magma chamber $(f = 0.1-0.8)$ is shown for four different bulk partition coefficients in Figure 4.14(a)–(d). The chief differences between *in situ* crystallization and Rayleigh fractionation may be observed by comparing Figures 4.14 and 4.13(a). The limiting case is where $f = 1.0$, which is Rayleigh fractionation. At low values of f the enrichment of incompatible elements and the depletion of compatible elements are not as extreme during *in situ* fractional crystallization as in Rayleigh fractionation.

Contamination   *(a) AFC processes* It was Bowen (1928) who first proposed that wallrock assimilation is driven by the latent heat of crystallization during fractional crystallization (see Section 3.3.1). DePaolo (1981b) and Powell (1984) have derived equations which describe the concentration of a trace element in a melt relative to the original magma composition in terms of AFC processes. It is worthy of note, however, that AFC processes are difficult to recognize and require a strong contrast in trace element concentrations between magma and wallrock before they can be detected (Powell, 1984). The equation for constant values of r and D is

$$C_L/C_0 = f' + \frac{r}{(r-1+D)} \cdot \frac{C_A}{C_0}(1-f') \qquad [4.22]$$

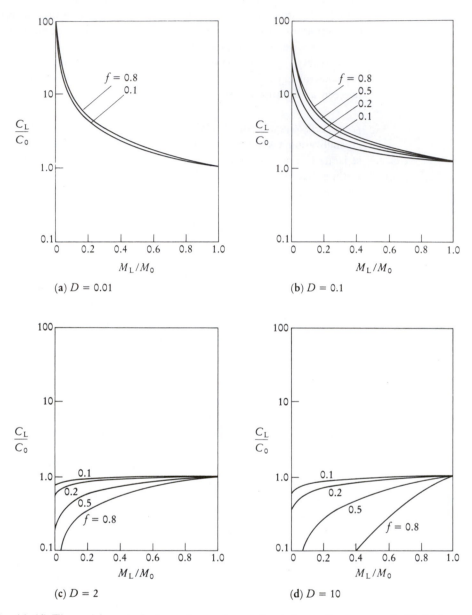

Figure 4.14 (a)–(d) The enrichment of a trace element in a melt relative to the parent melt (C_L/C_0) as a function of the fraction of remaining melt (M_L/M_0 — the ratio of the mass of the liquid to the mass of the magma chamber), bulk partition coefficient ($D = 0.01, 0.1, 2, 10$) and the fraction of magma in the solidification zone which is returned to the magma chamber ($f = 0.1, 0.2, 0.5, 0.8$) during *in situ* crystallization.

where r is the ratio of the assimilation rate to the fractional crystallization rate, C_A is the concentration of the trace element in the assimilated wallrock and f' is described by the relation

$$f' = F^{-(r-1+D)/(r-1)}$$

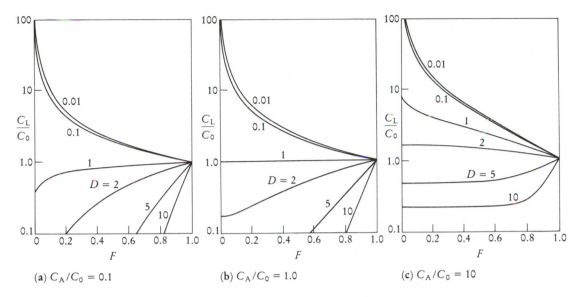

(a) $C_A/C_0 = 0.1$ (b) $C_A/C_0 = 1.0$ (c) $C_A/C_0 = 10$

Figure 4.15   (a)–(c) The enrichment of a trace element in a melt relative to the parental melt (C_L/C_0) as a function of the fraction of the remaining liquid (F) and the ratio of the concentration of the trace element in the assimilated country rock to the concentration in the parental liquid ($C_A/C_0 = 0.1$, 1, 10) and the bulk partition coefficient ($D = 0.01$, 0.1, 1, 2, 5, 10) during assimilation and fractional crystallization (AFC). The calculated curves are for the case where the ratio of the rate of assimilation to fractional crystallization (r) is 0.2.

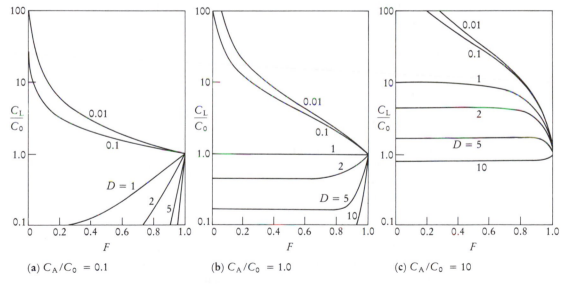

(a) $C_A/C_0 = 0.1$ (b) $C_A/C_0 = 1.0$ (c) $C_A/C_0 = 10$

Figure 4.16   (a)–(c) The enrichment of a trace element in a melt relative to the parental melt (C_L/C_0) as a function of the fraction of the remaining liquid (F) and the ratio of the concentration of the trace element in the assimilated country rock to the concentration in the parental liquid (C_A/C_0) and the bulk partition coefficient (D) during assimilation and fractional crystallization (AFC). The calculated curves are for the case where the ratio of the rate of assimilation to fractional crystallization (r) is 0.8.

where F is the fraction of magma remaining and D is the bulk distribution coefficient.

Figures 4.15 and 4.16 depict the enrichment of a trace element relative to its concentration in the parental magma with varying amounts of remaining melt for two different rates of assimilation relative to fractional crystallization ($r = 0.2$ and 0.8) and three different concentrations of the trace element in the assimilant relative to the parental magma (C_A/C_O), and different values of D.

Where the rate of assimilation to fractional crystallization is small ($r = 0.2$) incompatible elements behave in a similar manner to Rayleigh fractionation. Depletion in compatible elements is less dramatic, particularly when the concentration of the trace element in the assimilant is higher than in the primary magma (Figure 4.15c), although for very compatible elements concentrations level off after a small degree of fractionation.

Where the rate of assimilation is high ($r = 0.8$) and the concentration of the trace element relative to the parental magma is small (Figure 4.16a), incompatible elements are enriched and there is some separation between incompatible and strongly incompatible elements. Compatible elements are strongly depleted. As the trace element concentration in the assimilant increases relative to the parental melt, enrichment increases and even compatible elements are enriched (Figure 4.16c).

(b) Zone refining   In the section above on partial melting, it was suggested that batch melting and Rayleigh melting represent end-member models of natural melting processes. In reality it is likely that melts are neither instantaneously removed from the source nor do they remain totally immobile in their source. Rather they migrate at a finite rate and continuously react with the matrix through which they pass (Richter, 1986). During its passage through unmelted matrix it is possible that a melt will become further enriched in trace elements.

One possible mechanism for such a process is that of zone refining (Harris, 1974), analogous to a metallurgical industrial process, in which superheated magma consumes several times its own volume during melt migration and thus becomes enriched in incompatible elements. The equation for the enrichment of a trace element by zone refining is

$$C_L/C_0 = \frac{1}{D} - [\frac{1}{D} - 1]e^{-nD} \qquad [4.23]$$

where n is the number of equivalent rock volumes that have reacted with the liquid. Where n is very large the right-hand side reduces to $1/D$. The extent to which zone refining occurs in nature is the subject of debate, however, for a continual supply of superheated liquid is unlikely to occur on a large scale.

The numerical effects of zone refining are illustrated in Figure 4.17, which shows the enrichment of a trace element relative to its original concentration for the number of rock volumes consumed (n). For a compatible element with $D = 2.0$ the maximum enrichment ($1/D$) is reached when n is about 200, but for incompatible elements (0.1–0.01) the maximum enrichment is not reached even after 1000 rock volumes.

Dynamic models   *(a) Dynamic melting*   To improve upon earlier 'static' models of partial melting, Langmuir *et al.* (1977) proposed a model of partial melting in which a number of melting processes take place continuously and simultaneously. The main feature of

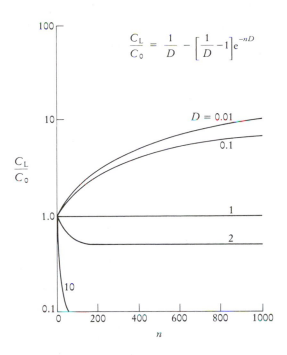

$$\frac{C_L}{C_0} = \frac{1}{D} - \left[\frac{1}{D} - 1\right]e^{-nD}$$

Figure 4.17   The enrichment of a trace element in a melt relative to the source (C_L/C_0) as a function of the number of rock volumes (n) and bulk partition coefficient (D) during zone refining.

the model is that there is incomplete melt extraction from the source which means that the source, always contains a mixture of melt and residue. Langmuir *et al.* (1977) used this approach to explain the variability of REE patterns from the Famous region of the Mid-Atlantic Ridge.

(b) The RTF magma chamber   In an attempt to view magma chamber processes in a more dynamic way O'Hara (1977) and O'Hara and Matthews (1981) proposed a model to describe the behaviour of trace elements in a periodically Replenished, periodically Tapped, continuously Fractionated magma chamber (abbreviated to RTF). They proposed that the life of a magma chamber comprises a series of cycles each of which has four stages — fractional crystallization, magma eruption, wallrock contamination and replenishment. The concentration of a trace element in a steady-state liquid produced after a large number of cycles relative to the concentration in the replenishing magma batch ($_{SS}C_B/C_0$) is given by the expression

$$_{SS}C_B/C_0 = \frac{(x + y)(1 - x)^{D-1}}{1 - (1 - x - y)(1 - x)} \qquad [4.24]$$

where x is the mass fraction of the liquid crystallized in each cycle, y is the mass fraction of the liquid escaping in each cycle and D is the bulk distribution coefficient, and where x, y and D do not vary in the life of the magma chamber.

The degree of enrichment $_{SS}C_B/C_0$ is plotted against x in Figure 4.18 for $y = 0.1$ and for several different values of D.

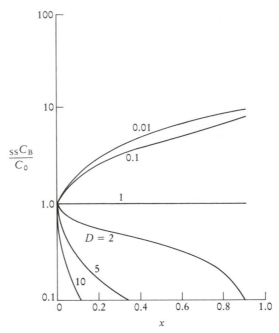

Figure 4.18   The enrichment of a trace element in a steady-state liquid relative to the concentration in the replenishing magma batch $_{ss}C_B/C_0$ as a function of the mass fraction of the liquid crystallized in each cycle (x) and the bulk partition coefficient (D) after a large number of cycles in a RTF magma chamber. The curves shown here are for the condition $y = 0.1$, i.e. the mass fraction of the liquid escaping in each cycle is 0.1. At higher values of y the degree of enrichment of incompatible elements is reduced.

In the special case where $x + y = 1$, i.e. when the magma chamber is emptied then

$$_{ss}C_B/C_0 = (1 - x)^{D-1} \qquad [4.25]$$

which is the Rayleigh fractionation equation.

When $D = 0$, i.e. in the case of a totally incompatible trace element,

$$_{ss}C_B/C_0 = 1 + \frac{x}{y} \qquad [4.26]$$

and is a measure of the maximum enrichment attainable.

Hagen and Neumann (1990) proposed that RTF processes are continuous rather than a series of cycles as suggested by O'Hara and Matthews (1981) and their paper usefully includes the FORTRAN 77 code to implement the two models.

Sedimentary processes   Trace element concentrations in sediments result from the competing influences of the provenance, weathering, diagenesis, sediment sorting and the aqueous geochemistry of the individual elements. The highest concentrations of trace elements are found in clay-rich sediments and most geochemical studies have concentrated on these lithologies. Selected trace elements may be used to identify particular geochemical processes and to identify the sedimentary provenance. The most important elements in this respect are the REE, Th, Sc and to a lesser extent Cr and Co. These elements have very low concentrations in sea and river waters, low residence times in the ocean and element ratios which are unaffected by diagenesis and metamorphism. Thus they are transported exclusively in the terrigenous component of a sediment and reflect the chemistry of their source.

Other elements are more soluble. For example, Fe, Mn, Pb and sometimes Cr are mobile during diagenesis. Cs, Rb and Ba are fixed during weathering but Sr is leached. Immobile elements such as Zr, Hf and Sn may be mechanically distributed according to grain size and may be controlled by the concentration of heavy minerals.

4.3   Rare earth elements (REE)

The rare earth elements (REE) are the most useful of all trace elements and REE studies have important applications in igneous, sedimentary and metamorphic petrology. The REE comprise the series of metals with atomic numbers 57 to 71 — La to Lu (Table 4.4). In addition, the element Y with an ionic radius similar to that of the REE Ho is sometimes included. Typically the low-atomic-number members of the series are termed the light rare earths (LREE), those with the higher atomic

Table 4.4   The rare earth elements

Atomic number	Name	Symbol	Ionic radius for eight-fold coordination *	
57	Lanthanum	La	La^{3+}	1.160
58	Cerium	Ce	Ce^{3+}	1.143
			Ce^{4+}	0.970
59	Praesodymium	Pr	Pr^{3+}	1.126
60	Neodymium	Nd	Nd^{3+}	1.109
61	Promethium	Pm	Not naturally occurring	
62	Samarium	Sm	Sm^{3+}	1.079
63	Europium	Eu	Eu^{3+}	1.066
			Eu^{2+}	1.250
64	Gadolinium	Gd	Gd^{3+}	1.053
65	Terbium	Tb	Tb^{3+}	1.040
66	Dysprosium	Dy	Dy^{3+}	1.027
67	Holmium	Ho	Ho^{3+}	1.015
68	Erbium	Er	Er^{3+}	1.004
69	Thulium	Tm	Tm^{3+}	0.994
70	Ytterbium	Yb	Yb^{3+}	0.985
71	Lutetium	Lu	Lu^{3+}	0.977
39	Yttrium	Y	Y^{3+}	1.019

* From Shannon (1976), in Ångstroms (10^{-10} m).

numbers the heavy rare earths (HREE) and less commonly the middle members of the group, Sm to Ho, are known as the middle REE (MREE).

4.3.1 The chemistry of the REE

The REE all have very similar chemical and physical properties. This arises from the fact that they all form stable 3+ ions of similar size. Such differences as there are in chemical behaviour are a consequence of the small but steady decrease in ionic size with increasing atomic number. This is illustrated for ions in eight-fold coordination state in Table 4.4. These small differences in size and behaviour are exploited by a number of petrological processes causing the REE series to become fractionated relative to each other. It is this phenomenon which is used in geochemistry to probe into the genesis of rock suites and unravel petrological processes.

A small number of the REE also exist in oxidation states other than 3+ but the only ions of geological importance are Ce^{4+} and Eu^{2+}. These form a smaller and a larger ion respectively, relative to the 3+ oxidation state.

Table 4.5 Chondrite values used in normalizing REE (concentrations in ppm)

	Wakita	Haskin	Masuda	Nakamura	Evensen	Boynton	T & M	Primitive mantle value
Analytical method	NAA	NAA	IDMS	IDMS	IDMS	IDMS	IDMS	
Chondrite(s) analysed: (Ref.)	Composite (1)	Composite (2)	Leedey (3)	Composite (4)	Avg.CI (5)	Avg.CI (6)	Avg. CI (7)	(8)
La	0.340	0.330	0.3780	0.3290	0.244 60	0.3100	0.3670	0.7080
Ce	0.910	0.880	0.9760	0.8650	0.637 90	0.8080	0.9570	1.8330
Pr	0.121	0.112			0.096 37	0.1220	0.1370	0.2780
Nd	0.640	0.600	0.7160	0.6300	0.473 80	0.6000	0.7110	1.3660
Sm	0.195	0.181	0.2300	0.2030	0.154 00	0.1950	0.2310	0.4440
Eu	0.073	0.069	0.0866	0.0770	0.058 02	0.0735	0.0870	0.1680
Gd	0.260	0.249	0.3110	0.2760	0.204 30	0.2590	0.3060	0.5950
Tb	0.047	0.047			0.037 45	0.0474	0.0580	0.1080
Dy	0.300		0.3900	0.3430	0.254 10	0.3220	0.3810	0.7370
Ho	0.078	0.070			0.056 70	0.0718	0.0851	0.1630
Er	0.200	0.200	0.2550	0.2250	0.166 00	0.2100	0.2490	0.4790
Tm	0.032	0.030			0.025 61	0.0324	0.0356	0.0740
Yb	0.220	0.200	0.2490	0.2200	0.165 10	0.2090	0.2480	0.4800
Lu	0.034	0.034	0.0387	0.0339	0.025 39	0.0322	0.0381	0.0737
Y							2.1000	

(1) Wakita *et al.* (1971): composite of 12 chondrites.
(2) Haskin *et al.* (1968): composite of nine chondrites.
(3) Masuda *et al.* (1973): Leedey chondrite.
(4) Nakamura (1974).
(5) Evensen *et al.* (1978): average of C1 chondrites.
(6) Boynton (1984).
(7) Taylor and McLennan (1985): 1.5 x values of Evensen [column (5)].
(8) McDonough *et al.* (1991).

4.3.2 Presenting REE data

Rare earth element concentrations in rocks are usually normalized to a common reference standard, which most commonly comprises the values for chondritic meteorites. Chondritic meteorites were chosen because they are thought to be relatively unfractionated samples of the solar system dating from the original nucleosynthesis. However, the concentrations of the REE in the solar system are very variable because of the different stabilities of the atomic nuclei. REE with even atomic numbers are more stable (and therefore more abundant) than REE with odd atomic numbers, producing a zig-zag pattern on a composition–abundance diagram (Figure 4.19). This pattern of abundances is also found in natural samples. Chondritic normalization therefore has two important functions. Firstly it eliminates the abundance variation between odd and even atomic number elements and secondly it allows any fractionation of the REE group relative to chondritic meteorites to be identified. Normalized values and ratios of normalized values are denoted with the subscript N — hence for example Ce_N, $(La/Ce)_N$.

The REE are normally presented on a concentration vs atomic number diagram on which concentrations are normalized to the chondritic reference value, expressed as the logarithm to the base 10 of the value. Concentrations at individual points on the graph are joined by straight lines (Figure 4.20). This is sometimes referred to as the Masuda–Coryell diagram after the original proponents of the diagram (Masuda, 1962; Coryell et al., 1963). Trends on REE diagrams are usually referred to as REE 'patterns' and the shape of an REE pattern is of considerable petrological interest.

Not infrequently the plotted position of Eu lies off the general trend defined by the other elements on an REE diagram (Figure 4.20) and may define a **europium anomaly**. If the plotted composition lies above the general trend then the anomaly is described as positive and if it lies below the trend then the anomaly is said to be negative. Europium anomalies may be quantified by comparing the measured concentration (Eu) with an expected concentration obtained by interpolating between the normalized values of Sm and Gd (Eu*). Thus the ratio Eu/Eu* is a measure of the europium anomaly and a value of greater than 1.0 indicates a positive anomaly whilst a value of less than 1.0 is a negative anomaly. Taylor and McLennan (1985) recommend using the geometric mean; in this case $Eu/Eu^* = Eu_N/\sqrt{[(Sm_N).(Gd_N)]}$.

(a) Difficulties with chondrite normalization   Unfortunately it has become apparent that chondritic meteorites are actually quite variable in composition and 'chondrites with "chondritic" REE abundances are the exception rather than the rule' (Boynton, 1984). This variability in chondritic composition has given rise to a large number of sets of normalizing vales for the REE (Table 4.5) and to date no standardized value has been adopted. The variability may be reduced to two factors — the analytical method and the precise type of chondrites analysed. Some authors use 'average chondrite' whilst others selected C1 chondrites as the most representative of the composition of the original solar nebula.

(b) Choosing a set of normalizing values   Figure 4.20 shows flat rare earth patterns typical of an Archaean tholeiite normalized to the range of chondritic values listed in Table 4.5. The patterns show both variety in shape and in concentration range. The consensus seems to favour values based upon average chondrite rather than C1

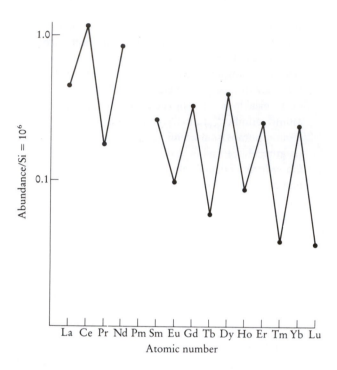

Figure 4.19   Rare earth element abundances (log scale) in the solar system plotted against atomic number. Data from Anders and Ebihara (1982), normalized to Si = 10^6 atoms. Elements with even atomic numbers have higher abundances than those with odd atomic numbers.

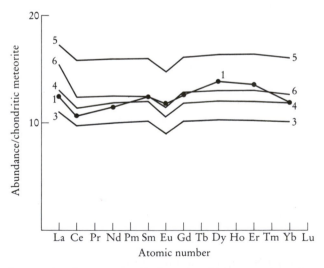

Figure 4.20   Rare earth element abundances normalized to chondritic meteorite values plotted against atomic number for an Archaean tholeiitic basalt (sample 11 — Rollinson, 1983). The same sample has been normalized using five different sets of values; the numbers refer to the columns of normalizing values listed in Table 4.5.

chondrites and either the values of Boynton (1984) based upon Evensen *et al.* (1978), or the values of Nakamura (1974), with additions from Haskin *et al.* (1968), seem to be satisfactory. In fact the two sets of values are very similar and lie in the middle of the range of values currently in use.

REE ratio diagrams

The degree of fractionation of a REE pattern can be expressed by the concentration of a light REE (La or Ce) ratioed to the concentration of a heavy REE (Yb or Y). Both elements are chondrite-normalized. The ratio $(La/Yb)_N$ is often plotted against either Ce_N or Yb_N on a bivariate graph and is a measure of the degree of REE fractionation with changing REE content. Similar diagrams may be constructed to measure the degree of light REE fractionation [$(La/Sm)_N$ vs Sm_N], heavy REE fractionation [$(Gd/Yb)_N$ vs Yb_N] and Eu anomaly [$(La/Sm)_N$ vs (Eu/Eu*)] in individual REE patterns.

NASC normal-ization for sediments

It has been observed that the concentration of many elements in fine-grained sedimentary rocks in continental platforms around the world is similar as a consequence of mixing through repeated cycles of erosion. This 'average sediment' is often used as the normalizing value for REE concentrations in sedimentary rocks. A frequently used composition is that of the North American Shale Composite (NASC) and the recommended values of Gromet *et al.* (1984) are given in Table 4.6 (column 5). Alternatives to NASC in current use are a composite European shale (Haskin and Haskin, 1966) and the post Archaean average Australian sedimentary rock (McLennan, 1989). Some authors have taken the average abundance of REE in sedimentary rocks as a measure of the REE content of the upper continental crust. This assumes that sedimentary processes homogenize the REE previously fractionated during the formation of igneous rocks. Thus an alternative to shale normalization is to use values for average upper continental crust (Table 4.6, column 8).

Relative to chondritic meteorites, NASC has about 100 times the light REE and about 10 times the heavy REE content and a small negative Eu anomaly (Figure 4.21). Normalization against NASC is a measure of how typical a sediment is, and may identify subtle enrichments and deficiencies in certain elements.

Rock normalization

Less commonly some authors normalize REE concentrations to a particular sample in a rock suite as a measure of relative change. This is also useful when the REE concentrations of the individual minerals in the rock have also been determined, for then they can be expressed relative to the concentration in the whole rock. A similar form of normalization is to express the concentration in a mineral relative to the composition of the groundmass; this is frequently used to display mineral/melt partition coefficients (Section 4.2.1).

4.3.3   Interpreting REE patterns

The REE are regarded as amongst the least soluble trace elements and are relatively immobile during low-grade metamorphism, weathering and hydrothermal altera-tion. Michard (1989), for example, shows that hydrothermal solutions have between 5×10^2 and 10^6 times less REE than the reservoir rock through which they have passed and therefore hydrothermal activity is not expected to have a major effect on

Table 4.6   Standard sedimentary compositions used for normalizing the REE concentrations in sedimentary rocks (concentrations in ppm)

(Ref.)	NASC (1)	NASC (2)	NASC (3)	NASC (4)	NASC (5)	ES (6)	PAAS (7)	Upper crust (8)	Typical seawater (9)	River water (10) $(10^{-12}\ mol\ kg^{-1})$
La	39.000	32.000	32.000	31.100	31.100	41.100	38.200	30.000	20.800	425.0
Ce	76.000	70.000	73.000	66.700	67.033	81.300	79.600	64.000	9.640	601.0
Pr	10.300	7.900	7.900			10.400	8.830	7.100		
Nd	37.000	31.000	33.000	27.400	30.400	40.100	33.900	26.000	21.100	365.0
Sm	7.000	5.700	5.700	5.590	5.980	7.300	5.550	4.500	4.320	80.4
Eu	2.000	1.240	1.240	1.180	1.253	1.520	1.080	0.880	0.823	20.7
Gd	6.100	5.210	5.200		5.500	6.030	4.660	3.800	5.200	83.5
Tb	1.300	0.850	0.850	0.850	0.850	1.050	0.774	0.640		
Dy					5.540		4.680	3.500	5.610	97.9
Ho	1.400	1.040	1.040			1.200	0.991	0.800		
Er	4.000	3.400	3.400		3.275	3.550	2.850	2.300	4.940	64.6
Tm	0.580	0.500	0.500			0.560	0.405	0.330		
Yb	3.400	3.100	3.100	3.060	3.113	3.290	2.820	2.200	4.660	51.7
Lu	0.600	0.480	0.480	0.456	0.456	0.580	0.433	0.320		
Y		27.000				31.800	27.000	22.000		

(1) North American shale composite (Haskin and Frey, 1966).
(2) North American shale composite (Haskin and Haskin, 1966).
(3) North American shale composite (Haskin *et al.*, 1968).
(4) North American shale composite (Gromet *et al.*, 1984) — INAA.
(5) North American shale composite (Gromet *et al.*, 1984) — recommended.
(6) Average European shale (Haskin and Haskin, 1966).
(7) Post-Archaean average Australian sedimentary rock (McLennan, 1989).
(8) Average upper continental crust (Taylor and McLennan, 1981).
(9) Elderfield and Greaves (1982), Table 1, 900 m sample.
(10) Hoyle *et al.* (1984): River Luce, Scotland; 0.7 μm filter.

rock chemistry unless the water/rock ratio is very high. However, the REE are not totally immobile, as is emphasized in the review by Humphries (1984), and the reader should be cautious in interpreting the REE patterns of heavily altered or highly metamorphosed rocks. Nevertheless REE patterns, even in slightly altered rocks, can faithfully represent the original composition of the unaltered parent and a fair degree of confidence can be placed in the significance of peaks and troughs and the slope of an REE pattern.

REE patterns in igneous rocks   The REE pattern of an igneous rock is controlled by the REE chemistry of its source and the crystal–melt equilibria which have taken place during its evolution. Here we describe in a qualitative manner the way in which the roles of individual minerals may be identified during magmatic evolution, either during partial melting in the source region or in subsequent crystal fractionation. In Section 4.9 the quantitative aspects of this approach are considered and applied to trace elements in general. The reasoning here is based upon the partition coefficients for the REE in the major rock-forming minerals listed in Tables 4.1 to 4.3 and depicted in Figures 4.8 to 4.10.

Europium anomalies are chiefly controlled by feldspars, particularly in felsic magmas, for Eu (present in the divalent state) is compatible in plagiocase and

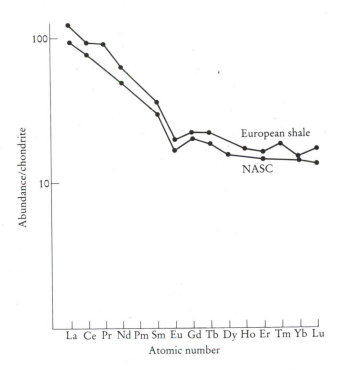

Figure 4.21   Rare earth element abundances in North American Shale Composite (NASC) and European shale, normalized to chondritic values. Data from Table 4.7, columns 5 and 6; normalizing values from Table 4.5, column 4.

potassium feldspar, in contrast to the trivalent REE which are incompatible. Thus the removal of feldspar from a felsic melt by crystal fractionation or the partial melting of a rock in which feldspar is retained in the source will give rise to a negative Eu anomaly in the melt. To a lesser extent hornblende, sphene, clinopyroxene, orthopyroxene and garnet may also contribute to a Eu anomaly in felsic melts, although in the opposite sense to that of the feldspars.

Enrichment in the middle REE relative to the light and heavy REE is chiefly controlled by hornblende. This is evident from the partition coefficients plotted in Figure 4.4. The REE are compatible in hornblende in felsic and intermediate liquids and the highest partition coefficients are between Dy and Er. Such large partition coefficients mean that even a moderate amount of hornblende (20–30 %) may dominate the bulk partition coefficient for this range of elements and influence the shape of the REE pattern. The same effect can also be observed with clinopyroxene, although the partition coefficients are not so high. Sphene also may also affect an REE pattern in a similar way although, because it is present usually in low concentrations, the effect may be masked by other phases.

Fractionation of the light REE relative to the heavy ones may be caused by the presence of olivine, orthopyroxene and clinopyroxene, for the partition coefficients increase by an order of magnitude from La to Lu in these minerals. In basaltic and andesitic liquids, however, the REE are all incompatible in each of these minerals and are only slightly fractionated.

Extreme depletion of the heavy REE relative to the light is most likely to indicate the presence of garnet in the source, for there is a large variation in the partition coefficients of the REE. In basaltic liquids the partition coefficient for Lu is more than 1000 times greater than that for La. The effect is less extreme, although still large, in felsic liquids. Hornblende in felsic liquids may also account for extreme enrichment of light REE relative to heavy, although the range of partition coeffients is not as great as in the case of garnet.

In felsic liquids accessory phases such as sphene, zircon, allanite, apatite and monazite may strongly influence an REE pattern for although they may be present in only small quantities (often much less than 1 % of the rock) their very high partition coefficients mean that they have a disproportionate influence on the REE pattern. Zircon will have an effect similar to that of garnet and will deplete in the heavy REE; sphene and apatite partition the middle REE relative to the light and heavy, and monazite and allanite cause depletion in the light REE.

REE patterns in sea and river water

The aqueous geochemistry of the REE is a function of the type of complexes that the REE may form, the length of time the REE remain in solution in the oceans (their residence time), and to a lesser extent the oxidizing potential of the water. The topic is well reviewed by Brookins (1989). The REE contents of rivers and seawater are extremely low (Table 4.6), for they are chiefly transported as particulate material. When normalized to a shale composite (Section 4.3.2), REE concentrations in seawater are between six and seven orders of magnitude smaller that the shale value. River waters are about an order of magnitude higher.

The REE in ocean waters provide information about oceanic input from rivers, hydrothermal vents and from aeolian sources (Elderfield, 1988). On a shale-normalized plot (Figure 4.22) seawater tends to show a gradual enrichment in REE concentrations from the light to heavy REE and often shows a prominent negative Ce anomaly (Elderfield and Greaves, 1982). This anomaly is expressed as Ce/Ce^* where Ce^* is an interpolated value for Ce based upon the concentrations of La and Pr or La and Nd. The Ce anomaly occurs in response to the oxidation of Ce^{3+} to Ce^{4+} and the precipitation of Ce^{4+} from solution as CeO_2. Eu anomalies in seawater reflect either aeolian or hydrothermal input. River water also shows a small negative Ce anomaly and an increase in REE concentrations from the light to heavy REE (Hoyle *et al.*, 1984) similar to that observed in seawater (Figure 4.22).

REE patterns in sediments

REE concentrations in sedimentary rocks are usually normalized to a sedimentary standard such as NASC, although this practice is not universal and some authors use chondritic normalization.

(a) Clastic sediments The single most important factor contributing to the REE content of a clastic sediment is its provenance (Fleet, 1984; McLennan, 1989). This is because the REE are insoluble and present in very low concentrations in sea and river water; thus the REE present in a sediment are chiefly transported as particulate matter and reflect the chemistry of their source. In comparison, the effects of weathering and diagenesis are minor. Studies such as those by Nesbitt (1979) show that whilst the REE are mobilized during weathering, they are reprecipitated at the site of weathering. A more recent study shows, however, that in the case of extreme weathering the degree of weathering of the source can be recognized in the REE chemistry of the derivative sediment (Nesbitt *et al.*, 1990).

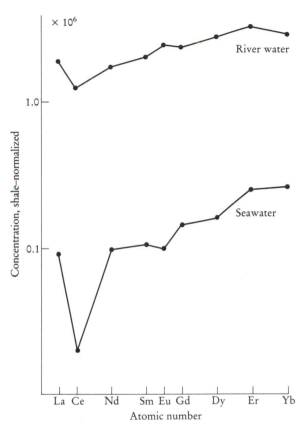

Figure 4.22   Concentrations of rare earth elements in average seawater and average river water normalized
to NASC. The data are given in Table 4.6, columns 9 and 10. Normalizing values are from
Table 4.6 column 5. Note that the concentrations in seawater and river water are quoted in
mol kg^{-1} and must in this case be multiplied by the atomic weight. Concentrations shown are
$\times 10^{-6}$.

Diagenesis has little influence on the redistribution of the REE, for very large
water/rock ratios are required to effect any change in sediment chemistry.

An important study by Cullers *et al.* (1987), on the effect of sedimentary
sorting on REE concentrations, found that the REE pattern of the source was most
faithfully represented in the clay-sized fraction of the sediment. Clay-bearing rocks
also have a much higher concentration of total REE than other sediments. It is for
this reason that many authors have used the REE content of the clay portion of a
sediment or clay-rich sediments in order to establish the sedimentary processes and
to identify the provenance. The presence of quartz has a diluting effect on REE
concentrations, as does carbonate. The presence of heavy minerals, particularly
zircon, monazite and allanite, may have a significant but erratic effect on the REE
pattern of an individual sample.

(b) Chemical sediments Chemical sediments are most likely to reflect the
composition of the seawater from which they were precipitated. This is seen in
ferromanganese nodules which show REE patterns that are the inverse of a typical

seawater pattern, i.e. they are enriched in light relative to heavy REE and show a positive Ce anomaly. This is not a universal feature of ferromanganese nodules, however, for the composition of chemical sediments also reflects local redox conditions and is strongly influenced by post-depositional changes (Elderfield and Greaves, 1981).

4.4   Normalized multi-element diagrams or incompatible element diagrams (spider diagrams)

Normalized multi-element diagrams are based upon a grouping of elements incompatible with respect to a typical mantle mineralogy. They are an extension of the more familiar chondrite-normalized REE diagrams in which other trace elements are added to the traditional REE diagram. They are a particularly useful way of depicting basalt chemistry although their use has been extended to all igneous and some sedimentary rocks. Mantle values or those of chondritic meteorites are used for normalization and they measure deviations from a primitive composition. The terms 'mantle (or chondrite)-normalized multi-element diagram' or 'incompatible element diagram' do not roll off the tongue with ease and the more colloquial **'spider diagram'** (or 'spidergram' for an individual pattern) is used here.

4.4.1   Multi-element diagrams for igneous rocks

There are three popular ways of normalizing trace element data for presentation as a spider diagram. These include an estimated primitive mantle composition and chondritic meteorites — two 'views' of the primitive undifferentiated earth. Others normalize their data to primitive MORB. Each version of the spider diagram has a slightly different array of elements with a slightly different order. In detail there are innumerable variations on each particular theme, usually dictated by the number of trace elements and the quality of their determinations in a particular data-set. This state of affairs is not satisfactory and some standardization is desirable. First, however, we consider the present 'state of the art'.

Primordial (primitive) mantle-normalized spider diagrams

The primitive mantle is the composition of the mantle before the continental crust formed. One of the most frequently used estimates of its composition is that of Wood *et al.* (1979a), who employed it as a means of comparing compositional variations between basic lavas. Nineteen elements are arranged in order of increasing compatibility with respect to a small percentage melt of the mantle. The values are given in Table 4.7, column 1. Element concentrations are plotted on a logarithmic scale (Figure 4.23) and average N-type MORB plots as a relatively smooth curve, depleted at the most incompatible end (Figure 4.23b).

There are a number of variations in the list of elements plotted. The most common alternative is a 13-element plot reflecting those elements whose concentrations are relatively high in basic igneous rocks and which are readily analysed by X-ray fluorescence. Normalizing values currently in use are given in Table 4.7, columns 1 to 5. Of these five sets of values, those of McDonough *et al.* (1992) (a slight revision of Sun and McDonough, 1989) — Table 4.7, column 4 — are becoming increasingly popular.

Chondrite-normalized spider diagrams Thompson (1982) proposed that normalization to chondritic values may be preferable to the primordial mantle composition since chondritic values are directly measured rather than estimated. The order of elements (Figure 4.24) is slightly different from that of Wood *et al.* (1979a) and is to some extent arbitrary but was chosen to give the smoothest overall fit to data for Icelandic lavas and North Atlantic ocean-floor basalts; it approximates to one of increasing compatibility from left to right. As a rule of thumb, concentrations below about ten times chondrite

Table 4.7 Normalizing values (in ppm) used in the calculation of 'spider diagrams' and listed in their plotting order

	Primordial mantle					Chondrite							MORB normalization			
(Ref.)	(1)	(2)	(3)	(4)	(5)		(6)		(7)		(8)	(9)		(10)		(11)
Cs	0.019	0.017		0.023	0.018					Cs	0.012	0.188				
Rb	0.860	0.660	0.810	0.635	0.550	Ba	6.900	Rb	1.880	Pb	0.120	2.470	Sr	120	Rb	1.00
Ba	7.560		6.900	6.990	5.100	Rb	0.350	K	850	Rb	0.350	2.320	K_2O (%)	0.15	Ba	12.00
Th	0.096		0.094	0.084	0.064	Th	0.042	Th	0.040	Ba	3.800	2.410	Rb	2.00	K_2O (%)	0.15
U	0.027		0.026	0.021	0.018	K	120	Ta	0.022	Th	0.050	0.029	Ba	20.00	Th	0.20
K	252.0	230.0	260.0	240.0	180	Nb	0.350	Nb	0.560	U	0.013	0.008	Th	0.20	Ta	0.17
Ta	0.043		0.040	0.041	0.040	Ta	0.020	Ba	3.600	Ta	0.020	0.014	Ta	0.18	Sr	136
Nb	0.620		0.900	0.713	0.560	La	0.329	La	0.328	Nb	0.350	0.246	Nb	3.50	La	3.00
La	0.710		0.630	0.708	0.551	Ce	0.865	Ce	0.865	K	120	545	Ce	10.00	Ce	10.00
Ce	1.900			1.833	1.436	Sr	11.800	Sr	10.500	La	0.315	0.237	P_2O_5 (%)	0.12	Nb	2.50
Sr	23.000		28.000	21.100	17.800	Nd	0.630	Hf	0.190	Ce	0.813	0.612	Zr	90.00	Nd	8.00
Nd	1.290			1.366	1.067	P	46.000	Zr	9.000	Sr	11.000	7.260	Hf	2.40	P_2O_5 (%)	0.12
P	90.400	92.000				Sm	0.203	Ti	610	Nd	0.597	0.467	Sm	3.30	Hf	2.50
Hf	0.350		0.350	0.309	0.270	Zr	6.840	Sm	0.203	P	46.000	1220	TiO_2 (%)	1.50	Zr	88.00
Zr	11.000		11.000	11.200	8.300	Hf	0.200	Y	2.000	Sm	0.192	0.153	Y	30.00	Eu	1.20
Sm	0.385		0.380	0.444	0.347	Ti	620	Lu	0.034	Zr	5.600	3.870	Yb	3.40	TiO_2 (%)	1.50
Ti	1200	1300	1300	1280	960	Tb	0.052	Sc	5.210	Ti	620	445	Sc	40.00	Tb	0.71
Tb	0.099			0.108	0.087	Y	2.000	V	49.000	Y	2.000	1.570	Cr	250.0	Y	35.00
Y	4.870		4.600	4.550	3.400	Tm	0.034	Mn	1720						Yb	3.50
Pb				0.071		Yb	0.220	Fe	265000						Ni	138
								Cr	2300						Cr	290
								Co	470							
								Ni	9500							

(1) Wood *et al.* (1979a); Ti from Wood *et al.* (1981).
(2) Sun (1980); Cs 0.017–0.008.
(3) Jagoutz *et al.* (1979).
(4) McDonough *et al.* (1992).
(5) Taylor and McLennan (1985).
(6) Thompson (1982); alternative value for Ba=3.85 (Hawkesworth *et al.*, 1984); Rb,K,P from primitive mantle values of Sun (1980).
(7) Wood *et al.* (1979b).
(8) Sun (1980): chondrite and undepleted mantle data.
(9) Sun and McDonough (1989): C1 chondrite.
(10) Pearce (1983); Sc and Cr from Pearce (1982).
(11) Bevins *et al.* (1984).

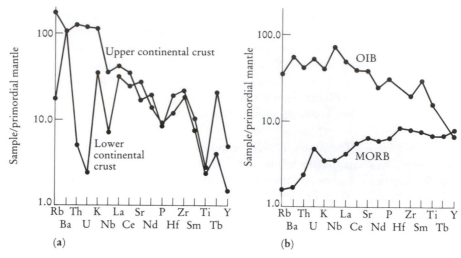

Figure 4.23   Trace element concentrations normalized to the composition of the primordial mantle and plotted from left to right in order of increasing compatibility in a small fraction melt of the mantle. The normalizing values are those of McDonough *et al.* (1992) — Table 4.7, column 4. (a) Upper and lower continental crust from Weaver and Tarney (1984) — data in Table 4.8; (b) Average N-type MORB from Saunders and Tarney (1984) and OIB from Sun (1980) — data in Table 4.8.

Table 4.8   Trace element data used in spider diagrams (Figures 4.23, 4.24, 4.26)

	Upper crust	Lower crust	Average crust	MORB	OIB
(Ref.)	(1)	(2)	(3)	(4)	(5)
Rb	110.00	11.00	61.00	1.00	22.00
Ba	700.00	757.00	707.00	12.00	380.00
Th	10.50	0.42	5.70	0.20	3.40
U	2.50	0.05	1.30	0.10	1.10
K	27393	8301	17430	830	9600
Nb	25.00	5.00	13.00	2.50	53.00
Ta				0.16	3.00
La	30.00	22.00	28.00	3.00	35.00
Ce	64.00	44.00	57.00	10.00	72.00
Sr	350.00	569.00	503.00	136.00	800.00
Nd	26.00	18.50	23.00	8.00	35.00
P	742.22	785.88	829.54	570.00	2760.00
Hf	5.80	3.60	4.70	2.50	
Zr	240.00	202.00	210.00	88.00	220.00
Sm	4.50	3.30	4.10	3.30	13.00
Ti	3597.00	2997.50	3597.00	8400.00	20000.00
Tb	2.20	0.43	0.24	0.71	
Y	22.00	7.00	14.00	35.00	30.00
Tm	0.33	0.19	0.24		
Yb	2.20	1.20	1.53		

(1) Upper continental crust (Taylor and McLennan, 1981).
(2) Lower continental crust (Weaver and Tarney, 1984).
(3) Average continental crust (Weaver and Tarney, 1984).
(4) Average N-type MORB (Saunders and Tarney, 1984; Sun, 1980).
(5) Average OIB (Sun, 1980).

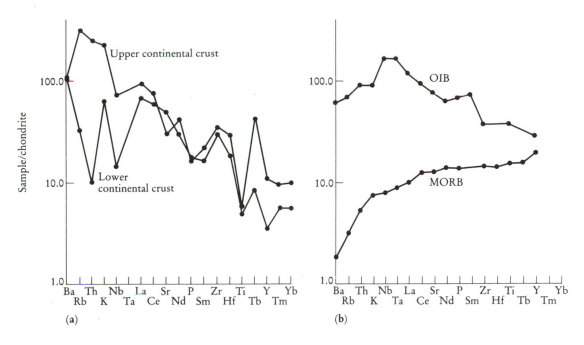

Figure 4.24   Trace element concentrations normalized to the composition of chondritic meteorites and plotted from left to right in order of increasing compatibility in a small fraction melt of the mantle. The normalizing values are those of Thompson (1982) — Table 4.7, column 6. (a) Upper and lower continental crust from Weaver and Tarney (1984) — data in Table 4.8; (b) Average N-type MORB from Saunders and Tarney (1984) and OIB from Sun (1980) — data in Table 4.8.

approach the detection limit for most techniques and so should be treated with caution (Thompson *et al.*, 1983). Thompson *et al.* (1983) point out that when basalts of varying MgO contents are compared on spider diagrams, there may be a very erratic set of patterns arising from the variable effects of fractional crystallization. To avoid this confusion they suggest the recalculation of the normalized data to make $(Yb)_N = 10.0$. This slight oversimplification of the data is justified because Yb values in lavas with the same MgO value vary only by a factor of about two and the net result is a much clearer set of patterns. Normalization values are given in Table 4.7, columns 6 to 9. The more obvious discrepancies between values in this list arise because some authors have used a mixture of chondritic and primordial mantle values.

MORB-normalized spider diagrams   MORB-normalized spidergrams are most appropriate for evolved basalts, andesites and crustal rocks — rocks to which MORB rather than primitive mantle could be parental. This form of spider diagram was proposed by Pearce (1983) and is based upon two parameters (Figure 4.25). Firstly, ionic potential (ionic charge for the element in its normal oxidation state, divided by ionic radius) is used as a measure of the mobility of an element in aqueous fluids. Elements with low (<3) and high (>12) ionic potentials are mobile and those with intermediate values are generally immobile. Secondly, the bulk distribution coefficient for the element between garnet lherzolite and melt is used as a measure of the incompatibility of an element in small-degree partial melts. The elements are ordered so that the most mobile elements (Sr, K, Rb and Ba) are placed at the left of the diagram and in order of increasing incompatibility. The immobile elements are arranged from right to left in order of increasing incompatibility (Figures 4.25 and 4.26). A slightly different version of this diagram is used by Saunders and Tarney (1984), who arrange the elements into a LIL-group (Rb, Ba, K, Th, Sr, La, Ce), followed by an HFS-group (Nb, Ta, Nd, P, Hf, Zr, Eu, Ti, Tb, Y, Yb), followed by the transition metals Ni and Cr. These authors proposed that data should be normalized to Zr = 10.0 in order to eliminate concentration differences arising from low-pressure crystal fractionation.

The normalizing values used by Pearce (1983) are taken from the average MORB of Pearce *et al.* (1981) and are given in Table 4.7, column 10. A longer list of elements arranged in a slightly different order is used by Bevins *et al.* (1984)

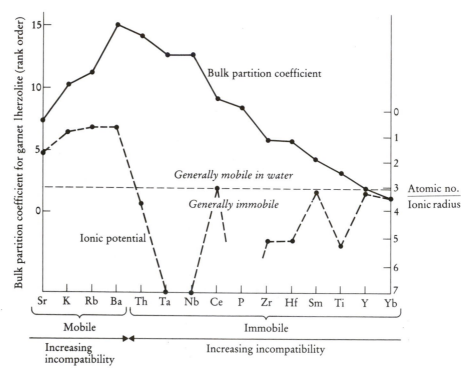

Figure 4.25   The petrogenetic variables that determine the order of elements in the MORB-normalized spider diagram of Pearce (1983). The ionic potential (broken line and right-hand scale) is the ratio of the atomic number to ionic radius and is a measure of the mobility of an element in aqueous fluids. Elements with a low ionic potential (<3, i.e. above the dotted line) are generally mobile in water whereas those with a higher ionic potential are generally immobile. (Note that P has an ionic potential which plots off-scale.) The bulk partition coefficient for an element between garnet lherzolite and a small percentage melt is a measure of incompatibility (left-hand scale). The immobile elements are plotted in rank order starting with the most compatible element (on the right). The mobile elements are ordered in a similar way so that element incompatibility increases from the edge to the centre of the diagram.

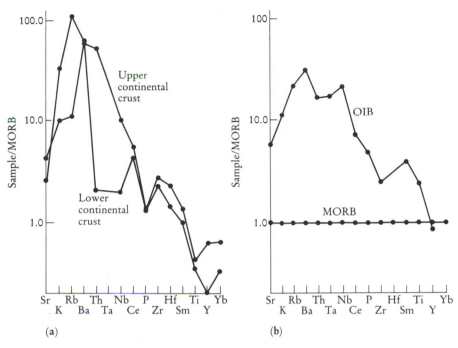

Figure 4.26 Trace element concentrations normalized to the composition of average MORB and plotted as in Figure 4.25. The normalizing values are those of Saunders and Tarney (1984) with additions from Sun (1980) — Table 4.7, column 11. (a) Upper and lower continental crust from Weaver and Tarney (1984) — data in Table 4.8; (b) Average N-type MORB from Saunders and Tarney (1984) and OIB from Sun (1980) — data in Table 4.8.

using values from Pearce *et al.* (1981) and Saunders and Tarney (1984) (Table 4.7, column 11).

Which spider diagrams do we use? There is a pressing need for the standardization of multi-element diagrams. Current practice is chaotic, some authors not even quoting the source of their normalizing values. This has led to a situation where it is impossible to compare one spidergram with another. Rock (1987b) has argued for a set of standard multi-element diagrams with a universal set of consistent normalizing values — not necessarily 'perfect values', but ones which are universally accepted. The price of standardization is that some authors may have to sacrifice some elements of personal preference from their private version of a multi-element diagram. For igneous rocks two multi-element diagrams are sufficient: one to compare rock chemistry with a 'mantle source' and another to compare rock chemistry with 'the most abundant volcanic rock' — MORB. To maintain consistency with REE plots and to avoid a measure of subjectivity implicit in models of the primordial mantle, the chondrite model is recommended and the element order and normalizing values of Thompson (1982) (Table 4.7, column 6) should be adopted. These values utilize the REE data of Nakamura (1974) and thus provide some consistency with REE plots. A condensed version of the diagram is permissible if the full range of trace elements have not been determined, although the reader should be aware of the possibility of induced anomalies resulting from missing elements. The standard diagram and normalizing values for the MORB source should be those of Pearce (1983) (Table 4.7, column 10), since this is the most widely used and the most objectively based (Rock, 1987b).

Interpreting multi-element diagrams for igneous rocks

Multi-element diagrams contain a more heterogeneous mix of trace elements than do REE diagrams. Consequently they often show a greater number of peaks and troughs reflecting the different behaviour of different groups of trace elements. For example, contrast the behaviour of the more mobile LIL elements (Cs, Rb, K, Ba, Sr, Eu) with the less mobile HFS elements (Y, Hf, Zr, Ti, Nb, Ta). On the one hand the LIL element concentrations may be a function of the behaviour of a fluid phase, whilst the HFS element concentrations are controlled by the chemistry of the source and the crystal/melt processes which have taken place during the evolution of the rock. Partition coefficients for the HFS elements in the major rock-forming minerals in a range of melt compositions are summarized in Tables 4.1 to 4.3.

Of the less mobile elements, mineralogical controls on the distribution of the REE have already been discussed. Other elements are often strongly controlled by individual minerals. For example, Zr concentrations may be controlled by zircon, P by apatite, Sr by plagioclase, Ti, Nb and Ta by ilmenite, rutile or sphene. Negative Nb anomalies are also characteristic of the continental crust and may be an indicator of crustal involvement in magma processes.

More mobile LIL element concentrations may be controlled by aqueous fluids but these elements are concentrated in the continental crust and can also be used as an indicator of crustal contamination of magmas.

4.4.2   Multi-element diagrams for sediments

The processes controlling the trace element composition of sedimentary rocks may be investigated using normalization diagrams similar to those for spidergrams discussed above, although they are not as widely used as their equivalents in igneous petrology. Different normalization values are employed for different types of sediment, each representing average Phanerozoic values for the particular sediment (see Table 4.9). The most commonly used normalizing values are those for average shale such as average post-Archaean shale and the North American shale composite (NASC), representing 'average crustal material', although average upper continental crust is also used. Wronkiewicz and Condie (1987), in a study of Archaean shales, plot 15 elements representing the LIL, HFS and transition metals, normalized to NASC. In clays which contain a variable amount of carbonate material, samples are better normalized against the most clay-rich, carbonate-free sample in the suite in order to emphasize the difference in trace element chemistry between the clay and carbonate-rich components (Norman and De Deckker, 1990).

The average trace element concentration in a Phanerozoic quartzite is given by Boryta and Condie (1990) for 18 elements; these values may be used in quartzite normalization (Table 4.9). Trace elements in marbles and calc–silicate rocks were normalized to average Phanerozoic limestone (Condie *et al.*, 1991). The normalizing values are listed in Table 4.9. To minimize the effect of varying amounts of carbonate, Condie *et al.* (1991) also normalized their elemental concentrations to Al_2O_3.

Table 4.9   Normalizing values for multi-element diagrams for sedimentary rocks (values in ppm)

(Ref.)	NASC (1)	Average upper crust (2)	Average Phanerozoic limestone (3)	Average Phanerozoic quartz arenite (4)
Na	7479	28200		
Al	89471	84700	5294	2647
K	31546	27400		
Ca	24303	25000		
Sc	14.90	10.00		1.00
Ti	4676	3600	1199	1499
V		60.00		10.00
Cr	124.50	35.00	15.00	30.00
Mn	4646	600	651	
Fe	39565	35000	3777	3777
Co	25.70	10.00		1.50
Ni	58.00	20.00	15.00	5.00
As	28.40			
Br	0.69			
Rb	125.00	110.00	20.00	
Sr	142.00	350.00	400.00	40.00
Y		22.00	5.00	5.00
Zr	200.00	240.00	20.00	200.00
Nb		25.00	1.50	20.00
Sb	2.09			
Cs	5.16	3.70		
Ba	636.00	700.00	85.00	350.00
La	31.10	30.00	5.00	4.00
Ce	66.70	64.00	10.00	
Nd	27.40	26.00		
Sm	5.59	4.50		
Eu	1.18	0.88	0.20	
Tb	0.85	0.64		
Yb	3.06	2.20		0.50
Lu	0.46	0.32		
Hf	6.30	5.80		3.50
Ta	1.12			2.00
W	2.10			
Pb		15.00	7.00	
Th	12.30	10.50		3.00
U	2.66	2.50		

(1) Gromet *et al.* (1984).
(2) Taylor and McLennan (1981).
(3) Condie *et al.* (1991).
(4) Boryta and Condie (1990).

Interpreting multi-element diagrams for sediments

Cullers (1988) showed that the silt fraction of a sediment most closely reflects the provenance of the sediment. The feldspars and sphene control concentrations of Ba, Na, Rb and Cs; ferromagnesian minerals control the concentrations of Ta, Fe, Co, Sc and Cr; Hf is controlled by zircon; and the REE and Th are controlled by sphene. The ratios La/Sc, Th/Sc, La/Co, Th/Co, Eu/Sm and La/Lu are also good indicators of provenance.

The effects of the extent of weathering and its influence on sediment composition were assessed by Wronkiewicz and Condie (1987) using the alkali and alkaline earth element content of sediments. They showed that large cations (Cs, Rb and Ba) are fixed in a weathering profile whilst smaller cations (Na, Ca and Sr) are more readily leached. In clay-bearing rocks with a carbonate content, the elements Mn, Pb and Sr are chiefly contained in the carbonate component and anomalous values of these elements may provide a signal for the presence of carbonate.

4.5 Platinum metal group element (PGE) plots

The platinum groups elements (PGEs) consist of Ru, Rh, Pd, Os, Ir and Pt (Table 4.10). The noble metal Au and the base metals Cu and Ni are often included on PGE plots. The PGEs can be divided on the basis of their associations into two sub-groups — the Ir-group (IPGEs — Os, Ir and Ru) and the Pd-group (PPGEs — Rh, Pt and Pd). Gold is often associated with the latter group.

   The platinum group elements are very strongly fractionated into a sulphide phase and are useful as a measure of sulphur saturation in a melt. They are also potentially useful as an indicator of partial melting in the mantle, although at present the appropriate partition coefficients are not sufficiently well known. PGEs occur at the ppb level (1 part in 10^9) in basic and ultrabasic igneous rocks but may be concentrated in coexisting chromitites and sulphides. The IPGEs are often associated with chromite as alloys or sulphides in dunites whilst the PPGEs and Au are often associated with the sulphides of Fe, Ni and Cu and are found in norites, gabbros and dunites (Barnes *et al.*, 1985).

Table 4.10   The platinum group elements (PGEs)*

Element	Symbol	Atomic no.	Charge	Ionic radius† (Å)	Melting point (°C)
Ruthenium	Ru	44	2+ 3+ 4+	0.74 0.68 0.62	2310
Rhodium	Rh	45	2+ 3+ 4+	0.72 0.66 0.60	1966
Palladium	Pd	46	2+ 3+ 4+	0.86 0.76 0.615	1552
Osmium	Os	76	2+ 4+	0.74 0.63	3045
Iridium	Ir	77	2+ 3+	0.74 0.68	2410
Platinum	Pt	78	2+ 4+	0.80 0.625	1722

* From compilation of Barnes *et al.* (1985).
† After Shannon (1976) (1Å = 1^{-10} m).

PGE concentrations are normally determined by instrumental neutron activation although the sensitivity of the method decreases in the order Ir> Au> Rh>> Pd> Pt> Os> Ru. In rocks such as ocean-floor basalts PGE concentrations are so low that some elements are below the limit of detection. In this case analysis may be limited to the elements Au, Pd and Ir.

4.5.1 Presenting PGE data

The platinum group elements may be presented in the same way as are the REE and incompatible elements and normalized to either chondrite meteorites or to the primitive mantle.

Chondrite normalization   Naldrett *et al.* (1979) showed that if PGE and Au concentrations are chondrite-normalized and plotted in order of decreasing melting point (see Table 4.10) they define smooth curves rather like REE patterns (Figure 4.27). Normalizing values currently in use are given in Table 4.11, columns 1 to 3. The analogy with the REE is not close, however, because the PGEs are not ordered from light to heavy in keeping with their order in the periodic table. Further, the two groups of PGEs, the Ir-group and the Pd-group, behave differently; the Ir-group tend to be compatible during mantle melting whereas the Pd-group are incompatible. In rocks with very low concentrations of PGEs only Au, Pd and Ir may be measurable. This allows the slope of the chondrite-normalized PGE pattern to be determined from the Ir/Pd ratio, although of course the detail of the pattern is unknown. Au is not used in this instance because it is much more mobile than Pd and is not a reliable indicator of the PGE slope.

Using the terminology developed for REE patterns, depletion and enrichment

Table 4.11   Normalizing values for PGEs, noble metals and associated transition metals (in ppb)

		Chondrite values			Primitive mantle		Fertile mantle	Archaean mantle
	(Ref.)	(1)	(2)	(3)	(4)	(5)	(6)	(7)
	Ni				2110000	2110000		
	Cu				28000	28000	28000	
	Os	510	514	700	3.3	(4.0–5.6)		
IPGEs	Ir	510	540	500	3.6	3.5	4	3.4
	Ru		690	1000	4.3			
	Rh		200					
PPGEs	Pt	1060	1020	1500		(7.0)	7	7.5
	Pd	510	545	1200	4.0		4	5
	Au	160	152	170	1.0	0.5	1	0.75–1.5
	Re	35				0.1		

(1) Sun (1982): C1 chondrites.
(2) Compilation of Naldrett and Duke (1980): C1 chondrites.
(3) Cocherie *et al.* (1989).
(4) Compilation of Brugmann *et al.* (1987): primitive mantle.
(5) Jagoutz *et al.* (1979): average of six primitive mantle nodules.
(6) Compilation of Sun *et al.* (1991): fertile mantle.
(7) Sun (1982): Archaean mantle based upon komatiites.

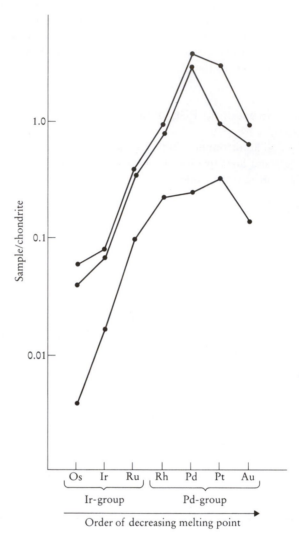

Figure 4.27   Chondrite-normalized platinum group element plot. The data are for komatiite-massive sulphide ores from Alexo Mine, northen Ontario (from Barnes and Naldrett, 1986). The chondrite-normalizing values are given in Table 4.11, column 2 and are taken from Naldrett and Duke (1980).

relative to the chondrite-normalized curve may be described as anomalous, although in this case there is no *a priori* reason why these variations do constitute anomalies. This is illustrated in the ultramafic and basaltic rocks described by Barnes *et al.* (1985) which show positive Pd anomalies and negative Pt and Au anomalies.

Primitive mantle   Brugmann *et al.* (1987) present a primitive mantle-normalized PGE (Pd, Ru, Os, Ir)
normalization   – Au – Cu – Ni plot (Figure 4.28). The elements are broadly arranged in order of increasing compatibility in the primitive mantle from left to right across the diagram. Primitive mantle values are listed in Table 4.11. Also shown are estimated values for the Archaean mantle (Sun, 1982) and the fertile mantle (Sun, *et al.*, 1991) to illustrate the observation of Brugmann *et al.* (1987) that the estimated

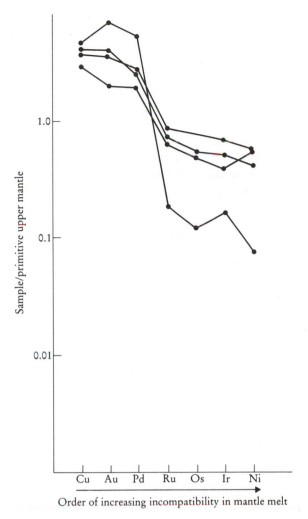

Figure 4.28   Primitive upper mantle-normalized platinum group element (plus Ni, Cu and Au) plot (after Brugmann *et al.*, 1987). The data are for Gorgona Island komatiites from Brugmann *et al.* (1987). The normalizing values are also taken from the compilation of Brugmann *et al.* (1987) and are given in Table 4.11, column 4.

concentrations of PGEs in the upper mantle have not changed since the late Archaean. There is some debate over the significance of the mantle concentrations of PGEs. They range between 0.005 and 0.01 times chondrite, indicating that the mantle may never have had chondritic abundances of PGEs. Garnet and spinel lherzolite xenoliths, on the other hand, show flat chondritic PGE patterns, indicating that the mantle does at least have chondritic PGE *ratios*.

4.5.2   Interpreting PGE patterns

The fractionation of the PGEs is principally governed by their partitioning between solid and melt phases. Unfortunately, partition coefficient data are not available for many minerals for the PGEs. One dominant control is that of a sulphide liquid, for whilst sulphide melt/silicate melt partition coefficients are not well known experimentally, they are thought to be very large. Campbell *et al.* (1983) suggest that *Kd* values for Pt and Pd between a sulphide melt and a silicate melt may be as high as 160 000 and 120 000 respectively. Barnes *et al.* (1985) state that sulphide/silicate melt *Kd* values for the PGEs are in the order Os> Ir> Ru> Rh> Pt> Pd. Thus PGE fractionation may be used as a measure of the level of sulphur-saturation of melt. The extreme fractionation of sulphur undersaturated basaltic melts may be due to the compatibility of the IPGEs in olivine (Brugmann *et al.*, 1987), indicating that they are retained in mantle melting or during fractional crystallization. Barnes *et al.* (1985) investigated the extent to which PGE patterns may be modified by alteration and concluded that, whilst Pt and Au may be mobilized, this is not the dominant process.

4.6 Transition metal plots

The elements of the first transition series (Sc, Ti, V, Cr, Mn, Fe, Co, Ni, Cu and Zn) vary in valence state and in geochemical behaviour. Quadrivalent Ti is a high field strength, incompatible element whilst divalent Mn, Co, Ni, Cu and Zn and trivalent V and Cr are compatible elements (see Figures 4.1 and 4.2). Transition element plots have been mainly used with basalts as a means of exploring the geochemical behaviour of the first transition series. There is no geochemical reason

Table 4.12 Normalizing values used for transition metals (ppm)

	Chondrite concentrations					Primitive mantle concentrations				
(Ref.)	(1)	(2)	(3)	(4)	(5)	(6)	(7)	(8)	(9)	(10)
Sc	5.8				5.21				17	17
Ti	410	720	440	660	610	1300	1300	1230		1300
V	49	94	42	50	49	82	87	59	97	77
Cr	2300	3460	2430	2700	2300	3140	3000	1020		3140
Mn	1720	2590	1700	2500	1720	1010	1100	1000		1010
Fe		219000	171000	250700	265000	61000	65000	67000		60800
Co	475	550	480	800	470	110	110	105		105
Ni	9500	12100	9900	13400	9500	2110	2000	2400		2110
Cu	115	140	110	100		28	30	26		28
Zn	350	460	300	50		50	56	53		50

(1) Langmuir *et al.* (1977).
(2) Kay and Hubbard (1978) from Mason (1971): chondrites, with Cu and Zn from carbonaceous chondrite.
(3) Sun (1982): Cl chondrites.
(4) Bougault *et al.* (1980).
(5) Wood *et al.* (1979b).
(6) Sun (1982): primitive mantle (nodules).
(7) Sun (1982): primitive mantle (partial melting model).
(8) Kay and Hubbard (1978): model mantle.
(9) Sun and Nesbitt (1977): Archaean mantle.
(10) Jagoutz *et al.* (1979): average of six primitive ultramafic nodules.

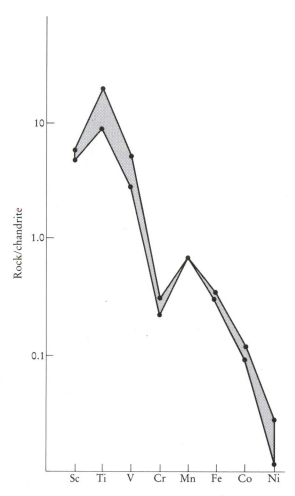

Figure 4.29   Transition metal concentrations in MORB (from Langmuir *et al.*, 1977) normalized to chondritic values. The normalizing values are taken from Langmuir *et al.* (1977) with the exception of Fe, which was taken from Wood *et al.* (1979b) — Table 4.12.

for them behaving as a coherent group nor is there any reason why they should have been present in the primitive earth in chondritic concentrations.

Mid-ocean ridge basalts from the Famous area of the Atlantic show consistent trends on chondrite-normalized plots (Table 4.12, Figure 4.29). They show progressive depletion from Ti to Ni, and have a positive Ti anomaly and a negative Cr anomaly. Cu and Zn are omitted because they are more variable and their concentrations may not reflect the original igneous values (Langmuir *et al.*, 1977). An alternative normalization scheme is to use estimated primitive mantle concentrations, (Figure 4.30). These values are given in Table 4.12, columns 6 to 10.

In summary, the elements Cu and Zn may be quite mobile during metamorphism and alteration and concentrations may diverge from expected smooth patterns. Anomalies in Ni and Cr concentrations may reflect the role of olivine (Ni) and clinopyroxene or spinel (Cr). Ni and Cu can also be concentrated into sulphide melts. Ti anomalies indicate the role of Fe–Ti oxides.

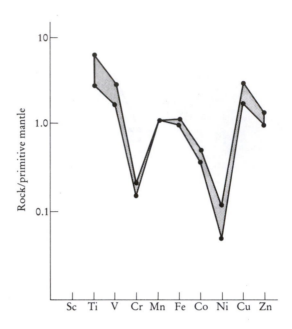

Figure 4.30   Transition metal concentrations in MORB (from Langmuir *et al.*, 1977) normalized to primitive mantle values. The normalizing values are taken from Sun (1982) — Table 4.12, column 6.

4.7   Bivariate trace element plots

So far in this chapter we have concentrated on the display of trace element data on multi-element plots. This approach has the advantage of utilizing a large number of elements and allowing broad conclusions to be drawn on their geochemical behaviour. Multi-element plots have one particular disadvantage, however, for only a few samples can be shown on a single diagram before it becomes becomes cluttered and confused. Thus the simple bivariate plot is superior to a multivariate plot when a large number of samples are plotted and in particular when geochemical trends are sought. Variation diagrams are discussed in detail in Section 3.3 and many of the principles outlined there and applied to major elements are equally valid for trace element plots.

4.7.1   The selection of trace elements in igneous rocks for plotting on bivariate graphs

Many igneous rocks have had a complex history of solid–liquid equilibria in their journey from the source region to their site of emplacement. In addition they may also have interacted with fluids during or after their solidification. The task for the igneous geochemist, therefore, is to work out which trace elements are indicative of which processes in this complex history. The clues come from a knowledge of mineral/melt (and the lesser known mineral/fluid) partition coefficients and the physical laws which govern the concentrations of trace elements in igneous rocks. Most fruitful are trace elements which show extreme behaviour, such as the highly incompatible and the highly compatible elements. Further clues may come from the inspection of multi-element plots, for this allows the identification of element pairs or ratios which can be used in trace element modelling and in subdividing rocks into similar suites.

Incompatible element plots Incompatible element concentrations are particularly sensitive to partial melting processes (see for example Figures 4.11a and 4.12a). The more highly incompatible an element is, the more sensitive it is to degrees of partial melting. This is true for both batch melting and fractional melting but is most extreme in fractional melting. Incompatible element concentrations also vary during fractional crystallization, although the effect is most strongly marked in AFC processes.

(a) Identification of igneous source characteristics from incompatible element plots The ratio of a pair of highly incompatible elements whose bulk partition coefficients are very similar will not vary in the course of fractional crystallization and will vary little during batch partial melting. Thus the slope of a correlation line on a bivariate plot of two such highly incompatible elements gives the ratio of the concentration of the elements in the source. In the case of mantle melting, the following groups of elements have almost identical bulk partition coefficients during mantle melting: Cs–Rb–Ba, U–Nb–Ta–K, Ce–Pb, Pr–Sr, P–Nd, Zr–Hf–Sm, Eu–Ti, Ho–Y. Bivariate plots of element pairs taken from within these groups can be expected to show to a first approximation the ratio of the elements in the source (Sun and McDonough, 1989). In addition the elemental pairs Y–Tb, La–Ta, La–Nb, Ta–Th, Ti–Zr and Ti–Y are also often assumed to have very similar bulk partition coefficients and are used in a similar way. Any variation in the ratio reflects heterogeneity in the source (Bougault *et al.*, 1980) resulting from source mixing or contamination.

This approach has been very fruitful in some areas of igneous petrology but is not universally applicable. Firstly, the relationship does not hold for very small degrees of melting for there is divergence between the enrichment paths of highly incompatible and moderately incompatible trace elements — see Figures 4.12(a) and (c). If a small degree of melting is suspected it may be recognized from the approximation $C_L/C_0 = 1/F$ derived from the batch melting equation for the case when D is very small (Section 4.2.2). Secondly, this method is difficult to apply to granitic rocks. This is because few trace elements are highly incompatible in granitic melts, chiefly because of the large number of possible minor phases incorporating many of the trace elements traditionally regarded as incompatible. The source region of granitic rocks is more easily characterized by the study of radiogenic isotopes (see Chapter 6).

(b) Identification of igneous source characteristics from incompatible element ratio–ratio plots Ratio–ratio plots of highly incompatible elements minimize the effects of fractionation and allow us to examine the character of the mantle source. Different mantle sources plot on different correlation lines (Figure 4.31). This approach was

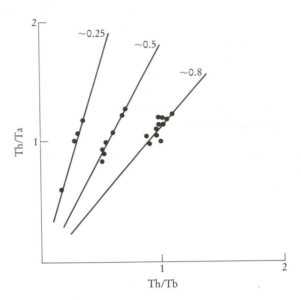

Figure 4.31   Bivariate ratio plot of Th/Ta vs Th/Tb for basalts from the Red Sea Rift (data from Eissen *et al.*, 1989). The three trends showing (Th/Tb)/(Th/Ta) ratios of approximately 0.25, 0.5 and 0.8 are thought to indicate three different mantle sources.

used by Saunders *et al.* (1988) who used the highly incompatible elements Th,U, Pb, K, Ba, Ce and Nb and the element pairs Th–Ce, K–Ce, U–Pb and Ba–Ce each ratioed to Nb, in order to explore the ratios Th/Ce, K/Ce and U/Pb in MORB. Where the three elements have identical bulk partion coefficients the ratio will not change during partial melting or fractional crystallization. Ratio–ratio plots do have some inherent problems and the reader should take note of possible spurious correlations arising from the common denominator effect discussed in Section 2.5.

(c) Calculation of partition coefficients from reciprocal concentration trace element plots   Minster and Allegre (1978) showed that, from a rearrangement of the batch melting equation, a bivariate plot of the reciprocals of incompatible elements can be used to obtain information about partition coefficients during melting. Provided the mass fractions of the minerals in the melt remain constant, a linear trend on such a diagram characterizes the batch melting process. Further, the slope and intercept of the trend can provide information about the difference in bulk partition coefficient between the two elements. If the samples are first normalized to the most enriched sample, elements with the same bulk partition coefficient will have a slope of 1 and an intercept of zero. If the bulk partition coefficients for the two elements change at different rates during melting then a curvilinear trend will be produced (Bender *et al.*, 1984).

Compatible element plots   Compatible trace element concentrations change dramatically in an igneous liquid during fractional crystallization (Figure 4.13). Thus bivariate plots of compatible elements, plotted against an index of fractionation (e.g. mg number) can be used to test for fractional crystallization. The effect is not so marked in *in situ* crystallization except when a large melt fraction is returned to the magma chamber (Figure 4.14d). In AFC processes when the rate of assimilation is high and the trace element concentration in the wallrock is less than in the melt, compatible elements are strongly depleted (Figure 4.16a and b)

During partial melting, however, highly compatible elements are buffered by the solid phase with respect to solid–melt equilibria during partial melting (Figures 4.11b, 4.12d). This means that their concentration in the source, even if it has undergone some partial melting, is largely unchanged.

4.7.2 Bivariate plots in sedimentary rocks

Bivariate trace element plots in sedimentary rocks are mostly used to detect mixing processes in sediments. Norman and De Deckker (1990) suggest that linear correlations amongst a diverse group of elements over a broad range of concentrations may be taken as indicative of mixing of two sedimentary components. Condie and Wronkiewicz (1990) and McLennan and Taylor (1991) have exploited the geochemical differences between elements such as Th and La (indicative of a felsic igneous source) and Sc and Cr (indicative of a mafic source) and have used plots such as Th/Sc vs Sc and Cr/Th vs Sc/Th as indicators of contrasting felsic and mafic provenance. Floyd *et al.* (1989) quantified such mixing processes with the general mixing equation of Langmuir *et al.* (1978) — see Section 4.9.3.

4.8 Enrichment–depletion diagrams

Enrichment–depletion diagrams are a convenient way of showing relative enrichment and depletion in trace (and major) elements. They can be useful, for example, in demonstrating the extent of elemental enrichment and depletion in an igneous suite by comparing the chemistries of early and late members of a series. Hildreth (1981) compares the relative concentrations of the early and late members of the Bishop's tuff (Figure 4.32). The *x*-axis of the graph shows the elements arranged by atomic number and the *y*-axis the concentration of an element in the latest erupted ejecta ratioed to that for the earliest erupted ejecta. Alternatively the values on the *y*-axis may be recorded on a logarithmic scale.

Enrichment–depletion diagrams are also useful as a way of displaying element mobility; this has been used particularly in alteration zones associated with hydrothermal mineralization. For example, Taylor and Fryer (1980) show the relative mobilities of trace and major elements in the zones of potassic and propylitic alteration associated with a porphyry copper deposit. In this case the enrichment/depletion is measured relative to the unaltered country rock.

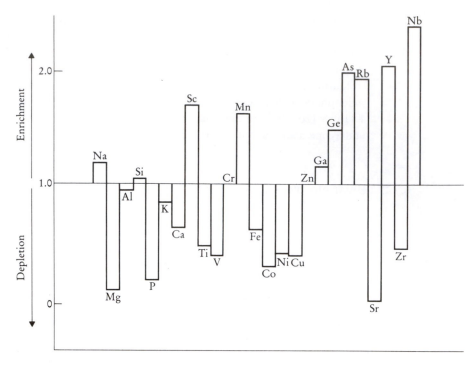

Figure 4.32   Enrichment–depletion diagram showing the enrichment factors for selected major and trace elements from the Bishop's tuff, arranged in order of increasing atomic number. The diagram compares the relative concentrations of the early and late members of the Bishop's tuff and is thought to be a measure of the zonation of the magma chamber. Data from Hildreth (1981).

4.9   Modelling trace element processes in igneous rocks

One of the most important uses of trace elements in modern igneous petrology is in the modelling of geochemical processes. Trace element modelling depends upon the mathematical expressions which describe the equilibrium partitioning of trace elements between minerals and melt during igneous processes (Section 4.2.2) and a precise knowledge of trace element partition coefficients (see Tables 4.1 to 4.3). For successful geochemical modelling, three conditions should be fulfilled. Firstly, trace element concentrations must be determined with great accuracy; otherwise it is impossible to discriminate between competing hypotheses (Arth, 1976). Secondly, partition coefficients must be known accurately for the conditions under which the process is being modelled and, thirdly, the starting composition must be known. This latter condition is not always fulfilled; and sometimes a reasonable assumption of the starting composition must be made which is later refined as the model is developed.

Calculated compositions are plotted on a bivariate graph and compared with an observed trend of rock compositions, or plotted on a multivariate diagram such as an REE diagram and the calculated composition compared with a measured composition. Below we describe these two modes of presentation in some detail.

4.9.1   Vector diagrams

Changes in trace element concentrations may be modelled on a bivariate plot using vectors to show the amount and direction of change which will take place as a consequence of a particular process. **Mineral vectors** show the trend of a fractionating mineral phase or mineral assemblage. These are illustrated in Figure 4.33 for the fractionation of plagioclase, clinopyroxene, orthopyroxene, hornblende, biotite and orthoclase from a granitic melt. **Partial melting** vectors are used to show changing melt and source compositions during the partial melting of a given source composition and mineralogy. The effects of different melting models, source

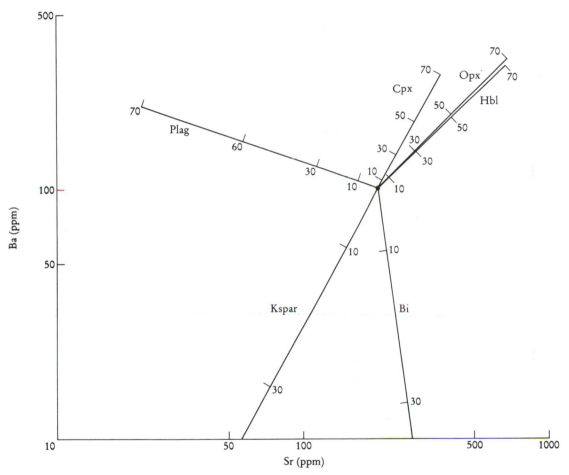

Figure 4.33   Mineral vector diagram showing fractional crystallization trends in a granitic source containing 200 ppm Sr and 100 ppm Ba. Fractionation trends are shown for 10, 30, 50 and 70 % fractional crystallization of the minerals clinopyroxene (Cpx), orthopyroxene (Opx), hornblende (Hbl), biotite (Bi), orthoclase (Kspar) and plagioclase (Plag). The direction of the lines shows the compositional change in the residual liquid when the specified phase is progressively removed during fractional crystallization. The partition coefficients for Ba and Sr are taken from Table 4.3 and the vectors calculated from the Rayleigh fractionation equation (Eqn [4.18]). Details of the calculation are given in Table 4.13. The logarithmic scale is used to produce a straight-line trend.

Table 4.13   Data used in the calculation of mineral vectors shown in Figure 4.33

Partition coefficients for rhyolitic liquids (from Table 4.3)

	Opx	Cpx	Hbl	Bi	Plag	Ksp
Ba	0.003	0.131	0.044	6.360	0.360	6.120
Sr	0.009	0.516	0.022	0.120	2.840	3.870

Initial composition

Ba	100
Sr	200

Calculated compositions (from Eqn [4.18])

	Opx	Cpx	Hbl	Bi	Plag	Ksp
10 % fractional crystallization						
Ba	111.1	109.6	110.6	56.9	107.0	58.3
Sr	222.0	210.5	221.7	219.4	164.8	147.8
30 % fractional crystallization						
Ba	142.7	136.3	140.6	14.8	125.6	16.1
Sr	284.8	237.7	283.5	273.7	103.8	71.9
50 % fractional crystallization						
Ba	199.6	182.6	194.0	2.4	155.8	2.9
Sr	397.5	279.7	393.9	368.1	55.9	27.4
70 % fractional crystallization						
Ba	332.1	284.7	316.1	0.2	216.1	0.2
Sr	659.5	358.2	649.2	577.0	21.8	6.3

compositions and mineralogies may all be explored in this way. Figure 4.34, for example, illustrates the differing melting paths during fractional and batch melting and the different degrees of enrichment caused by the two processes. Whilst vector diagrams select only two out of a vast array of possible elements, they have the advantage of being able to display data from a large number of samples. This means that it is possible to view trends in the data. Hence both mineral and partial melting vectors are compared with observed trends on bivariate plots in order to test the validity of a particular model.

4.9.2   Modelling on multivariate diagrams

Multivariate diagrams such as REE plots and spider diagrams are also used in petrogenetic modelling, although by their nature they are unable to show more than a few samples clearly on a single diagram. In this case the same operation is performed on each element in the plot and the resultant data array is compared with a measured rock composition. The process is illustrated in Figures 4.35 and 4.36 which show respectively the effect of olivine fractionation on a komatiite liquid and the partial melting of a primitive mantle source. Repetitive calculations such as

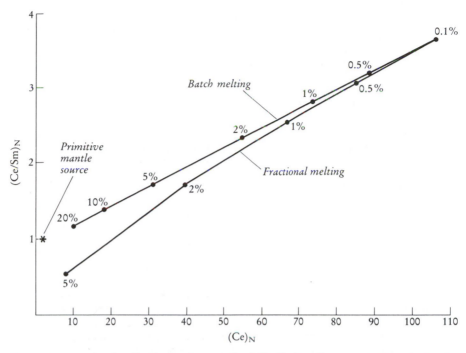

Figure 4.34   Vector diagram showing the change in normalized (Ce/Sm) vs Ce concentrations during the partial melting of a primitive mantle source with the mineralogy: olivine 55 %, orthopyroxene 25 %, clinopyroxene 11 % and garnet 9 %. The initial normalized concentrations of Ce and Sm in the primitive mantle were taken as 2.12 and 2.19 respectively. The vector for modal batch melting between 0.1 % and 20 % melting was calculated from Eqn [4.10]. The vector for single melt increments formed during modal fractional melting between 0.1 and 5 % melting was calculated from Eqn [4.11]. The partition coefficients were taken from Table 4.1. Details of the calculation are given in Table 4.14 and the same data are also illustrated in Figure 4.36.

Table 4.14   Data used in the calculation of partial melting vector diagram (Figure 4.34) and REE diagram (Figure 4.36)

| | Primitive mantle composition | | | Basaltic *Kd* values from Table 4.1 | | | | Weighted mean of bulk partition coeff. |
	Concn. (ppm)	Nakamura values	Normalized values	Ol	Opx	Cpx	Gt	(Ol 55 %, Opx 25%, Cpx 11 %, Gt 9 %)
La	0.708	0.329	2.15	0.007	0.020	0.056	0.001	0.015
Ce	1.833	0.865	2.12	0.006	0.020	0.092	0.007	0.019
Nd	1.366	0.630	2.17	0.006	0.030	0.230	0.026	0.038
Sm	0.444	0.203	2.19	0.007	0.050	0.445	0.102	0.074
Eu	0.168	0.077	2.18	0.007	0.050	0.474	0.243	0.091
Gd	0.595	0.276	2.16	0.010	0.090	0.556	0.680	0.150
Dy	0.737	0.343	2.15	0.013	0.150	0.582	1.940	0.283
Er	0.479	0.225	2.13	0.026	0.230	0.583	4.700	0.559
Yb	0.481	0.220	2.19	0.049	0.340	0.542	6.167	0.727
Lu	0.074	0.034	2.17	0.045	0.420	0.506	6.950	0.811

(Continued)

Table 4.14   (*Continued*)

	Assumed mantle composition (normalized)		Batch modal partial melting — Eqn [4.10]						
	Source	Calculated bulk partition coeff. for mineral assemblage (Ol 55 %, Opx 25 %, Cpx 11 %, Gt 9 %)	0.1 %	0.5 %	1 %	2 %	5 %	10 %	20 %
La	2.15	0.015	135.17	108.36	86.82	62.13	33.53	18.97	10.15
Ce	2.12	0.019	105.79	88.46	73.43	54.80	31.12	18.09	9.85
Nd	2.17	0.038	55.11	50.20	45.17	37.63	25.08	16.12	9.40
Sm	2.19	0.074	29.01	27.65	26.12	23.52	18.11	13.09	8.43
Eu	2.18	0.091	23.85	22.94	21.89	20.06	16.04	12.02	8.01
Gd	2.16	0.150	14.26	13.94	13.57	12.88	11.18	9.16	6.73
Dy	2.15	0.283	7.57	7.49	7.40	7.22	6.73	6.05	5.04
Er	2.13	0.559	3.81	3.80	3.78	3.75	3.67	3.53	3.29
Yb	2.19	0.727	3.01	3.00	3.00	2.99	2.95	2.90	2.80
Lu	2.17	0.811	2.68	2.68	2.67	2.67	2.65	2.62	2.56
Ce	2.12		105.70	88.46	73.43	54.80	31.12	18.09	9.85
Ce/Sm	0.97		3.65	3.20	2.81	2.33	1.72	1.38	1.17

	Assumed mantle composition (normalized)		Calculated composition assuming fractional modal melting — Eqn [4.11]				
	Source	Calculated bulk partition coeff. for mineral assemblage (Ol 55 %, Opx 25 %, Cpx 11 %, Gt 9 %)	0.1 %	0.5 %	1 %	2 %	5 %
La	2.15	0.015	134.89	103.53	74.26	38.01	4.89
Ce	2.12	0.019	105.65	85.93	66.30	39.31	7.93
Nd	2.17	0.038	55.09	49.82	43.91	34.05	15.63
Sm	2.19	0.074	29.00	27.59	25.92	22.85	15.52
Eu	2.18	0.091	23.85	22.90	21.78	19.67	14.39
Gd	2.16	0.150	14.26	13.94	13.55	12.79	10.73
Dy	2.15	0.283	7.57	7.49	7.39	7.21	6.66
Er	2.13	0.559	3.81	3.80	3.78	3.75	3.66
Yb	2.19	0.727	3.01	3.00	3.00	2.99	2.95
Lu	2.17	0.811	2.68	2.68	2.67	2.67	2.65
Ce	2.12		105.65	85.93	66.30	39.31	7.93
Ce/Sm	0.97		3.64	3.11	2.56	1.72	0.51

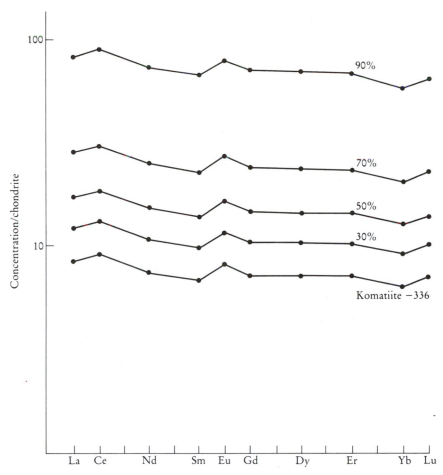

Figure 4.35   Chondrite-normalized REE patterns calculated for olivine fractionation from a komatiite melt (komatiite–336) at 30 %, 50 %, 70 % and 90 % fractional crystallization. The partition coefficients were taken from Table 4.1. The details of the calculations made for each of the ten REE shown using Eqn [4.18] are given in Table 4.15.

Table 4.15   Data used in the calculation of the REE diagrams shown in Figure 4.35

	Starting composition of Sula Mountains komatiite 336				Calculated values (from Eqn [4.18]) at a given % fractional crystallization				
	Concn (ppm)	Nakamura values	Normalized values	*Kd* for olivine (Table 4.1)	10 %	30 %	50 %	70 %	90 %
La	2.79	0.3290	8.48	0.0067	9.4	12.1	16.9	28.0	83.5
Ce	7.93	0.8650	9.17	0.0060	10.2	13.1	18.3	30.3	90.4
Nd	4.73	0.6300	7.51	0.0059	8.3	10.7	15.0	24.8	74.1
Sm	1.40	0.2030	6.90	0.0070	7.7	9.8	13.7	22.8	67.9
Eu	0.63	0.0770	8.18	0.0074	9.1	11.7	16.3	27.0	80.4
Gd	2.01	0.2760	7.28	0.0100	8.1	10.4	14.5	24.0	71.2
Dy	2.46	0.3430	7.17	0.0130	8.0	10.2	14.2	23.5	69.6
Er	1.61	0.2250	7.16	0.0256	7.9	10.1	14.1	23.1	67.5
Yb	1.41	0.2200	6.41	0.0491	7.1	9.0	12.4	20.1	57.2
Lu	0.24	0.0339	7.08	0.0454	7.8	10.0	13.7	22.3	63.8

these are done most easily by computer and the results displayed, ideally, on a graphics screen. Holm (1990) describes such a method using a spreadsheet.

4.9.3 Petrogenetic modelling — examples

In this section we illustrate the way in which trace element modelling has been used to identify differing petrogenetic processes in igneous rocks.

Partial melting   A numerical example showing the effects of between 0.1 % and 20 % partial melting on the REE concentrations of a primitive mantle source is given in Table 4.14. The results are displayed graphically on a vector diagram (Figure 4.34) and on a multivariate plot (Figure 4.36). Both diagrams show the extreme enrichment of the light REE relative to the heavy REE in the partial melts — a property which increases as the percentage of melting decreases. This type of modelling has been used extensively and many examples can be found in the geochemical literature. The results of partial melting calculations may equally well be presented on spider diagrams. For example, Thompson *et al.* (1984) calculated chondrite-normalized spidergrams for the dynamic melting of mantle lherzolite for comparison with measured oceanic basalt compositions.

Crystal   The modelling of fractional crystallization on vector diagrams and REE plots is
fractionation   described above in Sections 4.9.1 and 4.9.2 respectively and is illustrated in Figures 4.33 and 4.35. Details of the calculations are given in Tables 4.13 and 4.15.

Vector diagrams can be used to identify a fractionating phase on a bivariate plot. If, for example, on a plot of Ba vs Sr, the rock compositions define a liquid trend which could have been produced by crystal fractionation, then the slope of the trend can be compared with a mineral vector diagram such as Figure 4.33 and the phase responsible for the fractional crystallization trend can be identified. In addition, it is possible, from the compositional range in the two elements, to make an estimate of the amount of fractional crystallization that has taken place. When there is more than one fractionating phase present a composite vector must be calculated, although when this is the case it is not always possible to find a unique composition for the fractionating mineral assemblage.

Multivariate diagrams are used to compare calculated and measured rock compositions. Provided the composition of a parental melt is known, the composition of derivative liquids can be calculated and compared with the composition of liquids which are thought to have been derived from the parental melt. This is the approach used by Arth (1981), who modelled the REE chemistry of andesites and dacites in New Britain by fractional crystallization from basaltic and basaltic andesites. Arth estimated the mineralogy of the fractionating assemblage from the proportions of phenocrysts present in the lavas and calculated REE patterns which show excellent agreement with the observed REE patterns in the rocks.

Crystal   Komatiitic magmas are thought to have had exceptionally high liquidus
contamination   temperatures and to have been frequently contaminated with continental crust.
and AFC   Trace element data in support of this hypothesis are illustrated qualitatively by
processes   Arndt and Jenner (1986) on a REE diagram. Similarly, Condie and Crow (1990)

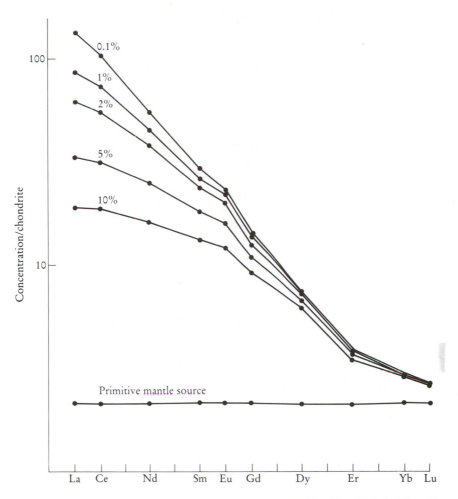

Figure 4.36 Chondrite-normalized REE paterns calculated for modal batch partial melting of a primitive mantle source with *ca* 2.15 chondritic concentration of REE and with the mineralogy — olivine 55 %, orthopyroxene 25 %, clinopyroxene 11 % and garnet 9 %. Curves are shown for 0.1 to 10 % partial melting. The calculations were made for the ten REE shown using Eqn [4.10] — see Table 4.14 for the details.

proposed that Archaean and early Proterozoic basalts erupted on the Kaapvaal craton in southern Africa were komatiitic in origin but contaminated with continental crust. They support this hypothesis by showing that AFC vectors, calculated from the equations of DePaolo (1981b) (see Section 4.2.2), coincide with observed trends on bivariate Zr–Ni and Hf–Th diagrams. They also use MORB-normalized spidergrams to show conformity between measured basaltic compositions and calculated spidergrams for varying degrees of assimilation and fractional crystallization from a komatiitic liquid.

Open system processes The behaviour of the REE in a periodically replenished, periodically tapped, continuously fractionated (RTF) magma chamber (Section 4.2.2) is described by O'Hara and Matthews (1981). Using chondrite-normalized values of Ce/Yb to represent the slope of the REE diagram, and Sm values as a measure of REE concentrations, they show on a bivariate plot of normalized Ce/Yb vs Sm the effects of the RTF process on partial melts of a variety of mantle sources coupled with crystal fractionation.

Recognizing RTF processes in a lava sequence requires a very detailed and complete geochemical section. Norman and Leeman (1990), in a study of Oligocene andesites and basalts from southwest Idaho, USA recognized a cyclicity in the trace and major element chemistry up the stratigraphic section. On bivariate Ba–Sc and Th–Sc plots they show that the scatter in the data cannot be accommodated within calculated fractional crystallization and AFC trends alone, and requires the recharge of the magma chamber by the addition of more mafic magmas during evolution of the magma chamber.

Magma and source mixing
A set of general mixing equations is given by Langmuir *et al.* (1978) which may be used to identify magma mixing and mixing in an igneous source region. The equations of Langmuir *et al.* (1978) predict that mixing between two elements produces a straight line whereas mixing between an element and a element ratio or between two ratios produces an asymptotic curve.

The two types of mixing, magma mixing and source mixing, can be differentiated by using differences in behaviour between compatible and incompatible elements. For example, since highly incompatible element ratios do not change during partial melting or fractional crystallization, a mixing curve based upon incompatible element ratios is ambiguous and can indicate either magma mixing or source mixing. Compatible element ratios, on the other hand, are strongly fractionated during partial melting and will not reflect the ratios of the source region. Thus, if mixing is in the source region a compatible element plot will show a scattered trend whereas the mixing of two melts will produce a simple mixing line.

Demonstrating element mobility
Two types of trace element plot lend themselves to the investigation of element mobility. Firstly, the MORB-normalized multi-element diagram of Pearce (1983) (Figure 4.23, Section 4.4.1) was constructed to show the difference in behaviour between elements which are mobile and those which are not. Brewer and Atkin (1989) found that this diagram successfully differentiated between the behaviour of mobile elements such as Sr, K, Rb and Ba and the immobile elements Nb, Ce, P, Zr, Ti and Y during the greenschist facies metamorphism of basalts. A second approach is to use the enrichment–depletion diagram described in Section 4.8 (Figure 4.32), although this type of diagram can only be used when the composition of the unaltered rock is known.

4.9.4 Inversion techniques using trace elements

Trace element inversion methods make use of the variability in elemental concentrations in a suite of cogenetic igneous rocks to determine unknowns such as the composition and mineralogy of the source, the physical process causing the variation — crystal fractionation, partial melting or other process — and the degree of partial melting, crystal fractionation or other process. Thus, the inverse method, with its emphasis on constraining the model from the trace element data, offers a much greater possibility of a unique solution to a geochemical problem. Inverse methods can in principle be applied to all petrological processes, although only fractional crystallization and partial melting are illustrated here. The methods outlined here are discussed in detail by Allegre and Minster (1978).

The first step in using the inverse approach to the study of trace elements is to identify the likely physical process which accounts for the variation in the data. This may be done by plotting selected trace elements on bivariate plots. For example, elements which are compatible will vary drastically in concentration during fractional crystallizsation, whilst highly incompatible elements will vary most in abundance during partial melting (Minster and Allegre, 1978).

Constraining fractional crystallization using an inversion method

A worked example of the inverse approach to fractional crystallization is given by Minster et al. (1977) for a suite of basaltic lavas from the Azores. In this case the unknowns are (1) the initial concentration of the trace elements in the parent magma, (2) the bulk partition coefficients for the elements and (3) the degree of crystallization corresponding to each sample. The initial concentrations of trace elements in the melt were estimated using Ni. The likely Ni concentration in a parental melt was calculated by melting a model mantle. On this basis a parental melt from the lava suite was thus identified. Bulk partition coefficients were calculated using the method of Allegre et al. (1977) as follows. It can be shown from the Rayleigh Fractionation Law (Eqn [4.18]) that the slope of a log–log plot of a highly incompatible trace element against any other trace element is proportional to the bulk partition coefficient D (see Figure 4.37). Where $D < 1$ the slope is $(1 - D)$ and where $D > 1$ the slope is $(D - 1)$. Minster et al. (1977) calculated bulk partition coefficients by assuming that the highly incompatible element Ta had a bulk partition coefficient of zero. The slopes of log–log plots of the data against Ta were used to estimate bulk partition coefficients for other elements. The degree of

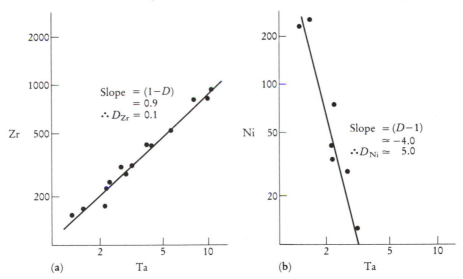

Figure 4.37   Log–log trace element plots showing the calculation of the bulk partition coefficient D during crystal fractionation, after the method of Allegre et al. (1977). (a) Plot of moderately incompatible element Zr against highly incompatible element Ta. The slope of the logarithmic plot is 0.9; hence the bulk partition coefficient $D = 0.1$. (b) Plot of compatible element Ni against highly incompatible element Ta. The slope of the logarithmic plot is *ca* −4.0, hence the bulk partition coefficient D is *ca* 5.

crystallization for each sample was calculated from a knowledge of the composition of the primary melt and the bulk partition coefficient.

Constraining partial melting using an inversion method

In the case of partial melting the unknowns are (1) the chemistry of the source, (2) the bulk partition coefficient for each element considered and (3) the degree of partial melting for each sample. There are too many unknowns for a direct solution and so a number of initial assumptions have to be made.

Bender *et al.* (1984), in a study of ocean-floor basalts, used the modal batch melting equation (Eqn [4.10]) and made two initial assumptions. Firstly, they selected a reference element (in this case the most incompatible element) and assigned it a bulk distribution coefficient (D). Secondly, an initial value was assumed for the degree of partial melting. This value is subsequently checked and refined by the geological plausibility of the end result. From these two assumptions the concentration in the source (C_0) may be calculated for the reference element: this is usually done for the most enriched sample. From the values obtained for D and C_0 for the reference element, values for the percentage melting (F) can be calculated for all the other samples. The batch melting equation now has two known parameters for each element — the concentration in the melt (C_L) and the percentage melting (F) — and two unknowns, C_0 and D. These unknowns may be found by solving simultaneous equations for pairs of parental magmas and by assuming that in each case a pair of magmas has values of C_0 and D which are constant.

The results of a preliminary calculation are inspected and refined as necessary. A comparison of calculated D values for each pair of parental magmas gives a test of the assumed batch melting model for these rocks. The calculated inverted partition coefficients may be compared with experimentally determined values to see if they are geologically plausible. The relative values of the inverted distribution coefficients are a robust feature of this inversion method over a range D and F values. Values for the composition of the source are more sensitive to values of D and F but solutions are restricted to a small range of geologically plausible possibilities.

4.9.5   A final comment on geochemical modelling

The development of trace element modelling in petrology represents a major advance in geochemistry. However, this approach must be used with some caution for two reasons.

(1)   Most of the models developed in Section 4.2.2 of this chapter are idealized and do not always conform to the actual physical conditions of the process being modelled.

(2)   Rarely can petrological modelling produce a unique solution. Often problems of this type contain too many unknowns.

For this reason, trace element modelling must be regarded as a means of hypothesis testing and should be used to verify an already constrained model. Wherever possible, trace element studies should be part of a broader approach to a geochemical problem which includes the constraints of the major element and isotopic chemistry.

Chapter 5

Discriminating between tectonic environments using geochemical data

5.1 Introduction

The idea of trying to fingerprint magmas from different tectonic settings chemically is probably best attributed to Pearce and Cann (1971, 1973). In two very important papers these authors showed that it was possible to use geochemistry to distinguish between basalts produced in different, known tectonic settings. They produced what have become known as tectono-magmatic discrimination diagrams — geochemical variation diagrams on which magmas produced in different tectonic settings may be distinguished from one another on the basis of their chemistry. The relatively simple approach and the wide applicability of their results meant that the environment of eruption of both ancient and modern basalts could be defined by the analysis of a rock for just a few readily determined elements. This led quickly to a plethora of papers purporting to show the tectonic setting of ancient volcanic rocks whose state of preservation and poor exposure had previously precluded the identification of the palaeotectonic environment. More recently, however, geochemists have moved away from the 'cookbook' type of approach to tectono-magmatic discrimination diagrams to a closer examination of the reason why different tectonic environments have variable geochemical signatures.

The work pioneered by Pearce and Cann (1971, 1973) brought together three separate strands of thinking. These are the statistical technique of discriminant analysis, the rapid and accurate analysis of trace elements thought to be normally immobile under hydrothermal conditions and the identification of a number of distinct tectonic environments. Each of these is now briefly considered.

5.1.1 Discriminant analysis

Discriminant analysis is a statistical technique used for classifying samples into predefined groups (Section 2.9). A large number of variables are examined in order to isolate the ones which most effectively classify the samples. The most powerful discriminants are used to define the axes on bivariate and triangular diagrams. The separate groups of samples are plotted either as elemental concentrations or as calculated discriminant functions based upon the elemental concentrations, and boundaries are drawn between the different groups of samples. Unknown samples are then classified according to the defined fields.

Most discrimination diagrams described here use individual elements as the discriminant functions rather than a composite of several elements, making the diagrams easy both to use and to understand. In the case of the study by Pearce and Cann (1971, 1973) the elements Ti, Zr, Y, Nb and Sr were found to be the most effective discriminants between basalts erupted in differing tectonic environments.

5.1.2   Immobile trace elements

A major step in the development of tectono–magmatic discrimination diagrams was the development of a method of rapid and accurate analysis of trace elements present at low concentrations in silicate materials. This work was initially carried out by X-ray fluorescence analysis although some trace elements are now analysed by neutron activation analysis. Of particular importance were trace elements thought to be immobile under most forms of hydrothermal activity. Not all discrimination diagrams use trace elements and not all use immobile trace elements, but the diagrams which have the widest applicability are based upon immobile trace elements as these can be used with altered and metamorphosed rocks. Much of the ensuing debate over the use of diagrams to discriminate between different tectonic settings has focused on the question of the mobility or otherwise of the so-called 'immobile elements'.

Many of the diagrams use high field strength elements such as Ti, Zr, Y, Nb and P which are thought to be relatively immobile in aqueous fluids unless there are high activities of F^-. This means that these elements will be stable under conditions of hydrothermal, sea-floor weathering and up to medium metamorphic grades (mid-amphibolite facies). Little is known about the stability of these elements at higher metamorphic grades.

In summary, a good tectonic discrimination diagram must be constructed with elements which are insensitive to secondary processes, and which are easy to measure with good precision even at low concentration levels, ideally by a relatively simple and rapid method of analysis.

5.1.3   Tectonic environments

The number of tectonic environments recognized today is much greater than 20 years ago. This reflects the advances made in understanding both earth processes and the chemistry of igneous rocks. Pearce and Cann (1971, 1973) originally identified the geochemical signature of rocks from volcanic-arcs, from the ocean floor and from within plates. Today the chemical discrimination of tectonic environments has expanded to include granitic rocks and sediments.

A summary of the tectonic environments discussed in this chapter is given in Box 5.1. Different types of ocean ridge are best distinguished using basalt chemistry, whereas discrimination between the different types of collision zone is better done using granite geochemistry. Passive continental margins are charac-terized by their absence of igneous activity and can only be recognized using the chemistry of sedimentary rocks. An intraplate setting can be recognized from the chemistry of both basalts and granites, and volcanic-arcs may be recognizable using all three types of discriminant analysis.

Box 5.1

Tectonic environments recognizable using geochemical criteria

Ocean ridge

Normal ocean ridge (characterized by N-type MORB)
Anomalous ocean ridge (characterized by E-type MORB)
Incipient spreading centre

Back-arc basin ridge
Fore-arc basin ridge (located above a subduction zone)

Volcanic-arc

Oceanic-arc — dominated by tholeiitic basalts
Oceanic-arc — dominated by calc–alkali basalts
Active continental margin

Collisional setting

Continent–continent collision
Continent–arc collision

Intraplate setting

Intracontinental — normal crust
Intracontinental — attenuated crust
Ocean-island

Passive continental margin

5.1.4   Using discrimination diagrams

Discrimination diagrams seldom provide unequivocal confirmation of a former tectonic environment. At best they can be used to suggest an affiliation. They should never be used as proof. This is all the more important the further back in time we go and the further away we move from the control set of samples used in the construction of the diagram. For example, using a discrimination diagram constructed from modern volcanic rocks to postulate an Archaean tectonic setting is likely to produce equivocal results. Furthermore, discrimination diagrams were never intended to be used for single samples, but rather with a suite of samples. This simple precaution will eliminate the occasional spurious result and highlight data-sets from mixed or multiple environments.

In this chapter we review the current use of discrimination diagrams and their application to igneous and sedimentary rocks. In each case a diagram is described and then evaluated in the light of its usage. Clearly, diagrams produced in the 1970s are more fully tested than those published recently and so in general they obtain the poorer reviews. These are sometimes justified but often reflect an increased level of understanding of both tectonic environments and geochemical processes since the publication of the diagram.

Factors that should be considered when evaluating discrimination diagrams are:

(1)   the number of samples used in constructing the diagram and in defining the boundaries;
(2)   the degree of overlap between the proposed fields;
(3)   the effects of element mobility on the usefulness of the diagram;
(4)   the range of tectonic environments represented.

Finally, it should be remembered that most discrimination diagrams are empirically derived. There is no harm in this, but for a diagram to be most useful we also need to understand how it works. Thus, where possible, the rationale behind a particular diagram is also given in the discussion below.

5.2   Discrimination diagrams for rocks of basaltic to andesitic composition

There are a large number of discrimination diagrams applicable to basalts and basaltic andesites which use trace elements, major and minor elements and the mineral clinopyroxene. These are considered in turn. Table 5.1 classifies the different types of basalt according to tectonic setting and shows which diagrams might be useful in their identification.

5.2.1   Trace element discrimination diagrams

Trace element discrimination diagrams are the largest group of discrimination diagrams which can be used to suggest the former tectonic environment of a suite of basalts. They can also be used to identify basaltic magma series. Both of these applications are discussed below. First, however, there is a full discussion of the Ti–Zr–Y diagrams proposed by Pearce and Cann (1973) to illustrate both the use and the pitfalls of tectono-magmatic discrimination diagrams. This particular set of diagrams was chosen not so much for its current usefulness but because it has been thoroughly evaluated and discussed in the literature.

The Ti–Zr,
Ti–Zr–Y and
Ti–Zr–Sr
diagrams (Pearce
and Cann, 1973)

The discrimination diagrams of Pearce and Cann (1973) apply to tholeiitic basalts in the compositional range 20 % > CaO + MgO > 12 %. Alkali basalts should be screened using their low Y/Nb ratio (Y/Nb < 1.0 for alkali ocean-island and alkali continental basalts; Y/Nb < 2.0 for ocean-floor alkali basalts). Care should also be taken in plotting rocks with a high cumulate content, for absolute concentrations of Ti, Zr, Y, Nb and Sr will be reduced by dilution, although this will not affect the relative proportions plotted on triangular diagrams. Rocks containing cumulus Ti-bearing phases such as titanomagnetite and clinopyroxene will also give erroneous results and should be avoided.

(a)   The Ti–Zr–Y diagram   The diagram in Figure 5.1 most effectively discriminates between within-plate basalts, i.e. ocean-island or continental flood basalts (field D) and other basalt types. Island-arc tholeiites plot in field A and

Table 5.1 Discrimination diagrams which may be used to determine the tectonic setting of basalts

MORB	Figure
Ti–Zr	5.2(a,b)
Ti–Zr–Sr	5.3
Zr/Y–Zr	5.5
Ti–V	5.10
Cr–Y	5.12
Cr–Ce/Sr	5.12
TiO_2–Y/Nb	5.15
FeO–MgO–Al_2O_3	5.20
MnO–TiO_2–P_2O_5	5.23
Discr. function (Ti–Zr–Y–Sr)	5.4
Discr. function (majors)	5.19

N-type MORB

Zr–Nb–Y	5.8
Th–Hf–Ta	5.9
La–Y–Nb	5.11

E-type MORB

Zr–Nb–Y	5.8
Th–Hf–Ta	5.9
La–Y–Nb	5.11
FeO–MgO–Al_2O_3	5.20

Transitional MORB

K_2O/Yb–Ta/Yb	5.13

Back-arc basin tholeiites	Figure
Ti–V	5.10
La–Y–Nb	5.11
K_2O–H_2O	5.23

Within-plate basalts	Figure
Ti–Zr–Y	5.1
Ti–Zr	5.2(b)
Discr. function	5.4
Zr/Y–Zr	5.5(a)
Zr/Y–Ti/Y	5.6
Cr–Y	5.12(a)
Cr–Ce/Sr	5.12
Discr. function (majors)	5.19

Alkali

Zr/Y–Nb/Y	5.7
Zr–Nb–Y	5.8
Th–Hf–Ta	5.9
Ti–V	5.10
K_2O/Yb–Ta/Yb	5.13
TiO_2–Y/Nb	5.15

Tholeiitic

Ti/Y–Nb/Y	5.7
Zr–Nb–Y	5.8
K_2O/Yb–Ta/Yb	5.13

Transitional

Ti/Y–Nb/Y	5.7
K_2O/Yb–Ta/Yb	5.13

Within-plate basalts	Figure
Ocean-island tholeiites	
Ti–V	5.10
FeO–MgO–Al_2O_3	5.20
MnO–TiO_2–P_2O_5	5.22
K_2O–H_2O	5.23
Ocean-island alkali basalts	
MnO–TiO_2–P_2O_5	5.22
Continental tholeiites	
Ti–V	5.10
La–Y–Nb	5.11
TiO_2–Y/Nb	5.15
FeO–MgO–Al_2O_3	5.20
K_2O–TiO_2–P_2O_5	5.21
Continental alkali basalts	
La–Y–Nb	5.11

Volcanic-arc basalts	Figure
Ti–Zr	5.2(b)
Zr/Y–Zr	5.5(a)
Cr–Y	5.12(a)
Cr–Ce/Sr	5.12(b)
FeO–MgO–Al_2O_3	5.20
K_2O–H_2O	5.23

Island-arc tholeiites

Ti–Zr–Y	5.1
Ti–Zr	5.2
Ti–Zr–Sr	5.3
Discr. function (Ti–Zr–Y–Sr)	5.4
Th–Hf–Ta	5.9
Ti–V	5.10
La–Y–Nb	5.11
K_2O/Yb–Ta/Yb	5.13
Discr. function (majors)	5.19
MnO–TiO_2–P_2O_5	5.22

Continental-arc

Zr/Y–Zr	5.5(b)

Oceanic-arc

Zr/Y–Zr	5.5(b)

Calc–alkaline basalts

Ti–Zr–Y	5.1
Ti–Zr	5.2(a)
Ti–Zr–Sr	5.3
Discr. function	5.4
Th–Hf–Ta	5.9
La–Y–Nb	5.11
K_2O/Yb–Ta/Yb	5.13
Discr. function (majors)	5.19
MnO–TiO_2–P_2O_5	5.22

Shoshonitic basalts

K_2O/Yb–Ta/Yb	5.13
Discr. function (majors)	5.19

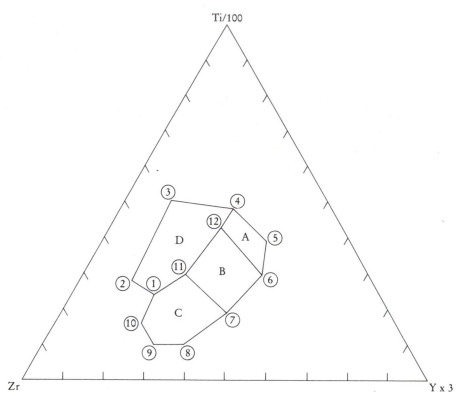

Figure 5.1   The Ti–Zr–Y discrimination diagram for basalts (after Pearce and Cann, 1973). A is the field of island-arc tholeiites, C the field of calc–alkali basalts, D is the field of within-plate basalts and B is the field of MORB, island-arc tholeiites and calc–alkali basalts. Rocks which plot in field B give an ambiguous result but can be separated by plotting on a Ti–Zr diagram (Figure 5.2) or if unaltered on a Ti–Zr–Sr diagram (Figure 5.3). The plotting coordinates, extracted from Pearce and Cann (1973 — Figure 3) are:

Point	Ti/100	Zr	Y × 3	Point	Ti/100	Zr	Y × 3
1	24	55.5	20.5	7	19	40	41
2	28	59	13	8	10	55	35
3	50	38.5	11.5	9	10	62.5	27.5
4	48	24	28	10	16	63	21
5	39	20.5	40.5	11	29.5	45	25.5
6	30	26	44	12	42.5	30	37.5

calc–alkaline basalts in field C. MORB, island-arc tholeiites and calc–alkaline basalts all plot in field B. Data points are calculated according to their assigned weightings [Ti/100 (ppm), Zr (ppm), Y x 3 (ppm)], recast to 100 %, and plotted on the triangular diagram in the manner described in Section 3.3.2.

(b) The Ti–Zr diagram   Figure 5.2(a) has four fields. Fields A, C and D contain island-arc tholeiites, calc–alkali basalts and MORB respectively and field B contains

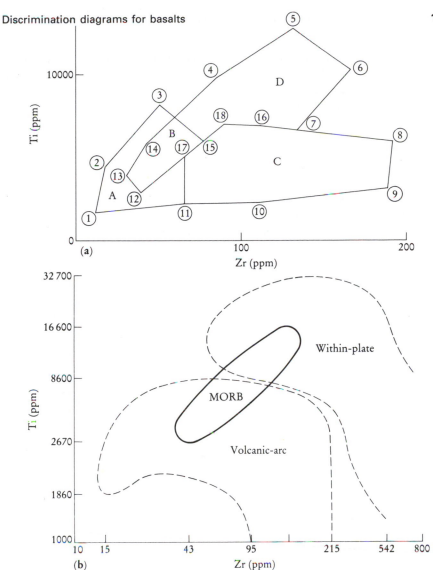

Figure 5.2   Discrimination diagrams for basalts based upon Ti–Zr variations. (a) Linear scale (after Pearce and Cann, 1973); the fields are as follows: A, island-arc tholeiites; B, MORB, calc–alkali basalts and island-arc tholeiites; C, calc–alkali basalt; D, MORB. The plotting coordinates (extracted from Pearce and Cann, 1973 — Figure 2) are as follows:

Point	Zr	Ti	Point	Zr	Ti
1	12	1 700	10	113	2 400
2	18	4 600	11	65	2 400
3	52	8 500	12	39	3 100
4	84	10 400	13	31	4 100
5	131	13 000	14	42	6 000
6	167	10 900	15	78	6 400
7	135	7 100	16	113	7 400
8	192	6 600	17	65	5 400
9	190	3 400	18	89	7 400

(b) Log scale (after Pearce, 1982) showing the fields of volcanic-arc basalts, MORB and within-plate basalts. The values given on the ordinate and abscissa can be used as a guide in drawing the field boundaries (data extracted from Pearce, 1982).

all three types. A modified version of this diagram, extended in compositional range to include within-plate lavas, is given by Pearce (1982) (Figure 5.2b). The same diagram presented in the form TiO$_2$–Zr, expressed as wt % and ppm respectively, is used by Floyd and Winchester (1975) to illustrate the compositional differences between alkali and tholeiitic basalts. On this diagram (not shown) there is considerable overlap between the tholeiites and alkali basalts from ocean-ridge and ocean-island settings.

(c) The Ti–Zr–Sr diagram   Figure 5.3 can only be used for fresh samples because Sr is a relatively mobile element in hydrothermal fluids. The main function of this diagram is to subdivide the rocks which plot in field B of the Ti–Zr–Y diagram (Figure 5.1) into their different tectonic settings. Island-arc tholeiites plot in field A

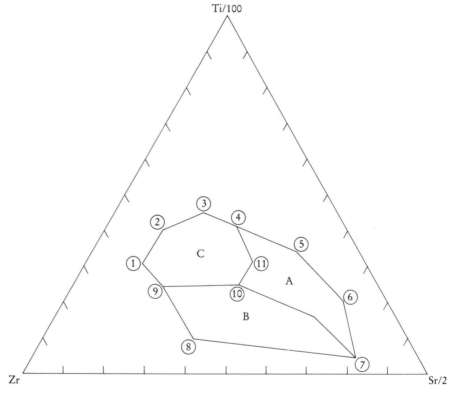

Figure 5.3   The Ti–Zr–Sr discrimination diagram for basalts (after Pearce and Cann, 1973). Island-arc tholeiites plot in field A, calc-alkaline basalts plot in field B and MORB plot in field C. The plotting coordinates for this diagram (extracted from Pearce and Cann, 1973 — Figure 4) are:

Point	Ti/100	Zr	Sr/2	Point	Ti/100	Zr	Sr/2
1	31	55	14	7	5	15	80
2	40	45.5	14.5	8	9.5	53	37.5
3	45	33	22	9	24	53.5	22.5
4	41	27	32	10	24	35	41
5	34	15.5	50.5	11	31	28	41
6	20.5	11	68.5				

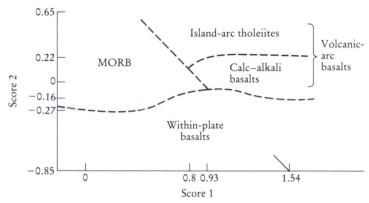

Figure 5.4 Plot of the first principal component score vs the second principal component score for 35 basalt trace element analyses.

Score 1 = −0.03707Ti − 0.0668Zr − 0.3987Y + 0.8362Sr
Score 2 = −0.3376Ti − 0.5602Zr + 0.7397Y + 0.1582Sr

The fields are those used by Pearce and Cann (1973). The approximate coordinates of the boundaries are extracted from Butler and Woronow (1986 — Figure 2).

of Figure 5.3, calc–alkaline basalts in field B and MORB in field C. Individual data points are calculated according to the weighting Ti/100 (ppm), Zr (ppm), Sr/2 (ppm), and recast to 100 %. Sr mobility has been demonstrated during zeolite and prehnite–pumpellyite facies metamorphism of basalts (Morrison, 1978; Smith and Smith, 1976).

(d) Discussion Pearce and Cann (1973) intended that the three diagrams described above should be used together and in the following way. The Ti–Zr–Y diagram (Figure 5.1) should be used first to identify within-plate basalts, then the Ti–Zr (Figure 5.2) and if the samples are unaltered the Ti–Zr–Sr diagram (Figure 5.3) to identify the other basalt types. This suite of diagrams has been widely used and a number of authors have reported inconsistencies in its use. These may be summarized as follows.

(1) Few samples of continental tholeiites were used in the construction of the Ti–Zr–Y diagram and continental flood basalts do not plot in the 'within-plate' setting — field D of Figure 5.1 — (Holm, 1982; Duncan, 1987).

(2) The effects of crustal contamination were not considered by Pearce and Cann (1973) and yet this is likely to have a significant effect on the chemistry of continental flood basalts and may be a contributory factor in their persistent misclassification (see Wood *et al.*, 1979c).

(3) Since the publication of the Ti–Zr–Y diagram we have learned a great deal about the chemistry of ocean-floor basalts and a number of types of MORB are now known. This point is made by Prestvick (1982), who showed that anomalous ocean-floor basalts from Iceland do not plot in the MORB field. More recent discrimination diagrams, however, can recognize different types of MORB (see Table 5.1).

(4) Recalculating results to 100 % prior to plotting on a triangular diagram causes samples to become grouped even when no real association exists (Section 2.5). Butler and Woronow (1986) discussed this problem and proposed an

alternative approach using principal component analysis (Section 2.9). Their results are plotted in a bivariate graph (Figure 5.4) which uses the scores of the first two principal components (functions of Ti, Zr, Y and Sr) as axes to discriminate between the four environments identified by Pearce and Cann (1973). However, whilst this graph may be more mathematically sound than the original Ti–Zr–Y diagram, the boundaries are drawn on the basis of only 35 analyses (averages) taken from the literature and so should be treated as preliminary.

Other discrimination diagrams using Ti–Zr–Y–Nb variations

A number of other tectono-magmatic discrimination diagrams have been proposed based upon Ti–Zr–Y variations and these are discussed below. In addition the HFS element Nb is used, although diagrams which introduce Nb should be used with care. Nb concentrations cannot be accurately determined by XRF below about 10 ppm and precise analyses must be made by INAA.

(a) The Zr/Y–Zr diagram for basalts (Pearce and Norry, 1979)   Pearce and Norry (1979) found that the ratio Zr/Y plotted against the fractionation index Zr proved an effective discriminant between basalts from ocean-island arcs, MORB and within–plate basalts (Figure 5.5a). Arc basalts plot in fields A and D, MORB in fields B, D and E and within–plate basalts in fields C and E. Pearce (1980) contoured the MORB field of this diagram for spreading rate.

The Zr/Y–Zr diagram can also be used to subdivide island-arc basalts into those belonging to oceanic arcs, where only oceanic crust is used in arc construction, and arcs developed at active continental margins (Pearce, 1983). Oceanic arcs plot in the island-arc field originally defined by Pearce and Norry (1979) whilst continental-arc basalts plot with higher Zr/Y and higher Zr (Figure 5.5b).

Pearce and Gale (1977) use a similar diagram based upon Zr/Y and Ti/Y variations to discriminate between within–plate basalts and other types of basalt, collectively termed 'plate margin basalts' (Figure 5.6). This diagram makes use of the enrichment in Ti and Zr but not Y in within–plate basalts.

(b) The Ti/Y–Nb/Y diagram (Pearce, 1982)   This diagram, shown in Figure 5.7, successfully separates the within–plate basalt group from MORB and volcanic-arc basalts, which overlap extensively on this plot. Within–plate basalts have higher Ti/Y and higher Nb/Y than the other types of basalt, differences which are thought to reflect an enriched mantle source relative to the sources of MORB and volcanic-arc basalts. Differences in Nb/Y ratio allow the within–plate basalt group to be further subdivided into tholeiitic, transitional and alkaline types. The broad linear array of fields on this diagram may result from the common denominator effect discussed in Section 2.5.

(c) The Zr–Nb–Y diagram (Meschede, 1986)   As our knowledge of oceanic basalt chemistry has expanded over the past 20 years, it has beome apparent that there is more than one type of MORB or 'ocean-floor basalt' as it was called by Pearce and Cann (1973). Meschede (1986) suggested that the immobile trace element Nb can be used to separate the different types of ocean-floor basalt and recognized two types of MORB. These are N-type MORB, basalt from a 'normal' mid-ocean ridge environment and depleted in incompatible trace elements, and E-type MORB (also known as P-type MORB) — ocean-floor basalts from plume-influenced regions such as Iceland which are generally enriched in incompatible trace elements.

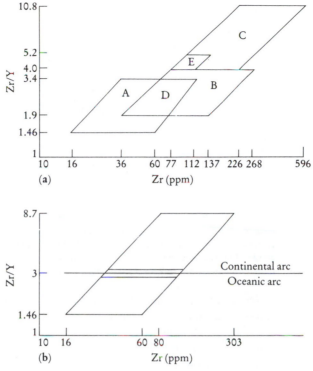

Figure 5.5 Discrimination diagrams for basalts based upon Zr/Y–Zr variations; both diagrams have a logarithmic scale. (a) The fields are A, volcanic-arc basalts; B, MORB; C, within-plate basalts; D, MORB and volcanic-arc basalts; E, MORB and within-plate basalts (after Pearce and Norry, 1979). The values along the ordinate and abscissa are on a logarithmic scale and are given to assist in constructing the field boundaries (data extracted from Pearce and Norry, 1979 — Figure 3). (b) Fields of continental and oceanic-arc basalts separated on the basis of a Zr/Y value of 3. The shaded area is the field of overlap between the two basalt types (after Pearce, 1983).

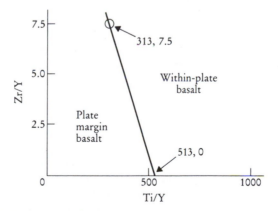

Figure 5.6 The Zr/Y–Ti/Y discrimination diagram for basalts showing the fields of within-plate basalts and plate margin basalts (i.e. all other basalt types). Coordinates are given for the bounding line between the two fields (after Pearce and Gale, 1977).

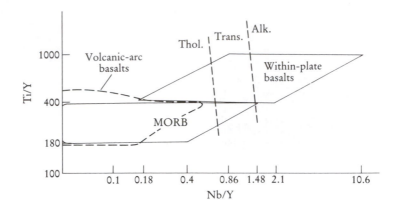

Figure 5.7 The Ti/Y–Nb/Y discrimination diagram for basalts showing the fields of within-plate basalts, MORB and volcanic-arc basalts (dashed line). The within-plate basalts may be divided into tholeiitic (Thol.), transitional (Trans.) and alkali (Alk.) basalt types (after Pearce, 1982). The values given along the ordinate and abscissa are to allow the construction of the field boundaries and are extracted from Pearce (1982 — Figure 9a).

On a triangular plot of Zr/4, 2 × Nb and Y (Figure 5.8) Meschede (1986) showed that four main basalt fields can be identified. The plotting coordinates are given in the caption to Figure 5.8 and Garcia and Frias (1990) have written a computer program in BASIC for plotting rock compositions on this diagram. The fields were defined on the basis of more than 1800 analyses of modern basalts selected from the composition range 20 wt % > CaO + MgO > 12 wt %.

Within-plate alkali basalts plot in field A; within-plate tholeiites plot in fields AII and C. E-type MORB plots in field B whilst N-type MORB plots in field D. Volcanic-arc basalts also plot in fields C and D. The several areas of overlap mean that only within-plate alkali basalts and E-type MORB can be identified without ambiguity.

(d) The causes of Ti–Zr–Y–Nb variations in basalts from different tectonic settings Pearce and Norry (1979) investigated the likely reasons for variation in Zr/Y and Ti/Y between basalt suites and concluded that they are most probably related to long-lived source heterogeneities. Certainly it seems to be true that the differences between within-plate basalts (high Zr/Y) and other basalt types reflects a difference in the mantle source regions. Differences in Zr/Nb and Y/Nb between alkali and tholeiitic basalts can be explained in a similar way. Island-arc basalts and MORB, however, may come from the same source, but in this case the lower absolute abundances of Ti, Zr, Y and Nb in island-arc basalts represent a greater degree of partial melting.

The Th–Hf–Ta diagram of Wood (1980) A discrimination diagram based upon the immobile HFS elements Th–Hf–Ta was proposed by Wood (1980). In order to expand and centre the fields of basalt types, concentrations are plotted (in ppm) as Th, Hf/3 and Ta. The elements Th, Hf and Ta are present in very low concentrations in basalts and cannot be accurately determined by XRF analysis and so must be determined by INAA. In cases where reliable Hf and Ta analyses are not given but Zr and Nb concentrations have been

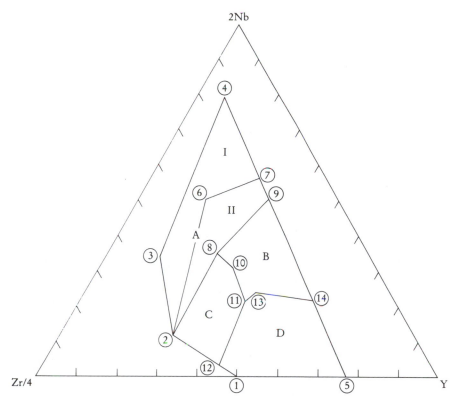

Figure 5.8   The Zr–Nb–Y discrimination diagram for basalts (after Meschede, 1986). The fields are defined as follows: AI, within-plate alkali basalts; AII, within-plate alkali basalts and within-plate tholeiites; B, E-type MORB; C, within-plate tholeiites and volcanic-arc basalts; D, N-type MORB and volcanic-arc basalts. The plotting coordinates for the boundary lines given by Meschede (1986) are:

Point	2Nb	Zr/4	Y	Point	2Nb	Zr/4	Y
1	0	50	50	8	35	37.5	27.5
2	12	60	28	9	50	17	33
3	34	52.5	13.5	10	31	36	33
4	79	14	7	11	22	37	41
5	0	23	77	12	3	53	44
6	50	33	17	13	24	33.5	42.5
7	56.5	16.5	27	14	22	21	57

measured, Hf concentrations may be obtained from a Zr/Hf ratio of 39 and Ta concentrations from a Nb/Ta ratio of 16 (Wood *et al.*, 1979c).

The chief features of this diagram are:

(1)   it recognizes different types of MORB;
(2)   it can be applied to intermediate and silicic lavas as well as to basalts; and
(3)   it is particularly good at identifying volcanic-arc basalts.

Figure 5.9 shows the fields for the different magma types. N-type MORB plots in field A whilst E-type MORB and tholeiitic within-plate basalts both plot in field B. It is not possible, therefore, to discriminate between E-type MORB and within-plate tholeiites on this diagram, although this separation can be made using the Ti–Zr–Y diagram of Pearce and Cann (1973). Within-plate alkali basalts plot on the Th–Hf–Ta diagram in field C and volcanic-arc basalts in field D. Field D may be subdivided into island-arc tholeiites (primitive arc tholeiites) — lavas with an Hf/Th ratio greater than 3 — and calc–alkaline basalts with an Hf/Th ratio less than 3.

There is some uncertainty over the mobility of Th in altered basalts and so ideally this diagram should only be used when samples do not contain a large proportion of altered glass. Crystal fractionation can also cause points to plot in the wrong fields. The fractionation of most silicate minerals tends to remove Ta and Hf from the melt and the residual liquid is pushed towards higher Th concentrations. Magnetite, on the other hand concentrates Ta relative to Hf and Th and so magnetite accumulation will displace liquid compositions towards the Ta apex of the diagram.

The Ti–V diagram of Shervais (1982)

Ti and V are adjacent members of the first transition series in the periodic table and yet in silicate systems they behave in different ways. This is the basis of the discrimination diagram of Shervais (1982), which is used to distinguish between volcanic-arc tholeiites, MORB and alkali basalts.

Partition coefficients for V are very variable and in minerals such as orthopyroxene, clinopyroxene and magnetite vary over several orders of magnitude as a function of oxygen activity. This is because V can exist in reduced (V^{3+}) or oxidized (V^{4+}, V^{5+}) states in natural magmas. In contrast Ti exists only as Ti^{4+}. Variations in concentrations of V, relative to Ti, act therefore as a measure of the oxygen activity of a magma and of the crystal fractionation processes which have taken place. These parameters, in turn, can be linked to the environment of eruption and are used as the basis for this discrimination diagram. Ti and V are immobile under conditions of hydrothermal alteration and at intermediate-to-high grades of metamorphism.

The different basalt fields are subdivided according to Ti/V ratio (Figure 5.10). MORB plots between Ti/V ratios of 20 and 50, although there is considerable overlap with the fields of continental flood basalt and back-arc basin basalts. Ocean-island and alkali basalts plot between Ti/V ratios of 50 and 100. Island-arc tholeiites plot between Ti/V ratios of 10 and 20 with a small overlap onto the field of MORB. Calc–alkali lavas plot with a near-vertical trend and with Ti/V ratios between 15 and 50.

The La–Y–Nb diagram of Cabanis and Lecolle (1989)

Using a comparatively small number of samples, Cabanis and Lecolle (1989) constructed a triangular diagram based upon La–Y–Nb concentrations which discriminates between volcanic-arc basalts, continental basalts and oceanic basalts. This diagram has not yet been widely tested but offers another means of discriminating between different types of MORB. Elemental concentrations are plotted in ppm as La/10, Y/15 and Nb/8 and the three main fields are further subdivided (Figure 5.11). Volcanic-arc basalts plot in field 1 and are subdivided into calc–alkali basalts (1A) and island-arc tholeiites (1C). Field 1B is where the two plot together. Field 2 characterizes continental basalts and field 2B may define continental back-arc tholeiites, although this subdivision is based upon a single

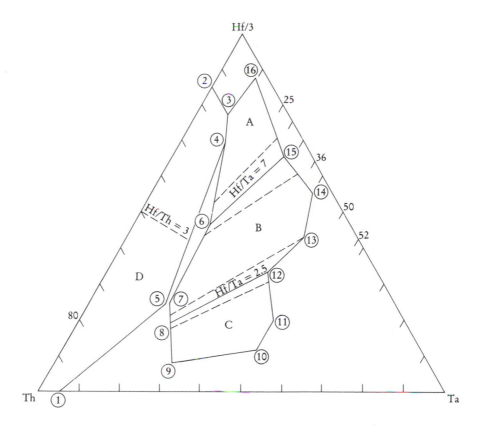

Figure 5.9   The Th–Hf–Ta discrimination diagram for basalts (after Wood, 1980). The fields are: A, N-type MORB; B, E-type MORB and within-plate tholeiites; C, alkaline within-plate basalts; D, volcanic-arc basalts. Island-arc tholeiites plot in field D where Hf/Th > 3.0 and calc–alkaline basalts where Hf/Th < 3.0. The broken lines indicate transitional zones between basalt types. The plotting coordinates for the boundary lines (extracted from Wood, 1980 — Figure 1) are:

Point	Th	Hf/3	Ta	Point	Th	Hf/3	Ta
1	95	0	5	9	63	8	29
2	15	85	0	10	40	12	48
3	15	77	8	11	32	20	48
4	20	69	11	12	27	33	40
5	56.5	24.5	19	13	13.5	42.5	44
6	35	46	19	14	5	55	40
7	55.5	24.5	20	15	7	65	28
8	57	20	23	16	3	87.5	9.5

value. Field 3 defines oceanic basalts and is subdivided as follows. Field 3D is N-type MORB; fields 3C and 3B E-type MORB (also known as P-type MORB) and field 3A is defined by alkali basalts from the Kenyan Rift.

There is some evidence that La is mobile under hydrothermal conditions and so highly altered and metamorphic rocks may show some distortion relative to the La apex.

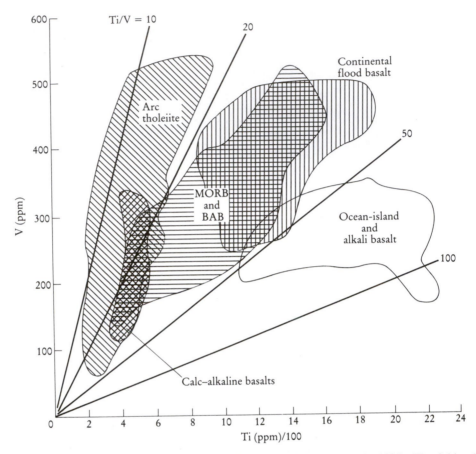

Figure 5.10   The Ti–V discrimination diagram for basalts (compiled from Shervais, 1982). The fields of arc tholeiite (diagonal hatching), MORB and back-arc basin basalts (BAB) (horizontal hatching), continental flood basalts (vertical hatching) and ocean-island and alkali basalt (unshaded) are recognized by their Ti/V ratio as shown. Calc–alkaline basalts (shaded) plot with low Ti concentrations with a wide range of Ti/V ratios.

Diagrams which preferentially select volcanic-arc basalts   Pearce (1982, 1983) used MORB-normalized spider diagrams (see Section 4.4.1) to identify the trace elements which best characterize magmas produced in an island-arc environment. From these studies three distinctive features of volcanic-arc basalts emerge (Pearce, 1982). Firstly, they are enriched in the elements Sr, K, Rb, Ba and Th relative to Ta and Cr. Secondly, island-arc tholeiites have low abundances of the high field strength elements relative to MORB and, thirdly, calc–alkaline basalts are enriched in Th, Ce, P and Sm relative to the other HFS elements. These features are used as basis for a number of discrimination diagrams which can be used to distinguish between MORB, within-plate basalts and volcanic-arc basalts.

(a) The Cr–Y diagram (Pearce, 1982)   The low concentrations of Cr in volcanic-arc basalts relative to other basalt types has been used in a number of discrimination diagrams to help characterize volcanic-arc basalts (Pearce and Gale, 1977; Garcia, 1978; Bloxham and Lewis, 1972). Cr is compatible in the minerals olivine,

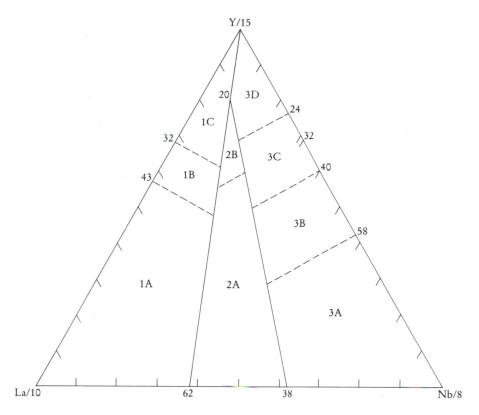

Figure 5.11   The La/10–Y/15–Nb/8 discrimination diagram for basalts (after Cabanis and Lecolle, 1989). The plotting coordinates are shown around the margin of the diagram; the broken lines are drawn normal to the sides of the triangular diagram. Field 1 contains volcanic-arc basalts, field 2 continental basalts and field 3 oceanic basalts. The subdivisions of the fields are as follows: 1A, calc–alkali basalts; 1C, volcanic-arc tholeiites; 1B is an area of overlap between 1A and 1C; 2A, continental basalts; 2B, back-arc basin basalts (although this is less well defined); 3A, alkali basalts from intercontinental rift; 3B, 3C, E-type MORB (3B enriched, 3C weakly enriched), 3D, N-type MORB.

orthopyroxene and clinopyroxene and the spinels in a basaltic melt. The low levels of Cr in volcanic-arc rocks therefore is either a function of a different amount of mantle melting from MORB and/or a difference in the fractionation history. The precise cause is difficult to define. Y is also depleted in island-arc basalts relative to other basalt types, for a given degree of fractionation. Thus a Cr vs Y plot (Figure 5.12a) discriminates effectively between MORB and volcanic-arc basalts, with only a small amount of overlap between the two fields. Within–plate basalts, on the other hand, overlap the fields of MORB and volcanic-arc basalts. The wide range of Cr values in the volcanic-arc basalt field is most efficiently obtained through crystal fractionation, indicating that Cr is a useful fractionation index in these rocks.

(b) The Cr–Ce/Sr diagram   Ce and Sr behave in a very similar way in MORB but, because of their differing mobilities in aqueous fluids, behave very differently in volcanic-arc basalts. When the data are displayed on a MORB-normalized spider diagram, volcanic-arc basalts show an enrichment in Sr relative to Ce, whereas in MORB Ce and Sr have very similar normalized concentrations. The ratio Ce/Sr is

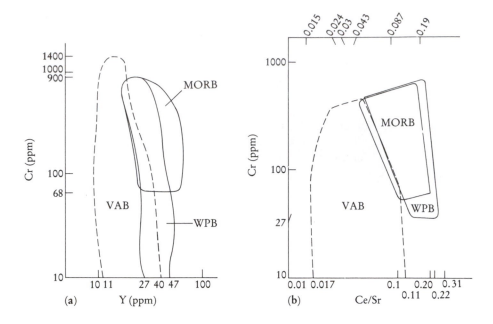

Figure 5.12 (a) The Cr–Y discrimination diagram for basalts (after Pearce, 1982), showing the fields for MORB, volcanic-arc basalts (VAB) and within-plate basalts (WPB). The plotting coordinates useful in constructing the diagram are extracted from Pearce (1982 — Figure 2). (b) The Cr–Ce/Sr discrimination diagram for basalts (after Pearce, 1982), showing the fields for MORB, volcanic-arc basalts (VAB) and within-plate basalts (WPB). The plotting coordinates useful in constructing the diagram are extracted from Pearce (1982 — Figure 5).

therefore a useful discriminant between MORB and volcanic-arc basalts. On a plot of Cr vs Ce/Sr (Figure 5.12b) volcanic-arc basalts have lower Ce/Sr ratios than MORB and within-plate basalts for the same Cr content.

Distinguishing between different types of volcanic-arc basalt

In contrast to the preceding diagrams in which the concentrations of trace elements vary chiefly as a consequence of different degrees of partial melting and crystal fractionation, the diagrams presented in this section show variations which are attributed to chemical differences in the source region. These differences might arise from mantle heterogeneity caused by melt movement or from the addition or removal of elements in a fluid phase.

Three element ratios are used — K/Ta, Ce/Ta and Th/Ta — each of which highlights the chemical difference between volcanic-arc basalts and MORB. In each case, however, the elements are ratioed to a common denominator (Yb) for the purposes of showing variations in the source region prior to subduction. Yb is assumed to be immobile in an aqueous fluid and to behave as an incompatible element and so the ratio of two incompatible elements such as Th/Yb will remain unchanged during partial melting and fractional crystallization (see Section 4.7.1). The same will be true for the Ta/Yb ratio and variations on a Th/Yb vs Ta/Yb diagram therefore reflect differences in source composition. The reader should beware, however, of the possible common denominator effect in these diagrams (Section 2.5) which could produce spurious linear correlations.

A number of subdivisions can be made of the conventional basaltic field on these diagrams. Volcanic-arc basalts are subdivided into tholeiitic, calc–alkali and shoshonitic varieties. [Shoshonites are a group of K-rich, near-saturated rocks ranging in composition from basalt to dacite with both calc–alkaline and alkaline affinities (Morrison, 1980)]. MORB and within–plate basalts are also subdivided into tholeiitic, alkali and transitional types.

The K_2O/Yb–Ta/Yb diagram   This diagram (after Pearce, 1982) is based upon the difference in behaviour between K and Ta in volcanic-arc basalts. In contrast to MORB, where the K/Ta ratio is almost constant, the greater mobility of K in an aqueous fluid relative to Ta means that in volcanic-arc basalts the K/Ta will always be high. The diagram (Figure 5.13) discriminates well between volcanic-arc basalts with high K_2O/Yb and MORB and within-plate basalts with a lower ratio. It also discriminates between tholeiitic, calc–alkaline and shoshonitic volcanic-arc basalts and between tholeiitic, transitional and alkaline MORB and within-plate basalts.

Source heterogeneity will affect both K and Ta equally on this diagram as they are both incompatible elements, and so mantle compositions will move to higher or lower K_2O/Yb and Ta/Yb ratios relative to primordial mantle along a slope of unity (Figure 5.13). Fluid enrichment, on the other hand, will enrich K but not Ta and will show as a trend parallel to the K_2O/Yb axis (Pearce, 1982). This diagram

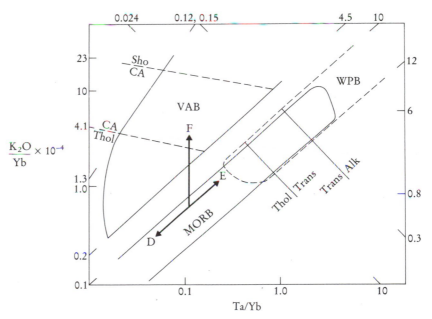

Figure 5.13   Discrimination diagram for basalts based upon K–Ta covariations and using Yb as a normalizing factor (after Pearce, 1982). The diagram shows the fields of volcanic-arc basalts (VAB), MORB and within-plate basalts (WPB). Volcanic-arc basalts are subdivided into tholeiitic (Thol), calc–alkaline (CA) and shoshonitic (Sho) varieties. MORB and within-plate basalts are subdivided into tholeiitic (Thol), transitional (Trans) and alkaline (Alk) varieties. Alkaline volcanic-arc basalts also plot in the alkaline field. The plotting coordinates, shown at the margin of the diagram, are extracted from Pearce (1982 — Figure 6). The solid arrows indicate the direction of mantle depletion (D), mantle enrichment (E) and enrichment via a fluid phase (F).

cannot be used in altered volcanic rocks because the mobility of K in aqueous fluids will produce very unreliable results. Instead, the less mobile elements Ce or Th can be used in place of K in analogous plots (Pearce, 1982).

Diagrams which distinguish between different types of volcanic-arc andesite (Bailey, 1981)

Bailey (1981) recognized four types of andesite — low-K oceanic island-arc andesites, 'other' oceanic island-arc andesites, continental island-arc andesites and Andean (active continental margin) andesites. These may be distinguished on the basis of their La/Yb and Sc/Ni ratios (Figure 5.14) and their Th content. The La/Yb ratio may be taken as a measure of the extent to which continental crust is involved in the magma genesis. A further group of andesites — the 'anorogenic' group — plot in the field of arc-related volcanism (Arculus, 1987).

Diagrams which discriminate between the alkali basalt and tholeiitic magma series

Floyd and Winchester (1975) and Winchester and Floyd (1976) proposed a series of diagrams based upon immobile HFS elements which discriminate between tholeiitic and alkali basalts. These diagrams are most useful in altered volcanic rocks where standard chemical tests for the tholeiitic and alkali magma series such as the TAS diagram (Section 3.2.1) are inappropriate because of the mobility of the alkali metals in aqueous solutions. These diagrams effectively discriminate between magma series but they do not allow the identification of tectonic setting. The only exception is the TiO_2–Y/Nb diagram, which shows some separation between continental tholeiites and MORB.

(a) The TiO_2–Y/Nb diagram (Floyd and Winchester, 1975) Alkali basalts tend to

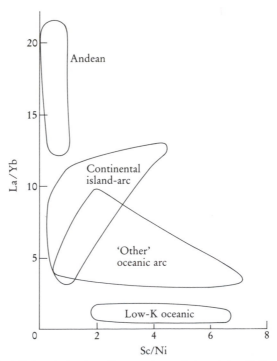

Figure 5.14   The La/Yb–Sc/Ni discrimination diagram for andesites (after Bailey, 1981) showing the fields of Andean-type (active continental margin), continental island-arc, low-K oceanic island-arc and 'other' oceanic island-arc andesites.

have low Y/Nb, a feature used by Pearce and Cann (1973) for eliminating them from their discrimination diagrams. Thus a plot of TiO_2 vs Y/Nb (Figure 5.15) shows three fields — MORB, alkali basalts (this includes continental and ocean-island alkali basalts) and the continental tholeiites. No field is unique but there is only a small amount of overlap. Winchester and Floyd (1976) report that the Y/Nb ratio is constant during metamorphism and alteration, the only exceptions being in the margins of altered pillow lavas and in metadolerites.

(b) The P_2O_5–Zr diagram (Floyd and Winchester, 1975)   Alkali basalts have higher P_2O_5 than tholeiitic basalts for a given Zr content and a straight line can be drawn separating the two fields (Figure 5.16). There is considerable overlap, however, between the fields of oceanic and continental tholeiites and between the fields of oceanic and continental alkali basalts.

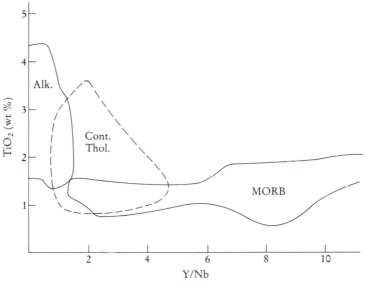

Figure 5.15   The TiO_2–Y/Nb discrimination diagram for basalts (adapted from Floyd and Winchester, 1975), showing the fields of alkali basalts (Alk.), continental tholeiites (Cont. thol.) and MORB.

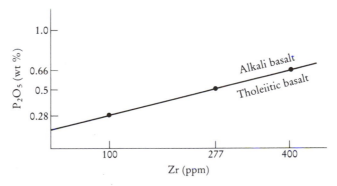

Figure 5.16   The P_2O_5–Zr discrimination diagram for basalts (after Winchester and Floyd, 1976) showing the fields of alkali basalts and tholeiitic basalts. The plotting coordinates of the boundary line are extracted from Winchester and Floyd (1976 — Figure 6).

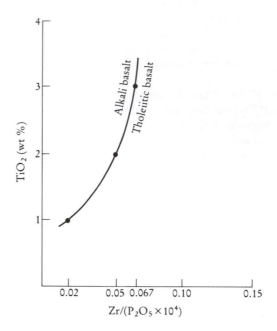

Figure 5.17   The TiO$_2$–Zr/(P$_2$O$_5$ × 10 000) discrimination diagram for basalts (after Winchester and Floyd, 1976) showing the fields of alkali basalt and tholeiitic basalt. The plotting coordinates are:

TiO$_2$	Zr/(P$_2$O$_5$ × 10 000)
1.00	0.020
2.00	0.054
3.00	0.067

(extracted from Winchester and Floyd, 1976 — Figure 2).

(c) The TiO$_2$–Zr/P$_2$O$_5$ diagram (Floyd and Winchester, 1975)   This diagram plots the TiO$_2$ content as wt % against the ratio Zr(ppm)/[P$_2$O$_5$(wt %) ×10 000]. It serves the same purpose as the Ti–Y/Nb diagram but has the advantage that Ti, Zr and P$_2$O$_5$ are more commonly determined in basalts, than are Y and Nb. There is almost complete separation between the fields of tholeiitic and alkali basalts with alkali basalts plotting in the field of low Zr/P$_2$O$_5$ and high TiO$_2$ (Figure 5.17). Winchester and Floyd (1976) found that the Zr/P$_2$O$_5$ ratio can change with progressive alteration, due to the mobility of P. This means that this diagram should be used with care when applied to strongly altered rocks.

(d) The Nb/Y–Zr/P$_2$O$_5$ diagram (Floyd and Winchester, 1975)   A plot of Nb/Y vs Zr/P$_2$O$_5$ produces the best discrimination between alkali and tholeiitic basalts. A boundary line separating the two fields from Winchester and Floyd (1976) is shown in Figure 5.18.

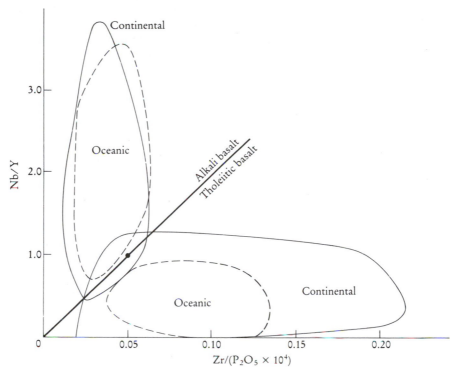

Figure 5.18   The Nb/Y–Zr/($P_2O_5 \times 10\,000$) discrimination diagram for basalts (adapted from Floyd and Winchester, 1975), showing the fields of continental and oceanic alkali basalts and tholeiitic basalts. The boundary line and plotting coordinates are taken from Winchester and Floyd (1976 — Figure 11).

(e) Discussion   A number of workers have used this suite of diagrams with varying degrees of success. Morrison (1978) found that the P_2O_5–Zr, TiO_2–Zr/P_2O_5 and the Nb/Y–Zr/P_2O_5 diagrams all failed to discriminate between hypersthene and nepheline normative basalts, metamorphosed at zeolite and greenschist facies. All the basalts plotted in the tholeiite field, whereas in reality they are probably transitional in chemistry between alkali and tholeiitic basalts. Clearly, the diagrams of Winchester and Floyd (1975) are unable to cope with such subtleties. Smith and Smith (1976), on the other hand, found that the Nb/Y–Zr/P_2O_5 diagram showed a close grouping of samples in the tholeiite field and successfully classified metamorphosed basalts from the prehnite–pumpellyite facies.

5.2.2   Discrimination diagrams for basalts based upon major elements

Discrimination diagrams which utilize the major elements are less likely to be successful than those for trace elements, for there is extensive overlap in major element chemistry between MORB, back-arc basin tholeiites and volcanic-arc basalts (Perfit *et al.*, 1980). This is because there is a large number of possible variables controlling the element concentrations and fewer elements to choose from. Thus it is difficult to identify elements which are either completely immobile or are unaffected by the effects of crystal fractionation.

Pearce (1976) calculated discriminant functions based upon the eight major element oxides SiO_2, TiO_2, Al_2O_3, FeO (recalculated from total Fe), MgO, CaO, Na_2O and K_2O and presented discriminant diagrams to identify MORB, within-plate basalts (ocean-island/continental basalts), calc–alkaline basalts, island-arc tholeiites and shoshonites. The method he employed is discussed in Section 2.9. The boundaries on the discriminant diagram were based upon fresh modern basalts (samples with $FeO/Fe_2O_3 < 0.5$ were rejected) in the compositional range 20 wt % > CaO +

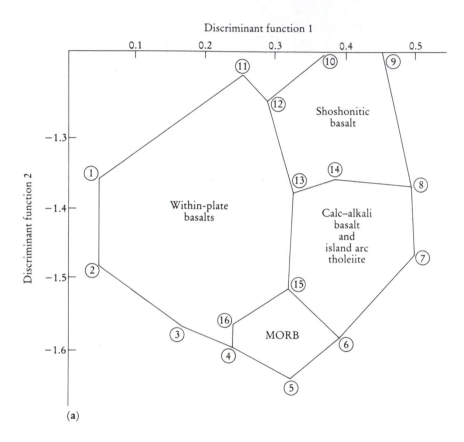

Figure 5.19   (a) Major element discrimination diagrams for basalts (after Pearce, 1976), showing the fields of within-plate basalts, MORB, calc–alkali basalts, island-arc tholeiites and shoshonitic basalts. The functions F1, F2 and F3 are defined in the text in terms of eight major element oxides. Discrimination diagram based upon functions F1 and F2; the plotting coordinates (extracted from Pearce, 1976 — Figure 1) are:

Point	Function 1	Function 2	Point	Function 1	Function 2
1	0.05	−1.36	9	0.455	−1.2
2	0.05	−1.49	10	0.37	−1.2
3	0.16	−1.58	11	0.255	−1.23
4	0.235	−1.61	12	0.29	−1.265
5	0.32	−1.655	13	0.325	−1.395
6	0.39	−1.6	14	0.38	−1.38
7	0.5	−1.485	15	0.32	−1.53
8	0.5	−1.385	16	0.24	−1.58

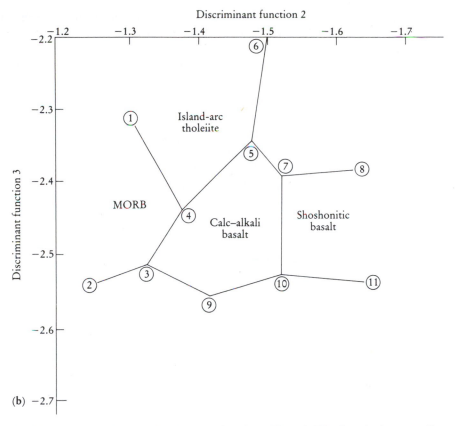

Figure 5.19 (b) Discrimination diagram based upon functions F2 and F3; the plotting coordinates (extracted from Pearce, 1976 — Figure 2) are:

Point	Function 2	Function 3	Point	Function 1	Function 2
1	−1.3	−2.3	7	−1.52	−2.39
2	−1.26	−2.53	8	−1.63	−2.38
3	−1.33	−2.51	9	−1.42	−2.55
4	−1.38	−2.44	10	−1.52	−2.52
5	−1.48	−2.34	11	−1.63	−2.53
6	−1.5	−2.2			

$MgO > 12$ wt %, with sums (including H_2O) between 99 and 101 %. Three discriminant functions, F1, F2 and F3 were identified

$$F1 = + 0.0088SiO_2 - 0.0774TiO_2 + 0.0102 \ Al_2O_3 + 0.0066FeO - 0.0017MgO$$
$$- 0.0143CaO - 0.0155Na_2O - 0.0007K_2O$$
$$F2 = - 0.0130SiO_2 - 0.0185TiO_2 - 0.0129 \ Al_2O_3 - 0.0134FeO - 0.0300MgO$$
$$- 0.0204CaO - 0.0481Na_2O - 0.0715K_2O$$
$$F3 = - 0.2210SiO_2 - 0.0532TiO_2 - 0.0361 \ Al_2O_3 - 0.0016FeO - 0.0310MgO$$
$$- 0.0237CaO - 0.0614Na_2O - 0.0289K_2O$$

A plot of F1 and F2 separates MORB, volcanic-arc basalts, shoshonites and within-plate basalts (ocean-island basalts and continental tholeiites) from each other (Figure 5.19a). A plot of F2 and F3 separates island-arc tholeiites, calc–alkali basalts and shoshonites from each other and from MORB (Figure 5.19b). Submarine weathering has a severe effect on these variables and only 46.7 % of weathered samples were correctly classified. The effects of greenschist facies metamorphism are less significant.

The MgO–FeO–Al₂O₃ diagram of T.H. Pearce et al. (1977)

The MgO–FeO– Al_2O_3 *diagram* *of T.H. Pearce* *et al. (1977)*

Pearce *et al.* (1977) used a data file of 8400 analyses of recent volcanic rocks to discriminate between basalts from different tectonic environments on the basis of their major element chemistry. This diagram is different from most others described in this chapter because it applies to rocks in the silica range 51–56 wt % (analyses recalculated dry), i.e. for subalkaline basalts and basaltic andesites. Pearce *et al.* (1977) found that the oxides MgO, Al_2O_3 and FeO (total Fe recalculated as FeO) were able to discriminate between the following tectonic environments: ocean-ridge and floor basalts (MORB); ocean-island basalts; continental basalts; volcanic-arc and active continental margin basalts (orogenic basalts in the terminology of Pearce *et al.*, 1977); spreading centre island basalts (e.g. Iceland, Galapagos). The boundaries between the different fields are shown in Figure 5.20.

The diagram works well for fresh, modern subalkaline volcanic rocks in the given silica range and the similar weight per cent of each of the oxides plotted means that there is minimal distortion in the projection. Alkali basalts from all environments plot in a single, but distinct, elongate field stretching from the continental basalt/spreading centre island-/volcanic-arc basalt triple point towards the Al_2O_3 corner and overlap a number of the subalkaline basalt fields. This means that alkali basalts cannot be plotted on this diagram and should be screened out using the TAS diagram of MacDonald and Katsura (1964) (Section 3.2.1).

The usefulness of the diagram is limited by the relative mobility of the major elements in basalts. Pearce (1976) shows for example that MgO and FeO are mobile in submarine weathering, and MgO and Al_2O_3 are mobile during greenschist facies metamorphism. A further factor which should be considered when using this diagram is the extent to which basalts might move across field boundaries during crystal fractionation.

5.2.3 Discrimination diagrams for basalts based upon minor elements

The advantage of using minor elements (Ti, Mn, P, K) as discriminants between basalt types is that the minor elements are more readily detected and more accurately determined than trace elements. Two of the methods outlined below can be used with elements whose concentrations can be adequately determined with XRF techniques. A third method requires the accurate determination of H_2O and applies only to volcanic glass.

The TiO₂–K₂O–P₂O₅ diagram of T.H. Pearce et al. (1975)

The chief merit of the TiO_2–K_2O–P_2O_5 discrimination diagram of Pearce *et al.* (1975) is that it claims to differentiate between oceanic and continental basalts. Most trace element discriminant plots allocate ocean-island and continental-flood basalts to the same field making it difficult to distinguish between the two. Pearce *et al.* (1975), in an empirical study, found that it was possible to discriminate between oceanic and non-oceanic basalts with a single straight boundary line on a TiO_2–K_2O–P_2O_5 triangular diagram (Figure 5.21). Their field of oceanic basalts includes MORB and the shield-building stage of ocean islands. The field of continental basalts successfully classified four out of six trial data-sets correctly

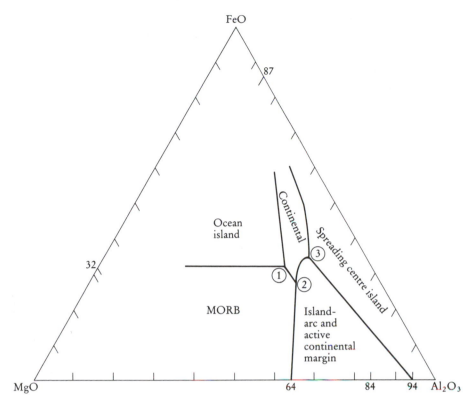

Figure 5.20 The MgO–FeO$_{(tot)}$–Al$_2$O$_3$ diagram (after Pearce *et al.*, 1977) showing the discriminant boundaries for a range of tectonic settings based upon the compositional range of recent volcanic rocks. The diagram can only be used with subalkaline rocks in the silica range SiO$_2$ = 51–56 wt % (calculated dry). The positions of the triple points are those given by Pearce *et al.* (1977). They are:

Point	MgO	FeO	Al$_2$O$_3$
1	21.5	32.0	46.5
2	21.0	27.5	51.5
3	14.0	34.5	51.5

The other plotting coordinates are extracted from Pearce *et al.* (1977 — Figure 1).

although Duncan (1987) found that continental-flood basalts from the Karoo igneous province are frequently mis-assigned.

The diagram does not work for alkali basalts and so samples must first be plotted on a weight per cent AFM diagram — (Fe$_2$O$_3$ + FeO)–MgO–(Na$_2$O + K$_2$O) — (Section 3.3.2) and values with 'A' greater than 20 % screened out. Another limitation is that K$_2$O is mobile and so concentrations may be variable in altered and metamorphosed rocks (Morrison, 1978; Smith and Smith, 1976). Pearce *et al.* (1975) assume that this will have the effect of increasing the K$_2$O value, in which case sample compositions will move towards the K$_2$O corner of the triangular diagram into the continental basalt field. Altered rocks which plot in the oceanic field therefore can be identified as oceanic with some degree of certainty although the result should be checked on another plot which uses immobile elements.

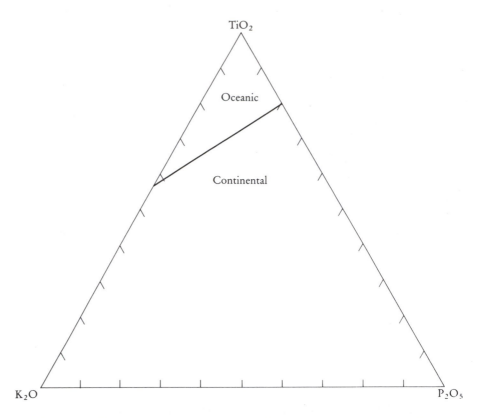

Figure 5.21   The TiO$_2$–K$_2$O–P$_2$O$_5$ discrimination diagram for basalts (after Pearce *et al.*, 1975). This
diagram is not suitable for fractionated and alkaline basalts. Oceanic basalts (MORB and
ocean-island basalts) plot near the TiO$_2$ apex; non-oceanic basalts plot below the boundary
line. The coordinates (given by the authors) of the boundary line separating oceanic and
continental basalts are K$_2$O 45.5 %, TiO$_2$ 54.5 %, P$_2$O$_5$ 0 % and K$_2$O 0 %, TiO$_2$ 79.6 %,
P$_2$O$_5$ 20.4 %.

The MnO–
TiO$_2$–P$_2$O$_5$
diagram of
Mullen (1983)

Basalts and basaltic andesites in the silica range 45–54 wt % SiO$_2$ can be subdivided
on the basis of their MnO, TiO$_2$ and P$_2$O$_5$ concentrations into the following types:
MORB; ocean-island tholeiites; ocean-island alkali basalts; island-arc tholeiites;
calc–alkali basalts (Mullen, 1983). The boundaries defined in Figure 5.22 are based
upon an empirical study of 507 published basalt analyses. MnO and P$_2$O$_5$ values are
multiplied by 10 in order to expand the plotted fields and although this also
amplifies the analytical errors for MnO and P$_2$O$_5$, the enhanced errors still do not
exceed the width of the fields. It should be noted, however, that the compositional
ranges for these elements are small — the mean values for all basalt types are in the
range MnO 0.16–0.24 wt %, P$_2$O$_5$ 0.14–0.74 wt % and TiO$_2$ 0.81–3.07 wt % —
and require accuracy of measurement.

Mn and Ti are readily accommodated in fractionating phases in basalts, Mn in
olivine, pyroxenes and titanomagnetite and Ti in titanomagnetite and pyroxenes.
Thus differences between volcanic-arc magmas and oceanic basalts may be
explained in terms of different patterns of fractional crystallization. P$_2$O$_5$
abundances, on the other hand, are related to either the magma source or the degree
of partial melting. The elements Mn, Ti and P are relatively immobile and

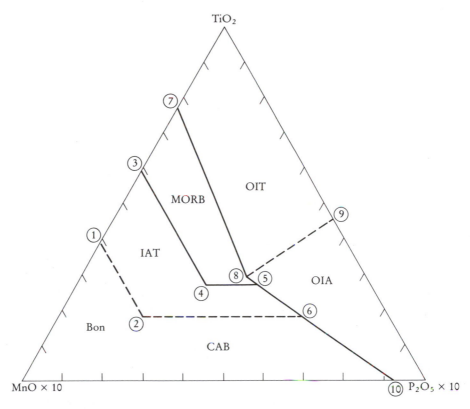

Figure 5.22 The MnO–TiO$_2$–P$_2$O$_5$ discrimination diagram for basalts and basaltic andesites (45–54 wt %
SiO$_2$) (after Mullen, 1983). The fields are MORB; OIT — ocean-island tholeiite or seamount
tholeiite; OIA — ocean-island alkali basalt or seamount alkali basalt; CAB — island-arc
calc–alkaline basalt; IAT — island-arc tholeiite; Bon — boninite. The boninite field occupies
the MnO-rich sector of the CAB field. The plotting coordinates (extracted from Figure 1 of
Mullen, 1983) for constructing the boundary lines on the triangular diagram are as follows:

Point	MnO × 10	TiO$_2$	P$_2$O$_5$ × 10	Point	MnO × 10	TiO$_2$	P$_2$O$_5$ × 10
1	61	39	0	6	21	18	61
2	61	18	21	7	23	77	0
3	41	59	0	8	30	29	41
4	41	27	32	9	0	45	55
5	28	27	45	10	8	0	92

insensitive to hydrothermal processes in the temperature range of the greenschist
facies, although in carbonated rocks Mn–Ti–P relationships are thought to be
unreliable.

*The K$_2$O–H$_2$O
diagram of
Muenow et al.
(1990)*

A diagram which identifies back-arc basin basalts has been proposed by Muenow *et
al.* (1990) on the basis of the K$_2$O, TiO$_2$ and H$_2$O content of volcanic glasses. Back-
arc basin basalts and MORB plot with a K$_2$O/H$_2$O ratio of < 0.70, and ocean-island
basalts and volcanic-arc basalts have K$_2$O/H$_2$O ratios of > 0.70 (Figure 5.23).
MORB and back-arc basin basaltic glasses have overlapping fields although the

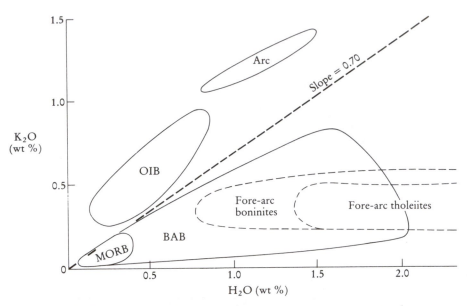

Figure 5.23   The K_2O–H_2O discrimination diagram for basalts (after Muenow *et al.*, 1990). The fields of MORB, OIB (ocean-island basalt), BAB (back-arc basin basalt) and Arc (volcanic-arc basalts) are from Muenow *et al.* (1990). The fields of fore-arc boninites and fore-arc tholeiites are taken from Bloomer and Stern (1990).

back-arc basin field is much more extensive than that of MORB. Water analyses were determined on fresh volcanic glasses by high-temperature mass spectrometry and represent water released above 600 °C and in many samples above 750 °C.

The small degree of overlap between MORB and back-arc basalts means that the diagram (Figure 5.23) can usefully separate the two. There is, however, extensive overlap between back-arc and fore-arc basalts (Bloomer and Stern, 1990).

5.2.4   Discrimination diagrams for basalts based upon clinopyroxene composition

The composition of clinopyroxenes varies according to the chemistry of their host lavas. This is particularly true for clinopyroxene phenocrysts, whose compositions reflect the chemical differences that exist between different basaltic magma types more closely than do groundmass compositions. This property has been used as a discriminant for basalts from different tectonic settings and offers much promise in altered basalts in which the cores of clinopyroxene phenocrysts may be chemically unchanged. A proposal by Nisbet and Pearce (1977) based upon the oxide concentrations MnO–TiO_2–Na_2O in clinopyroxenes has been superseded by the Ti–Cr–Ca–Al–Na plots of Leterrier *et al.* (1982).

Leterrier *et al.* (1982) criticized the diagram of Nisbet and Pearce (1977) on the grounds that it was based upon a small number of analyses, some of which were groundmass clinopyroxenes, and that two of the three chemical discriminants, MnO and Na_2O, were often at levels close to the detection limits in electron microprobe

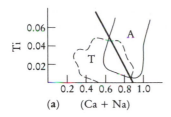
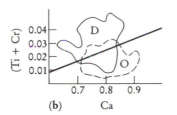
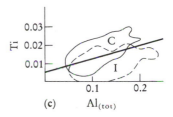

Figure 5.24 Discrimination diagrams for clinopyroxene phenocrysts in basalts (after Leterrier *et al.*, 1982). Clinopyroxene compositions are expressed in cations per six oxygens. (a) Ti vs (Ca+Na) plot showing the fields of alkali basalts (A) and tholeiitic and calc–alkali basalts (T). The equation of the line separating the two fields is $y = -0.4x + 0.38$; (b) (Ti+Cr) vs Ca plot showing the fields of MORB and other tholeiites from spreading zones (D) and volcanic-arc basalts (*O*). The equation for the line separating the two fields is $y = -0.08x - 0.04$; (c) Ti vs total-Al plot showing the fields of calc–alkali basalt (C) and island-arc tholeiite (I). The equation of the line separating the two fields is $y = 0.075x + 0.05$.

analysis. They proposed instead a series of diagrams based upon the elements Ti, Cr, Ca, Al and Na drawn from a larger analytical database (Figure 5.24). Three diagrams are presented which progressively discriminate between alkali basalts, spreading centre tholeiites and island-arc basalts. Clinopyroxene analyses are recalculated as cations to six oxygens and only analyses with more than 0.5 atoms of Ca per formula unit should be used. Fe^{3+} and Fe^{2+} are calculated to provide stoichiometric proportions (see Cameron and Papike, 1981). The first diagram (Figure 5.24a) uses a plot of Ti vs (Ca + Na) and discriminates between alkali basalts (A; ocean-island and continental alkali basalts) and tholeiitic and calc–alkali basalts T. The second diagram (Figure 5.24b), a plot of (Ti + Cr) vs Ca, separates clinopyroxenes from non-alkali basalts into non-orogenic basalts (D; MORB, ocean-island tholeiites and back-arc basin tholeiites) and volcanic-arc basalts (O). The final diagram (Figure 5.24c), a plot of Ti vs total Al, separates clinopyroxenes from volcanic-arc basalts into calc–alkali basalts (C) and island-arc tholeiites (I).

These diagrams should not be used with a single clinopyroxene analysis; rather, Leterrier *et al.* (1982) recommend that a minimum of 10 but if possible more than 20 analyses should be plotted before a meaningful result can be obtained. The method is thought to work in metabasites up to greenschist facies; thereafter clinopyroxene compositions are modified by metamorphic reactions.

5.3 Discrimination diagrams for rocks of granitic composition

The first systematic study of the geochemistry of granites from known tectonic settings was made by Pearce *et al.* (1984), who defined the term **granite** very loosely as 'any plutonic rock containing more than 5 per cent of modal quartz'. They classified granites into ocean-ridge, volcanic-arc, within-plate and collisional types, each category being further subdivided as shown in Box 5.2. A preliminary survey of trace element concentrations plotted against silica content from a suite of 600 selected granites revealed that the elements Y, Yb, Rb, Ba, K, Nb, Ta, Ce, Sm, Zr and Hf most effectively discriminate between granites from different tectonic settings. These variables are used by Pearce *et al.* (1984) in two suites of variation diagrams to classify granites according to their tectonic setting.

Box 5.2

Granite types classified according to tectonic setting (after Pearce *et al.*, 1984)

Ocean-ridge granites (ORG)

Granites associated with normal ocean ridges
Granites associated with anomalous oceanic ridges
Granites associated with back-arc basin ridges
Granites associated with fore-arc basin ridges

Volcanic-arc granites (VAG)

Granites in oceanic arcs dominated by tholeiitic basalt
Granites in oceanic arcs dominated by calc–alkali basalt
Granites in active continental margins

Within-plate granites (WPG)
Granites in intracontinent ring complexes
Granites in attenuated continental crust
Granites in oceanic islands

Collisional granites (COLG)

Syn-tectonic granites associated with continent–continent collision
Post-tectonic granites associated with continent–continent collision
Syn-tectonic granites associated with continent–arc collision

---

5.3.1 Discrimination diagrams for granites based upon Rb–Y–Nb and Rb–Yb–Ta variations (Pearce *et al.*, 1984)

From the list of elements given above, the elements Rb, Y (and its analogue Yb) and Nb (and its analogue Ta) were selected as the most efficient discriminants between most types of ocean-ridge granite (ORG), within-plate granites (WPG), volcanic-arc granites (VAG) and syn-collisional granites (syn-COLG). Post-orogenic granites and one group of ocean-ridge granites — the supra–subduction zone, fore-arc basin granites — give a more ambiguous result. Post-orogenic granites cannot be distinguished from volcanic-arc and syn-collisional granites on the diagrams below but they can be identified on the Hf–Rb–Ta diagram discussed in Section 5.3.2. Supra-subduction zone granites can only be identified successfully

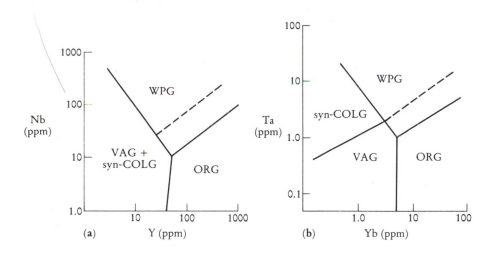

Figure 5.25   (a) The Nb–Y discrimination diagram for granites (after Pearce *et al.*, 1984), showing the fields of volcanic-arc granites (VAG), syn-collisional granites (syn-COLG), within-plate granites (WPG) and ocean-ridge granites (ORG). The broken line is the field boundary for ORG from anomalous ridges. The plotting coordinates (Y,Nb) are as follows (after Pearce *et al.*, 1984) 1,2000 to 50,10; 40,1 to 50,10; 50,10 to 1000,100; 25,25 to 1000,400; (b) The Ta–Yb discrimination diagram for granites (after Pearce *et al.*, 1984), showing the fields of volcanic-arc granites (VAG), syn-collisional granites (syn-COLG), within-plate granites (WPG) and ocean-ridge granites (ORG). The broken line is the field boundary for ORG from anomalous ridges. The plotting coordinates (Yb, Ta) are as follows (after Pearce *et al.*, 1984): 0.55,20 to 3,2; 0.1,0.35 to 3,2; 3,2 to 5,1; 5,0.5 to 5,1; 5,0.05 to 100,7; 3,2 to 100,20.

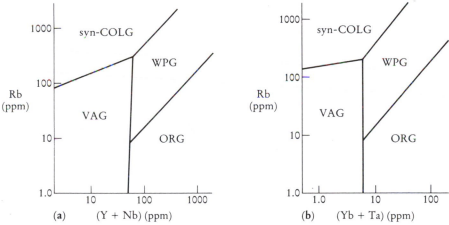

Figure 5.26   (a) The Rb–(Y + Nb) discrimination diagram for granites (after Pearce *et al.*, 1984) showing the fields of syn-collisional granites (syn-COLG), within-plate granites (WPG), volcanic-arc granites (VAG) and ocean-ridge granites (ORG). The plotting coordinates [(Y + Nb), Rb] for the field boundaries (taken from Pearce *et al.*, 1984) are: 2,80 to 55,300; 55,300 to 400,2000; 55,300 to 51.5,8; 51.5,8 to 50,1; 51.5,8 to 2000, 400. (b) The Rb–(Yb + Ta) discrimination diagram for granites (after Pearce *et al.*, 1984), showing the fields of syn-collisional granites (syn-COLG), within-plate granites (WPG), volcanic-arc granites (VAG) and ocean-ridge granites (ORG). The plotting coordinates [(Yb + Ta), Rb] for the field boundaries (taken from Pearce *et al.*, 1984) are: 0.5,140 to 6,200; 6,200 to 50,200; 6,200 to 6,8; 6,8 to 6,1; 6.8 to 200,400.

when there is geological evidence for an ocean setting. They may then be identified on an Nb–Y diagram from their lower Y content.

The Nb–Y and Ta–Yb discrimination diagrams

A bivariate plot of Nb and Y can be subdivided into three fields into which oceanic granites (ORG), within-plate granites (WPG) and volcanic-arc granites (VAG) together with syn-collisional granites (syn-COLG) plot (Figure 5.25a). A similar plot for Ta and Yb allows the fields of syn-collisional and volcanic-arc granites to be separated (Figure 5.25b). The plotting coordinates for the field boundaries are given in the caption to Figure 5.25.

The Rb– (Y + Nb) and Rb–(Yb + Ta) discrimination diagrams

A bivariate plot of Rb and (Y + Nb) more efficiently separates syn-collisional granites from volcanic-arc granites. There is also a clear division between within-plate and oceanic granites on this diagram (Figure 5.26a). The analogous plot using (Yb + Ta) along the *x*-axis of the bivariate plot produces a very similar set of fields (Figure 5.26b).

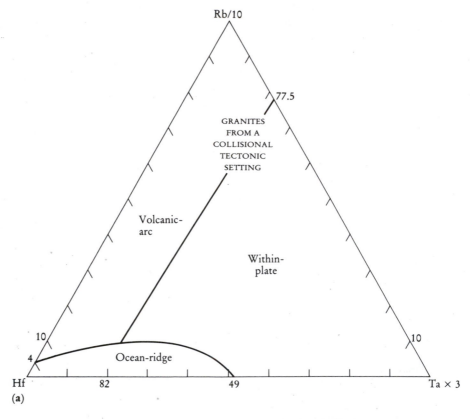

Figure 5.27 (a)   The Hf–Rb/10–Ta × 3 discrimination diagram for granites (after Harris *et al.*, 1986), showing the fields for volcanic-arc granites, within-plate granites and ocean-ridge granites. Granites from a collisional tectonic setting plot across the field boundary between volcanic-arc and within-plate granites. The plotting coordinates are extracted from Harris *et al.* (1986 – Figure 1).

5.3.2 Discrimination diagrams for granites based upon Hf–Rb–Ta variations

A trivariate plot of Hf–Rb/10–Ta × 3 (Figure 5.27a) discriminates between ocean-floor granites, volcanic-arc granites and within-plate granites, although collisional granites scatter across the volcanic-arc/within-plate boundary (Harris *et al.*, 1986). A modification of this diagram to expand the field of collisional granites using Hf, Rb/30 and Ta × 3 (Figure 5.27b) allows this particular group to become distinctive and permits their subdivision into syn-collisional granites and post-collisional granites. These two groups can also be recognized on a bivariate plot of the Rb/Zr ratio against SiO$_2$ although in this case there is extensive overlap between the post-collisional granites and volcanic-arc granites.

5.3.3 A measure of arc maturity for volcanic-arc granites

Brown *et al.* (1984) found that with increasing maturity, volcanic-arc granites are enriched in the elements Rb, Th, U, Ta, Nb, Hf and Y and depleted in the elements Ba, Sr, P, Zr and Ti. They showed that a bivariate plot of the ratio Rb/Zr

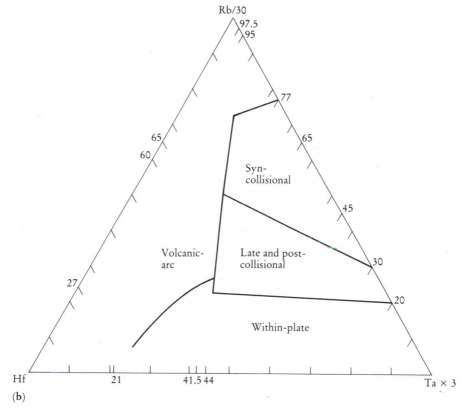

(b)

Figure 5.27 (b)　The Hf–Rb/30–Ta × 3 discrimination diagram for granites (after Harris *et al.*, 1986), showing the fields for volcanic-arc granites, within-plate granites, syn-collisional granites and late to post-collisional granites. The plotting coordinates are extracted from Harris *et al.* (1986 – Figure 6).

against either Nb or Y showed a positive correlation in which values increase with increasing arc maturity.

5.3.4   Discussion

As with all discrimination diagrams, the suite of diagrams for granitic rocks cannot be used without giving some careful thought to the effects of element mobility and crystal fractionation. For example, Rb is used in several of these diagrams and yet is a very mobile element in hydrothermal fluids. Rb was chosen on the assumption that the effects of element mobility are much less in granitic rocks than in basic rocks, chiefly because granitic rocks are generally less altered. However, examples of Rb mobility are known (see, for example, Mukasa and Henry, 1990), and will give erroneous results.

Pearce *et al.* (1984) consider the effects of crystal accumulation in granites and show that the dilution effect of plagioclase accumulation may shift granites from the within-plate and ocean-ridge fields into the volcanic-arc granite field on the Nb–Y and Rb–(Y + Nb) plots (Figures 5.25a and 5.26a). In a similar way, volcanic-arc and syn-collisional granites may be shifted into the within-plate or oceanic granite field through the accumulation of ferromagnesian and minor phases.

Twist and Harmer (1987) applied the Rb–(Y + Nb) and Nb–Y diagrams to the Proterozoic Bushveld granites and associated felsite lavas. Their study raises two rather different but important points. Firstly, these diagrams can be applied to felsitic lavas as well as to plutonic felsic rocks, provided the lavas are not contaminated. Secondly, they found that differences in trace element chemistry could be just as much a function of variable source chemistry as of differences in tectonic environment.

5.4   Discrimination diagrams for clastic sediments

Plate tectonic processes impart a distinctive geochemical signature to sediments in two separate ways. Firstly, different tectonic environments have distinctive provenance characteristics and, secondly, they are characterized by distinctive sedimentary processes. Sedimentary basins may be assigned to the following tectonic settings (Bhatia and Crook, 1986):

Oceanic island-arc — fore-arc or back-arc basins, adjacent to a volcanic-arc developed on oceanic or thin continental crust.

Continental island-arc — inter-arc, fore-arc or back-arc basins, adjacent to a volcanic-arc developed on thick continental crust or thin continental margins.

Active continental margin — Andean-type basins developed on or adjacent to thick continental margins. Strike–slip basins also develop in this environment.

Passive continental margin — rifted continental margins developed on thick continental crust on the edges of continents; sedimentary basins on the trailing edge of a continent.

Collisional setting — sedimentary basins developed on thick continental crust.
Rift setting — intercratonic basin developed on thick continental crust.

5.4.1　Discrimination diagrams for clastic sediments using major elements

Three types of discrimination diagram are described below for sedimentary rocks, based upon major element chemistry. A further discrimination diagram identifies sedimentary provenance.

The sandstone discriminant function diagram (Bhatia, 1983)　Bhatia (1983) proposed a discrimination diagram based upon a bivariate plot of first and second discriminant functions of major element analyses of 69 Palaeozoic sandstones (Figure 5.28). The sandstones were chosen to represent four different tectonic settings, assigned on the basis of comparisons with modern sediments. The four fields thus defined (Figure 5.28) were then tested with modern sediments collected from known tectonic settings. This approach is of course the inverse of that used for igneous rocks, for normally a discrimination diagram is produced from suites of samples whose tectonic setting is known, not the other way round.

When this diagram is used, samples with a high content of CaO as carbonate

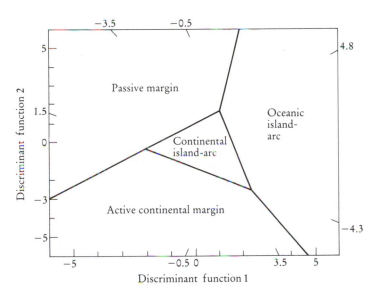

Figure 5.28　The discriminant function diagram for sandstones (after Bhatia, 1983), showing fields for sandstones from passive continental margins, oceanic island-arcs, continental island-arcs and active continental margins. The discriminant functions are as follows:

$$\text{Discriminant function 1} = -0.0447 \, SiO_2 - 0.972 \, TiO_2 + 0.008 \, Al_2O_3 - 0.267 \, Fe_2O_3$$
$$+ \, 0.208 \, FeO - 3.082 \, MnO + 0.140 \, MgO + 0.195 \, CaO$$
$$+ \, 0.719 \, Na_2O - 0.032 \, K_2O + 7.510 \, P_2O_5 + 0.303$$

$$\text{Discriminant function 2} = -0.421 \, SiO_2 + 1.988 \, TiO_2 - 0.526 \, Al_2O_3 - 0.551 \, Fe_2O_3$$
$$- \, 1.610 \, FeO + 2.720 \, MnO + 0.881 \, MgO - 0.907 \, CaO$$
$$- \, 0.177 \, Na_2O - 1.840 \, K_2O + 7.244 \, P_2O_5 + 43.57$$

(from Bhatia, 1983 — Table 3). The plotting coordinates are extracted from Bhatia (1983 — Figure 7).

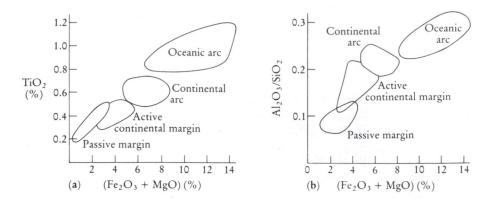

Figure 5.29   Discrimination diagrams for sandstones (after Bhatia, 1983), based upon (a) a bivariate plot of TiO_2 vs ($Fe_2O_{3(tot)}$ + MgO) and (b) a bivariate plot of Al_2O_3/SiO_2 vs ($Fe_2O_{3(tot)}$ + MgO). The fields are oceanic island-arc, continental island-arc, active continental margin and passive margin.

must be corrected for carbonate content. Haughton (1988), in a study of an ancient arc terrane from California, and Winchester and Max (1989), in a study of a Proterozoic ensialic basin, all found that this diagram misclassified their sediments.

Bivariate plots as sandstone discrimination diagrams (Bhatia, 1983)

Modern sandstones from oceanic and continental arcs and active and passive continental margins are variable in composition, particularly in their $Fe_2O_{3(tot)}$ + MgO, Al_2O_3/SiO_2, K_2O/Na_2O and $Al_2O_3/(CaO + Na_2O)$ contents. Bhatia (1983) used this chemical variability to discriminate between the different tectonic settings on a series of bivariate plots, two of which are shown in Figure 5.29.

The K_2O/Na_2O vs SiO_2 sandstone–mudstone discrimination diagram (Roser and Korsch, 1986)

Three tectonic settings — the Passive continental Margin (PM), the Active Continental Margin (ACM) and the oceanic island-ARC (ARC) — are recognized on the K_2O/Na_2O–SiO_2 discrimination diagram of Roser and Korsch (1986) (Figure 5.30). Chemical analyses of ancient sediments whose tectonic setting had been inferred were taken from the literature and recalculated dry to ensure comparability. The fields thus defined were then tested against modern sediments from a known geological setting. However, because the chemical composition of sediments is influenced in part by grain size, Roser and Korsch (1986) plot sand–mud couplets from single sites for modern sediments to test the validity of their diagram for muds as well as for sands. Generally sediments plot where expected, although fore-arc sands plot in the ARC field whilst the associated muds plot in the ACM field. Where sediments are rich in a carbonate component, the analyses were recalculated $CaCO_3$-free. Failure to do this will shift samples to lower SiO_2 values and from the passive margin field into the volcanic-arc field.

Provenance signatures of sandstone–mudstone suites using major

A discriminant function diagram has been proposed by Roser and Korsch (1988) to distinguish between sediments whose provenance is primarily mafic, intermediate or felsic igneous and quartzose sedimentary. Their analysis was based upon 248 chemical analyses in which Al_2O_3/SiO_2, K_2O/Na_2O and $Fe_2O_{3(tot)}$ + MgO proved the most valuable discriminants. A plot of the first two discriminant functions based

elements (Roser and Korsch, 1988) upon the oxides of Ti, Al, Fe, Mg, Ca, Na and K most effectively differentiates between the four provenances (Figure 5.31). The problem of biogenic CaO in $CaCO_3$ and also biogenic SiO_2 is circumvented by using ratio plots in which discriminant functions are based upon the ratios of TiO_2, $Fe_2O_{3(tot)}$, MgO, Na_2O and K_2O all to Al_2O_3 (Figure 5.32). The ratio discrimination diagram is not as effective as the one based upon the raw oxides.

5.4.2 Discrimination diagrams for clastic sediments using trace elements

Unlike igneous rocks, where most discriminant diagrams are based upon trace element chemistry, diagrams of this sort are in their infancy in sedimentary geochemistry. Instead, some use can be made of multi-element plots of the type described in Section 4.4.

Greywackes Bhatia and Crook (1986) identified the elements La, Th, Zr, Nb, Y, Sc, Co and Ti as the most useful in discriminating between greywackes from different tectonic environments. Distinctive fields for four environments — oceanic island-arc, continental island-arc, active continental margin and passive margin — are recognized on bivariate plots of La vs Th, La/Y vs Sc/Cr, Ti/Zr vs La/Sc and the trivariate plots La–Th–Sc, Th–Sc–Zr/10 and Th–Co–Zr/10. On a La–Th–Sc plot, the fields of active continental margin sediments and passive margin sediments overlap, but the Th–Sc–Zr/10 shows complete separation (Figure 5.33).

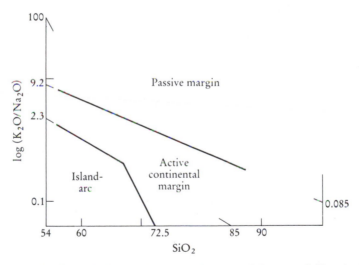

Figure 5.30 The log (K_2O/Na_2O) vs SiO_2 discrimination diagram of Roser and Korsch (1986) for sandstone–mudstone suites and showing the fields for a passive continental margin, an active continental margin and an island-arc. The plotting coordinates for the field boundaries have been extracted from Roser and Korsch (1986 — Figure 2).

Spider diagrams as discriminants of tectonic setting for shales Shale-normalized spider diagrams were shown to be distinctive for shales deposited in oceanic island-arcs, passive continental margins and destructive margins (continental island-arcs and active continental margins). Winchester and Max (1989) emphasize that their approach is preliminary for the results are based upon only a

(Winchester and Max, 1989)

small sample set. Oceanic arc profiles are characterized by depletion in many elements relative to post-Archaean continental shale (Table 4.6), whereas continental arcs and active margin samples have higher LIL concentrations and a broadly curving profile. Passive margin profiles are similar to post-Archaean continental shale and therefore have a flat trend.

Provenance studies (Cullers et al., 1988)

Cullers *et al.* (1988) showed that the immobile elements La and Th are more abundant in felsic than in basic rocks, but that the opposite is true for Sc and Co. Ratios such as La/Sc, Th/Sc, Th/Co, La/Co, Ba/Sc and Ba/Co, therefore, in sand-sized sediments allow a distinction to be made between a felsic and a mafic source. However, such ratios should be interpreted in terms of provenance with care, for the ratios can be fractionated during weathering and transport and may be strictly valid only for locally derived sediments. Recycled sediments and those from a mixed source are much more difficult to interpret.

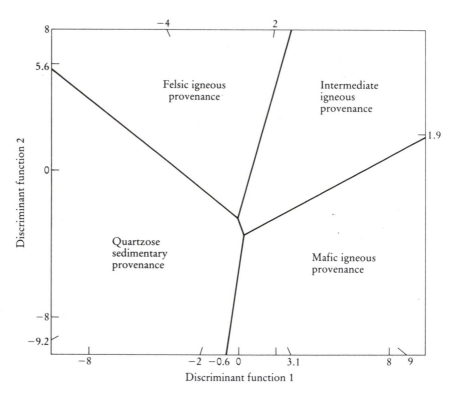

Figure 5.31   Discriminant function diagram for the provenance signatures of sandstone–mudstone suites using major elements (after Roser and Korsch, 1988). Fields for dominantly mafic, intermediate and felsic igneous provenances are shown with the field for a quartzose sedimentary provenance. The discriminant functions are

$$\text{discriminant function 1} = -1.773\text{TiO}_2 + 0.607\text{Al}_2\text{O}_3 + 0.76\text{Fe}_2\text{O}_{3(\text{total})}$$
$$- 1.5\text{MgO} + 0.616\text{CaO} + 0.509\text{Na}_2\text{O} - 1.224\text{K}_2\text{O} - 9.09$$
$$\text{discriminant function 2} = 0.445\text{TiO}_2 + 0.07\text{Al}_2\text{O}_3 - 0.25\text{Fe}_2\text{O}_{3(\text{total})}$$
$$- 1.142\text{MgO} + 0.438\text{CaO} + 1.475\text{Na}_2\text{O} + 1.426\text{K}_2\text{O} - 6.861$$

(Data from Roser and Korsch, 1988 — Table II; plotting coordinates extracted from Roser and Korsch, 1988 — Figure 4.)

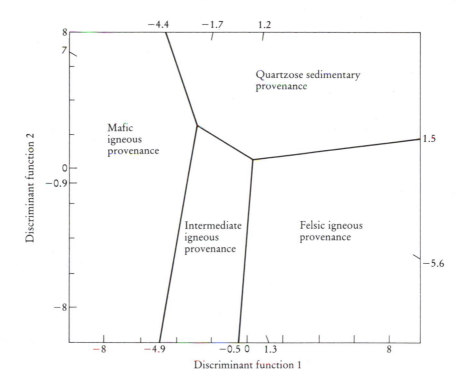

Figure 5.32 Discriminant function diagram for the provenance signatures of sandstone–mudstone suites using major element ratios (after Roser and Korsch, 1988). Fields for dominantly mafic, intermediate and felsic igneous provenances are shown with the field for a quartzose sedimentary provenance. The discriminant functions are

$$\text{discriminant function } 1 = 30.638 TiO_2/Al_2O_3 - 12.541 Fe_2O_{3(total)}/Al_2O_3$$
$$+ 7.329 MgO/Al_2O_3 + 12.031 Na_2O/Al_2O_3 + 35.402 K_2O/Al_2O_3 - 6.382$$
$$\text{discriminant function } 2 = 56.500 TiO_2/Al_2O_3 - 10.879 Fe_2O_{3(total)}/Al_2O_3$$
$$+ 30.875 MgO/Al_2O_3 - 5.404 Na_2O/Al_2O_3 + 11.112 K_2O/Al_2O_3 - 3.89$$

(Data from Roser and Korsch (1988 — Table IV), plotting coordinates extracted from Roser and Korsch, 1988 — Figure 9).

5.4.3   Discussion

The underlying assumption of geochemical discrimination diagrams for sedimentary rocks is that there is close link between plate tectonic setting and sediment provenance. Largely this is true and the chief successes of the technique are with immature sediments containing a significant volume of lithic fragments from which provenance and hence tectonic setting may be identified. However, there is one major area of uncertainty, for some sediments are transported from their tectonic setting of origin into a sedimentary basin in a different tectonic environment (McLennan *et al.*, 1990). This observation means that users of discrimination diagrams for sedimentary rocks must be cautious in their claims for the original tectonic setting of a sedimentary basin.

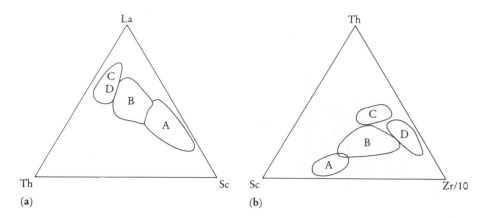

Figure 5.33   (a) La–Th–Sc discrimination diagram for greywackes; (b) Th–Sc–Zr/10 discrimination diagram for greywackes (after Bhatia and Crook, 1986). The fields are: A, oceanic island-arc; B, continental island-arc; C, active continental margin; D, passive margin.

5.5   Tectonic controls on magmatic and sedimentary geochemistry

The underlying tectonic controls on the chemistry of magmatic and sedimentary rocks is an issue much wider than that of discrimination diagrams, for a large part of modern geochemistry is ultimately concerned with this very problem. This then poses for us a question over the value of tectono-magmatic discrimination diagrams. Have they been superseded or are they still a valuable shorthand way of detecting a particular tectonic environment?

It can be argued that the more we learn about modern tectonic environments, in particular the controls on their trace element and isotope geochemistry, the more complex they appear to be and the less confidence we have in precisely fingerprinting them and then identifying them in the ancient record. Certainly the medium of a bivariate or trivariate variation diagram used to distinguish different tectonic environments might be thought too simplistic. Some would go as far as Morrison (1978) and argue that discrimination diagrams should be 'treated merely as a species of variation diagram; useful in identifying magma types and their evolutionary trends, but *not* as indicators of their palaeotectonic environment'.

On the other hand, different tectonic environments do have distinctive geochemical signatures and, if carefully used, geochemical data can be used to extract important environmental information from ancient rocks. Thus some type of geochemical variation diagram can be used to assign tectonic environment. Weaver (1987) has suggested that spider diagrams are the modern descendants of discrimination diagrams.

The quandary facing the user of discrimination diagrams is that tectonic environments do often carry a geochemical fingerprint but some of the fingerprints are not unique. Two possible solutions may be found and they are outlined here in this final section. Firstly, all available information may be put together to give a probable view of the former tectonic environment. This is the expert system approach described in Section 5.5.1. The other alternative is to take a hard and critical look at discrimination diagrams and only use the ones which are likely to provide a meaningful result. This approach is discussed in Section 5.5.2.

5.5.1 An expert system for identifying the tectonic environment of ancient volcanic rocks (Pearce, 1987)

An ideal way in which to incorporate all the information embodied in the discrimination diagrams described in this chapter is to use a computer-based expert system. An expert system is a computer system which is able, on the basis of a predefined knowledge base, to make intelligent decisions about that particular field of knowledge. Pearce (1987) described the prototype of such a system — known as ESCORT — developed to identify the tectonic environment of eruption of ancient volcanic rocks.

The method is multidisciplinary and requires information about the field relationships, petrology and mineralogy as well as the geochemistry of a suite of volcanic rocks. This integrated approach obviates some of the errors which can arise from the use of geochemical data alone. From the knowledge base the computer assigns each rock to a magma-type (basic, intermediate or evolved) and to a particular tectonic environment of eruption. The expert system as designed by Pearce (1987) also assigns probabilities to the decisions made in terms of the likely, most probable, possible and unlikely tectonic setting. Thus, in a suite of volcanic rocks each sample is assessed and designated to a tectonic environment but then a cumulative assessment is made of the whole suite and its probable tectonic setting.

The knowledge base contains information on rock-type associations drawn from the Phanerozoic record, phenocryst assemblages in different magma types and the association between trace element chemistry and tectonic setting. Field evidence uses information from sediments associated with the volcanic assemblage, but also uses estimates of the proportions of basic, intermediate and evolved lavas erupted in each environment since the Phanerozoic, and a measure of the probability of their preservation. The petrological and geochemical parts of the knowledge base uses the presence or absence of an iron enrichment trend and trace element data for TiO_2, Zr, Y, Nb, P_2O_5 and Cr.

This approach, albeit only a prototype, has the potential for expansion to include more elements (Hf, Th, Ta, V, Ni, Sc and the REE) and to include the information from sedimentary as well as volcanic rocks.

5.5.2 Do tectonic discrimination diagrams still have a function?

Practising geochemists will differ in their answer to this question. Many, however, including this author, will opt for the compromise of 'sometimes'. In seeking an answer to this question a number of factors should be considered.

(1) There is no doubt that there are tectonic controls on magma chemistry.
(2) Discrimination diagrams can work, although they may not always work very

well. For example, there are 'mixed signals' — data which 'straddle' the boundaries on discrimination diagrams. There may be geological reasons for this, as in the case of continental flood basalts which may erupt in a variety of within-plate tectonic settings. There may also be geochemical reasons. For example, the interaction of magma and a fluid phase may take place in a number of different tectonic settings, resulting in a similar trace element signal for different tectonic environments (Arculus, 1987).

(3)   Discrimination diagrams must never be used uncritically. The possible effects of element mobility and crystal–liquid fractionation processes must always be considered.

(4)   Discrimination diagrams must be used carefully with very old rocks. It is likely that the trace element composition of mantle source regions has changed with time and the mantle may have been less fractionated early in Earth history. Thus field boundaries will be in different places. With higher mantle temperatures in the Archaean, higher degrees of mantle melting and a greater probability of crustal melting should be considered (Pearce *et al.*, 1984).

(5)   Discrimination diagrams may be telling us more about process than environment. Trace element concentrations in igneous rocks, for example, are a function of initial mantle concentrations, the percentage melt fraction, fractional crystallization processes and the effects of crustal contamination. Thus it is the process that is characterized by the trace element concentrations in the first instance. Where processes can be linked to tectonic environment, then the discrimination diagram is useful. Where there is ambiguity some careful thought must be given to interpreting the results.

Chapter 6

Using radiogenic isotope data

6.1 Introduction

Radiogenic isotopes are used in geochemistry in two principal ways. Historically they were first used to determine the age of rocks and minerals. More recently they have been used in petrogenetic studies to identify geological processes and sources. The former application is normally described as geochronology, the latter as isotope geology or isotope geochemistry. There are a number of excellent texts which deal with these two disciplines (Faure, 1986; Jager and Hunziker, 1977; DePaolo, 1988) and the reader is referred to these for more detailed treatments of the topics covered here.

In the first part of this chapter the main principles of geochronology are briefly described and the interpretation of geochronological results are reviewed. The second half of the chapter describes the use of radiogenic isotopes in petrogenesis and takes us into one of the most exciting and fast-moving developments in geochemistry over the past two decades. The use of radiogenic isotopes as tracers of petrogenetic processes has allowed geochemists to sample the deep interior of the Earth, previously solely the domain of geophysicists. The results of such studies have led to important geochemical constraints on the nature of the continental crust and the Earth's mantle which now may be combined with our physical knowledge of these domains to help provide a unified chemical–physical model of the deep Earth.

6.2 Radiogenic isotopes in geochronology

The foundations of modern geochronology were laid at the turn of the century in the work of Rutherford and Soddy (1903) on natural radioactivity. They showed that the process of radioactive decay is exponential and independent of chemical or physical conditions. Thus rates of radioactive decay may be used for measuring geological time. Isotopic systems used in age calculations are listed in Table 6.1 and Box 6.1. In this section we discuss two of the most common techniques used in geochronological calculations — isochron diagrams and model age calculations. This is followed by a discussion of the significance of the calculated ages.

Box 6.1

Constants used in lead isotope geochronology

(a) Decay constants (Steiger and Jager, 1977)

$^{238}U \rightarrow {}^{206}Pb$ 0.155125×10^{-9} yr^{-1} (λ_1)
$^{235}U \rightarrow {}^{207}Pb$ 0.98485×10^{-9} yr^{-1} (λ_2)
$^{232}Th \rightarrow {}^{208}Pb$ 0.049475×10^{-9} yr^{-1} (λ_3)

(b) Isotope ratios of primeval lead (from the Canyon Diablo troilite (Tatsumoto et al., 1973))

$(^{206}Pb/^{204}Pb)_0 = 9.307$
$(^{207}Pb/^{204}Pb)_0 = 10.294$
$(^{208}Pb/^{204}Pb)_0 = 29.474$

(c) Age of the earth derived from the meteoritic isochron (Tatsumoto et al., 1973; Tilton, 1973)

Slope of $^{207}Pb/^{204}Pb$ vs $^{206}Pb/^{204}Pb$ isochron = 0.626208.
Defines age of earth as 4.57 Ga.

(d) Present-day ratio $^{238}U/^{235}U$

$^{238}U/^{235}U = 137.88$

(e) Symbols used in Pb isotope geochemistry

$\mu = {}^{238}U/^{204}Pb$ $\kappa = {}^{232}Th/^{238}U$

(f) Ratios used for plotting the U–Pb concordia curve

Age (Ga)	$^{206}Pb/^{238}U$	$^{207}Pb/^{235}U$
0.0	0.00000	0.00000
0.4	0.06402	0.48281
1.0	0.16780	1.67741
1.4	0.24256	2.97009
1.8	0.32210	4.88690
2.2	0.40674	7.72917
2.6	0.49679	11.94371
3.0	0.59261	18.19308
3.4	0.69456	27.45973
3.6	0.74796	33.65562
3.8	0.80304	41.20041
4.0	0.85986	50.38776
4.2	0.91846	61.57526
4.4	0.97892	75.19836
4.6	1.04128	91.78732

6.2.1 Isochron calculations

An **isochron diagram** is a bivariate plot of measured parent–daughter isotope ratios for a suite of cogenetic samples. Where the sample suite defines a linear array, this is said to be an **isochron** and the slope of the line is proportional to the age of the sample suite. Consider as an example the Rb–Sr system. The total number of

Table 6.1   Isotopic systems which are used in age calculations

Technique	Decay scheme	Decay constant, $\lambda(\text{yr}^{-1})$		Ratios plotted on the isochron diagram	
				x-axis	*y*-axis
Rb–Sr	$^{87}\text{Rb} \rightarrow ^{87}\text{Sr} + \beta$	1.42×10^{-11}	(1)†‡	$^{87}\text{Rb}/^{86}\text{Sr}$	$^{87}\text{Sr}/^{86}\text{Sr}$
Sm–Nd	$^{147}\text{Sm} \rightarrow ^{143}\text{Nd} + \text{He}$	6.54×10^{-12}	(2)	$^{147}\text{Sm}/^{144}\text{Nd}$	$^{143}\text{Nd}/^{144}\text{Nd}$
Lu–Hf*	$^{176}\text{Lu} \rightarrow ^{176}\text{Hf} + \beta$	1.96×10^{-12}	(3)	$^{176}\text{Lu}/^{177}\text{Hf}$	$^{176}\text{Hf}/^{177}\text{Hf}$
		1.94×10^{-12}	(4)		
Re–Os*	$^{187}\text{Re} \rightarrow ^{187}\text{Os} + \beta$	1.61×10^{-11}	(5)	$^{187}\text{Re}/^{186}\text{Os}$	$^{187}\text{Os}/^{186}\text{Os}$
K–Ar	$^{40}\text{K} \rightarrow ^{40}\text{Ar} - \beta$	0.581×10^{-10}	(1)	$^{40}\text{K}/^{36}\text{Ar}$	$^{40}\text{Ar}/^{36}\text{Ar}$
K–Ca*	$^{40}\text{K} \rightarrow ^{40}\text{Ca} + \beta$	4.962×10^{-10}	(1, 6)	$^{40}\text{K}/^{42}\text{Ca}$	$^{40}\text{Ca}/^{42}\text{Ca}$
	K total	5.543×10^{-10}	(1)		
La–Ce*	$^{138}\text{La} \rightarrow ^{138}\text{Ce} + \beta$	2.30×10^{-12}	(7)	$^{138}\text{La}/^{136}\text{Ce}$	$^{138}\text{Ce}/^{136}\text{Ce}$
La–Ba*	$^{138}\text{La} \rightarrow ^{138}\text{Ba} - \beta$	4.44×10^{-12}	(8)	$^{138}\text{La}/^{137}\text{Ba}$	$^{138}\text{Ba}/^{137}\text{Ba}$

*   These techniques are not routine and are carried out only in a few laboratories.
†   Older determinations use 1.39×10^{-11}. Ages which have used this decay constant may be recalculated using the factor (1.39/1.42).
‡   References: (1) Steiger and Jager (1977); (2) Lugmair and Marti (1978); (3) Patchett and Tatsumoto (1980); (4) DePaolo (1988); (5) Hirt *et al.* (1963); (6) Marshall and DePaolo (1982); (7) Dickin (1987); (8) Nakai *et al.* (1986).

^{87}Sr atoms in a rock which has been a closed system for *t* years is given by the equation

$$^{87}\text{Sr}_\text{m} = {}^{87}\text{Sr}_0 + {}^{87}\text{Rb}_\text{m}(e^{\lambda t} - 1) \qquad [6.1]$$

where $^{87}\text{Sr}_\text{m}$ is the total number of atoms of ^{87}Sr present today, $^{87}\text{Sr}_0$ is the number of atoms of ^{87}Sr present when the sample first formed, $^{87}\text{Rb}_\text{m}$ is the number of atoms of ^{87}Rb present today and λ is the decay constant (given in Table 6.1). The precise measurement of absolute isotope concentrations is difficult; instead, isotope ratios are normally determined. An isotope not involved in the radioactive decay scheme is used as the ratioing isotope. In the case of the Rb–Sr isotope system the ratioing isotope is ^{86}Sr and Eqn [6.1] may therefore be rewritten in the form

$$\left[\frac{^{87}\text{Sr}}{^{86}\text{Sr}}\right]_\text{m} = \left[\frac{^{87}\text{Sr}}{^{86}\text{Sr}}\right]_0 + \left[\frac{^{87}\text{Rb}}{^{86}\text{Sr}}\right]_\text{m}(e^{\lambda t} - 1) \qquad [6.2]$$

The ratios $(^{87}\text{Sr}/^{86}\text{Sr})_\text{m}$ and $(^{87}\text{Rb}/^{86}\text{Sr})_\text{m}$ can be measured by mass spectrometry, leaving $(^{87}\text{Sr}/^{86}\text{Sr})_0$ (the initial ratio) and *t* (the age of the rock) as the unknowns. Since Eqn [6.2] is the equation of a straight line, the age and intercept can be calculated from a plot of measured $(^{87}\text{Sr}/^{86}\text{Sr})_\text{m}$ and $(^{87}\text{Rb}/^{86}\text{Sr})_\text{m}$ ratios for a suite of cogenetic samples. The methodology is illustrated in Figure 6.1. The age is calculated from the slope of the line from the equation

$$t = 1/\lambda \, \ln(\text{ slope} + 1) \qquad [6.3]$$

where *t* is the age and λ is the decay constant. Time is measured from the present and is expressed as either Ma (10^6 years) or as Ga (10^9 years). The intercept, the initial ratio, has considerable petrogenetic importance and will be discussed fully in the latter half of this chapter.

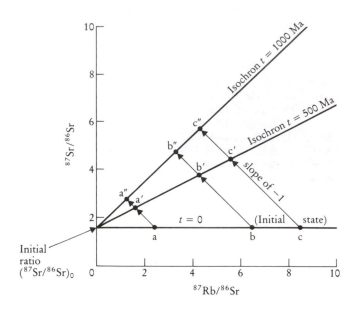

Figure 6.1   A schematic isochron diagram showing the evolution of a suite of igneous rocks (a, b, c) over 500 Ma and 1000 Ma. The samples formed at $t = 0$ from the same batch of magma which subsequently differentiated. At t = 0 each member of the rock suite has the same initial ratio $(^{87}Sr/^{86}Sr)_0$ but because the magma suite is chemically differentiated each rock has a different concentration of Rb and Sr and so a different $^{87}Rb/^{86}Sr$ ratio. Each sample plots as a separate point on the $^{87}Rb/^{86}Sr$ vs $^{87}Sr/^{86}Sr$ isochron diagram. From $t = 0$ to $t = 500$ Ma or t = 1000 Ma individual samples evolve along a straight line with a slope of −1 (for example a–a'–a''), reflecting the decay of a single atom of ^{87}Rb to a single atom of ^{87}Sr. [In practice the resulting change in $^{87}Sr/^{86}Sr$ is small and so the vertical scale is normally exaggerated and the path taken by points is therefore much closer to a vertical line.] The amount of ^{87}Sr produced in a given sample is proportional to the amount of ^{87}Rb present. The slopes of the isochrons ($t = 500$ Ma, $t = 1000$ Ma) are proportional to the ages of the sample suite. The intercept on $^{87}Sr/^{86}Sr$ is the initial ratio.

Thus isochron calculations require a suite of cogenetic samples, formed from the same parental material, and assume that there has been no exchange of parent–daughter isotopes other than through radioactive decay. An example of an isochron drawn for the Sm–Nd system from data in Table 6.2 is shown in Figure 6.2.

Pb isotope isochrons   A Pb–Pb whole-rock isochron is constructed by plotting the isotope ratios $^{206}Pb/^{204}Pb$ on the x-axis and $^{207}Pb/^{204}Pb$ on the y-axis of an isochron diagram. However, the interpretation of linear arrays on such diagrams is much more complex than for systems such as Rb–Sr because ^{206}Pb and ^{207}Pb are the product, of two separate radioactive decay schemes with two different decay constants (Box 6.1). The logic behind the construction of a Pb–Pb isochron is outlined in Figure 6.18. In this case the isochron equation cannot be directly solved and must be solved iteratively (see Harmer and Eglington, 1987).

Fitting an isochron   A simple but approximate way in which to fit an isochron is to draw a best-fit straight line by eye through the plotted points. Provided a suitable scale is chosen,

Table 6.2 Samarium–neodymium isotopic ratios for the lower ultramafic unit of the Onverwacht volcanics. South Africa (from Hamilton *et al.*, 1979b) used in the construction of the isochron in Figure 6.2

Sample	Rock type	$^{147}Sm/^{144}Nd$*	$^{143}Nd/^{144}Nd$†
HSS-74	Sodic porphyry	0.1030	0.510487 ± 36
HSS-161	Acid tuff	0.1054	0.510570 ± 32
HSS-52B	Felsic pillow lava	0.1653	0.511950 ± 22
HSS-56	Basaltic lava	0.2040	0.512875 ± 32
R-14	Basaltic komatiite	0.1888	0.512504 ± 34
HSS-32	Basaltic komatiite	0.1649	0.511957 ± 22
HSS-88A	Peridotitic komatiite	0.1792	0.512292 ± 34
HSS-92	Peridotitic komatiite	0.1858	0.512439 ± 34
HSS-95	Peridotitic komatiite	0.1902	0.512541 ± 28
HSS-523	Peridotitic komatiite	0.1701	0.512084 ± 20

* $^{147}Sm/^{144}Nd$ ratios determined to a precision of 0.2 % at the 2 σ level.
† Errors on $^{143}Nd/^{144}Nd$ are $\pm$ 2 σ.

both the slope of the line and the intercept can be determined with reasonable accuracy and these may be used to make preliminary estimates of the age and initial ratio. Precise results are obtained from statistical line-fitting procedures which estimate the slope and intercept of the isochron. These normally use a version of weighted least squares regression (see Section 2.4.3). Faure (1977 — p.425–33) gives the FORTRAN IV code for a computer program for the analysis of isochron data based upon equation 6 of York (1969).

A most important aspect of isochron regression treatment is that it provides a realistic measure of the uncertainty in the age and initial ratios. A measure of the 'goodness of fit' of an isochron is the Mean Squares of Weighted Deviates (MSWD). This is a measure of the fit of the line to the data within the limits of analytical error. Ideally an isochron should have an MSWD of 1.0 or less; anything greater than this is not strictly an isochron because the scatter in the points cannot be explained solely by experimental error. Brooks *et al.* (1972) suggest, however, that the ideal of an MSWD of 1.0 or less is only applicable to isochrons involving infinitely large numbers of samples and that for normal data-sets the higher value of 2.5 is an acceptable cut-off for the definition of an isochron.

Errorchrons The origin of the scatter on an isochron is one of the most important interpretive aspects of geochronology. If the scatter results in an MSWD of 2.5 or less it is deemed analytical. If the MSWD is greater than 2.5 it is geological. Brooks *et al.* (1972) proposed the term 'errorchron' for the situation where a straight line cannot be fitted to a suite of samples within the limits of analytical error. An errorchron implies that the scatter of the points is a consequence of geological error and indicates that one or more of the initial assumptions of the isochron has not been fulfilled.

The geochron The geochron represents an isochron for $t = 0$ drawn through the composition of the appropriate isotope ratio at the time of the formation of the Earth — the primordial composition. In principle a geochron can be defined for any isotopic system although in practice it is most commonly used in the interpretation of lead

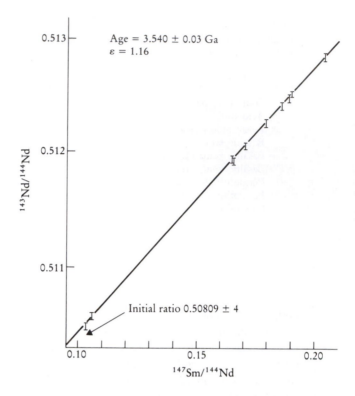

Figure 6.2   An isochron plot of ^{143}Nd/^{144}Nd vs ^{147}Sm/^{144}Nd for volcanic rocks from the Onverwacht group, South Africa (after Hamilton *et al.*, 1979b). The data are given in Table 6.2. Error bars (2σ) are shown for ^{143}Nd/^{144}Nd but 2σ errors on ^{147}Sm/^{144}Nd are too small to show. The slope of the best-fit line is proportional to an age of 3.54 ± 0.03 Ga; the intercept of this line on the ^{143}Nd/^{144}Nd axis where ^{147}Sm/^{144}Nd = 0 is the initial ^{143}Nd/^{144}Nd ratio and is 0.50809 ± 0.00004. The epsilon value (ε_{Nd} = +1.16) is a measure of the difference between the initial ratio and a chondritic model for the Earth's mantle at 3.54 Ga expressed in parts per 10 000 (see Section 6.3.4 and Box 6.2). The positive value for epsilon indicates that the volcanic rocks were derived from a (slightly) depleted mantle source at 3.54 Ga, although there are large errors on this estimate.

isotopes. Since there is some uncertainty in the age of the Earth, the geochron may be calculated for a range of ages between 4.4 and 4.6 Ga; its precise position is a function of the value adopted for the age of the Earth. Thus, although a geochron represents a zero-age isochron, the age of the Earth used in its computation may be specified so that the geochron is, for example, for 4.57 Ga. The initial values for the geochron are given in Box 6.1.

6.2.2   Model ages

A model age is a measure of the length of time a sample has been separated from the mantle from which it was originally derived. Model ages are most commonly quoted for the Sm–Nd system and are valuable because they can be calculated for an individual rock from a single pair of parent–daughter isotopic ratios. They must, however, be interpreted with care. The basis of all model age calculations is an assumption about the isotopic composition of the mantle source region from which the samples were originally derived. In the case of neodymium isotopes there are two frequently quoted models for the mantle reservoir — CHUR (the CHondritic Uniform Reservoir) and Depleted Mantle (DM).

T–CHUR model ages   The CHUR model assumes that the Earth's primitive mantle had the same isotopic composition as the average chondritic meteorite at the formation of the Earth, which in this case is taken to be 4.6 Ga. For neodymium isotopes CHUR is synonymous with the composition of the bulk Earth. A model age calculated relative to CHUR, therefore, is the time in the past at which the sample suite separated from the mantle reservoir and acquired a different Sm/Nd ratio. It is also the time at which the sample had the same ^{143}Nd/^{144}Nd ratio as CHUR. This is illustrated in Figure 6.3, in which the present-day ^{143}Nd/^{144}Nd composition of the sample is extrapolated back in time until it intersects the CHUR evolution line. This gives the *T–CHUR* model age. The evolution curve for the sample is constructed from the present day ^{143}Nd/^{144}Nd ratio and the ^{143}Nd/^{144}Nd ratio for that sample at some time in the past (t), calculated from the equation

$$\left[\frac{^{143}\text{Nd}}{^{144}\text{Nd}}\right]_{\text{sample}} = \left[\frac{^{143}\text{Nd}}{^{144}\text{Nd}}\right]_{t} + \left[\frac{^{147}\text{Sm}}{^{144}\text{Nd}}\right]_{\text{sample}}(e^{\lambda t} - 1) \qquad [6.4]$$

where λ is the decay constant, t is an arbitrarily selected time in the past, used for constructing the curve, and ^{147}Sm/^{144}Nd is the present-day ratio in the sample. Alternatively, a *T–CHUR* model age is calculated from the present-day ^{147}Sm/^{144}Nd and ^{143}Nd/^{144}Nd isotope ratios of a single sample using the equation

$$T^{\text{Nd}}_{\text{CHUR}} = \frac{1}{\lambda} \ln \left[\frac{(^{143}\text{Nd}/^{144}\text{Nd})_{\text{sample,today}} - (^{143}\text{Nd}/^{144}\text{Nd})_{\text{CHUR,today}}}{(^{147}\text{Sm}/^{144}\text{Nd})_{\text{sample, today}} - (^{147}\text{Sm}/^{144}\text{Nd})_{\text{CHUR,today}}} + 1 \right] \qquad [6.5]$$

where λ is the decay constant for ^{147}Sm to ^{143}Nd (Table 6.1) and where the values for CHUR are given in Table 6.3. Care must be taken to choose the correct CHUR value because different laboratories use different normalizing values. Model ages are also sensitive to the difference in Sm/Nd between the sample and CHUR and only fractionated samples with Sm/Nd ratios which are sufficiently different from the chondritic value will yield precise ages.

T–depleted mantle (DM) model ages   Studies of the initial ^{143}Nd/^{144}Nd ratios from Precambrian terrains suggest that the mantle which supplied the continental crust has evolved since earliest times with an Sm/Nd ratio greater than that of CHUR. For this reason model ages for the continental crust are usually calculated with reference to the depleted mantle (DM) reservoir rather than CHUR. Depleted mantle model ages are calculated by substituting the appropriate DM values in place of $(^{143}\text{Nd}/^{144}\text{Nd})_{\text{CHUR}}$ and $(^{147}\text{Sm}/^{144}\text{Nd})_{\text{CHUR}}$ in Eqn [6.5]. These values are given in Table 6.3, and a range of possible depleted mantle growth curves are shown in Figure 6.17, using the epsilon notation (depletion relative to CHUR). A graphical solution of a *T–DM* model age calculation is illustrated Figure 6.3.

Assumptions made in the calculation of model ages   It is important when calculating model ages to remember the assumptions upon which they are based, for these are not always fulfilled. Firstly, assumptions are made about the isotopic composition of the reservoir which is being sampled — either CHUR or depleted mantle. This aspect of model age calculations in itself raises three further problems.

Table 6.3   Parameters used in the normalization of neodymium isotope data for the calculation of model ages and epsilon values†

	Authors: DePaolo; Wasserburg*				Authors: O'Nions; Allegre; Hawkesworth*	
Normalizing factor	$\dfrac{^{146}Nd}{^{142}Nd} = \begin{cases} 0.636151\ (1)\ \text{or} \\ 0.63613\ (3) \end{cases}$		$\dfrac{^{150}Nd}{^{142}Nd} = \begin{cases} 0.2096\ (1) \\ 0.209627\ (2) \end{cases}$		$\dfrac{^{146}Nd}{^{142}Nd} = 0.63223\ (4)$ $\qquad \dfrac{^{146}Nd}{^{144}Nd} = 0.7219\ (5)$	

Chondritic uniform reservoir

			Preferred values
		Sm/Nd=0.31 (6)	Sm/Nd=0.325 (2)
$\dfrac{^{143}Nd}{^{144}Nd}_{CHUR,4.6\ Ga}$	0.505829 (1)	0.50677±10 (7)	0.50663 (8)
$\dfrac{^{143}Nd}{^{144}Nd}_{CHUR,today}$	0.511847 (2) 0.511836 (10)	0.51262 (8)	0.512638 (9)
$\dfrac{^{147}Sm}{^{144}Nd}_{CHUR,today}$	0.1936 (now superseded) (7) 0.1967 (1)	0.1966 (9) 0.1967 (12) 0.19637–0.1968 (18)	

Depleted mantle

$\dfrac{^{143}Nd}{^{144}Nd}_{DM,today}$	0.51235 (10) 0.512245 (11)	0.51315(12) 0.51316 (9) 0.513114 (17) 0.51317–0.51330 (18)	
$\dfrac{^{147}Sm}{^{144}Nd}_{DM,today}$	0.214 (10) 0.217 (13) 0.225 (14) 0.230 (11)	0.2137 (12) 0.222 (17) 0.233–0.251 (18)	

Standard rock BCR-1

$\dfrac{^{143}Nd}{^{144}Nd}_{BCR-1}$	0.51184 (15)	0.512669 ± 8 (16)	

* References: (1) Jacobsen and Wasserburg (1980); (2) Wasserburg *et al.* (1981); (3) DePaolo (1981a); (4),(5) O'Nions *et al.* (1977); (6) O'Nions *et al.* (1979); (7) Lugmair *et al.* (1975); (8) Hawkesworth and van Calsteren (1984); (9) Goldstein *et al.* (1984); (10) McCulloch and Black (1984); (11) McCulloch and Chappell (1982); (12) Peucat *et al.* (1988); (13) Taylor and McLennan (1985); (14) McCulloch *et al.* (1983); (15) DePaolo and Wasserburg (1976); (16) Hooker *et al.* (1981); (17) Michard et al. (1985); (18) Allegre *et al.* (1983a).

† $\lambda = 6.54 \times 10^{-12}\ yr^{-1}$.

(1) The correct choice of reservoir is important for, as Figure 6.3 shows, there can be as much as 300 Ma difference between Nd CHUR and DM model ages.

(2) There is a variety of possible models for the depleted mantle. These are shown for Nd in terms of the epsilon notation (deviation from CHUR) in Figure 6.17.

(3) There is great confusion over the actual numbers that should be used for mantle values. The difficulty has arisen because the Sm–Nd technique evolved very rapidly and different laboratories developed different normalization schemes in parallel. These are summarized in Table 6.3. In practical terms however, provided the same set of values is consistently used, the overall effect on calculated model ages is not great. Care must be taken, however, when

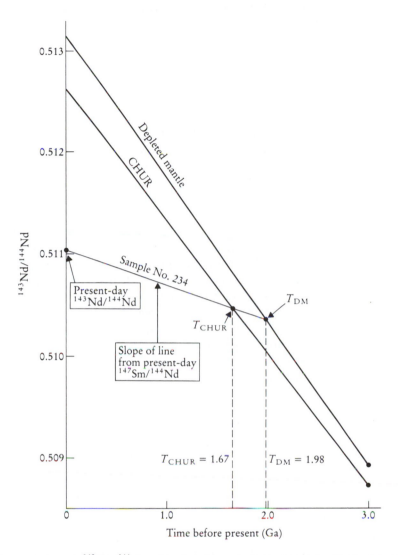

Figure 6.3 The evolution of $^{143}Nd/^{144}Nd$ with time in an individual sample (Table 6.4, No. 234) compared with two models of the mantle — CHUR (the chondritic uniform reservoir) and depleted mantle. A model Nd age is the time at which the sample had the same $^{143}Nd/^{144}Nd$ ratio as its mantle source. In this case there are two possible solutions, depending upon which mantle model is preferred. The T_{CHUR} model age is approximately 1.67 Ga and the T_{DM} model age is approximately 1.98 Ga. A similar diagram can be constructed using the epsilon notation (see Section 6.3.4) in place of $^{143}Nd/^{144}Nd$.

plotting data from different sources to normalize the isotope ratios to the same value of unradiogenic isotopes, and the normalizing factor used should be stated.

A second assumption used in model age calculations is that the Sm/Nd ratio of the sample has not been modified by fractionation after its separation from the mantle source. In the case of Nd isotopes this is a reasonable assumption to make.

Thirdly, it is assumed that all material came from the mantle in a single event. Table 6.4 shows model Nd–CHUR and depleted mantle ages for granulites and gneisses from central Australia, calculated using Eqn [6.5].

6.2.3 Interpreting geochronological data

Any detailed geochronological study of slowly cooled rocks will tend to yield discordant results when several different methods of age determination are used. In this section therefore we look closely at the different isotopic systems in an attempt to discover what the different forms of geochronological data mean and what it is that we are actually measuring when we claim to have calculated the geological age of a rock. In this respect one of the most fruitful areas of research has been the application of the concept of blocking temperatures to geochronological systems.

Blocking temperatures Many geochronologists now agree that the principal control on the retentivity of radiogenic daughter products in minerals is temperature. In other words, the point at which a mineral becomes a closed system with respect to a particular daughter isotope is temperature-controlled. Different minerals will close at different temperatures and different isotopic systems in the same mineral will also close at different temperatures. The concept of closure temperature, or blocking temperature as it is also called, is defined by Dodson (1973, 1979) as 'the temperature of a system at the time of its apparent age'. It is the temperature below which the isotopic clock is switched on. This is illustrated in Figure 6.4. Blocking temperatures are most commonly used to describe minerals although the principle can equally be applied to slowly cooled rocks. Hofmann and Hart (1978) have

Table 6.4 Model Nd ages calculated for CHUR and depleted mantle for granulites and gneisses from the Strangeways range, central Australia. An Sm–Nd whole-rock isochron for this suite gives an age of 2.07 ± 0.125 Ga (from Windrim and McCulloch, 1986)

Sample no.	$\dfrac{^{147}\text{Sm}}{^{144}\text{Nd}}$	$\dfrac{^{143}\text{Nd}}{^{144}\text{Nd}}$	$T^{\text{Nd}}_{\text{(CHUR)}}(\text{Ga})$	$T^{\text{Nd}}_{\text{DM}}(\text{Ga})$
Mafic granulites				
226	0.1853	0.511772±36	0.856	2.203
234	0.1251	0.511049±30	1.672	1.975
551	0.2134	0.512122±36	2.596	2.950
868	0.2046	0.512013±24	3.388	2.491
Quartzofeldspathic and calc-silicate rocks				
501A	0.1248	0.511006±28	1.755	2.034
507	0.1310	0.511039±24	1.844	2.115
503	0.1248	0.510967±32	1.837	2.093
512	0.1150	0.510794±32	1.938	2.145

Notes

1. Nd isotopes normalized to $^{146}\text{Nd}/^{142}\text{Nd} = 0.636151$.
2. DM values for Nd are $(^{143}\text{Nd}/^{144}\text{Nd})_{today} = 0.51235$, $(^{147}\text{Sm}/^{144}\text{Nd})_{today} = 0.225$.
3. Calculations made using Eqn [6.5].

Figure 6.4   The definition of blocking temperature: (a) the cooling curve; (b) the accumulation curve of the concentration of the daughter to parent isotope, both as a function of time (after Dodson, 1979). When a mineral is close to its temperature of crystallization (1), the daughter isotope will diffuse out of the mineral as fast as it is produced and so cannot accumulate. On cooling the mineral enters a transition stage in which some of the daughter isotope is lost and some is retained until finally at low temperatures the rate of escape is negligible and all of the daughter isotope is retained (2). The 'age' for the system (t_B) is effectively an extrapolation of the accumulation curve [broken line in (b)] back onto the time axis. The value of the blocking temperature (T_B) will depend upon the exact cooling history of a particular system but will be independent of the starting temperature if the latter is sufficiently high.

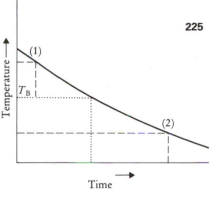

(a) Cooling curve

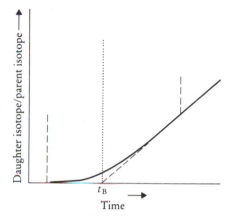

(b) Accumulation curve

shown that element exchange in slowly cooled rocks is chiefly controlled by volume diffusion and that in different whole-rock systems diffusion ceases at slightly different temperatures.

A distinction must be made however between blocking temperature and the resetting of an isotopic system. The latter will take place in whole-rock systems which have been thermally overprinted or subject to fluid movement. It is likely that in the case of resetting whole-rock systems the amount of fluid and the chemistry of a fluid are more important controls than temperature. Choosing between blocking temperature (mainly controlled by volume diffusion) and the resetting of an isochron (most commonly a function of fluid movement) as an explanation of discordant ages requires the consideration of the scale of isotopic homogenization. Rocks which have been isotopically reset may yield ages which vary from the time of crust formation to the most recent metamorphic event, depending upon the scale of sampling. Very large samples may have escaped the isotopic rehomogenization and will preserve old ages whereas individual minerals may have recrystallized and yield young ages.

In the discussion below we shall look more closely at the term 'age' in the light of the blocking temperature concept, and evaluate the different ways in which it has been used in both mineral and whole-rock systems.

Concepts of   As the discussion above has shown, the term 'age', whether applied to a model age,
geological age   or a whole-rock or mineral isochron age, may have a number of different meanings in geochronology and must be defined carefully by using an additional qualifying term.

(a) Cooling age In a metamorphic rock the term 'cooling age' is conventionally used to describe the time, after the main peak of metamorphism, at which a mineral which had crystallized at the metamorphic peak passes through its blocking temperature. In an igneous rock, cooling age is the time after the solidification of the melt at which a mineral passes through its blocking temperature.

(b) Crystallization age The crystallization age of a mineral or rock records the time at which it crystallized. In the case of a metamorphic mineral, for example, if the temperature of crystallization is lower than the blocking temperature, the instant the mineral forms the isotopic clock is switched on and the age of crystallization is recorded. In an igneous rock, the crystallization age of a mineral records the **magmatic age** of the rock.

(c) Metamorphic age The term 'metamorphic age' is often confused with cooling age but it means the time of the peak of metamorphism. The determination of metamorphic age depends greatly upon the grade of metamorphism. In low grades of metamorphism the metamorphic maximum may be determined from the blocking temperature of a specific mineral. In the case of high grades of metamorphism, the timing of the peak of metamorphism is normally inferred from the resetting of a whole-rock system such as Rb–Sr or Pb–Pb.

(d) Crust formation age This is the time of formation of a new segment of continental crust by fractionation of material from the mantle (O'Nions *et al.*, 1983). Whether or not it is possible to determine directly the crust formation age will depend upon the geological history of the segment of the crust. In many areas of ancient continental crust the formation of the crust was followed by deformation, metamorphism and melting and it may be possible to determine only a *cratonization age* rather than the age of formation.

(e) Crust residence age Sediment eroded from a segment of continental crust will possess a crust residence age, which may reflect the crust formation age. A neodymium crustal residence (T_{cr}) age is calculated from Eqn [6.5] by substituting the appropriate values for depleted mantle in place of the CHUR values. Some authors use the term **provenance age** as a synonym, although this does not signify a specific historical event such as a stratigraphic age but is the average crustal residence time of all the components of the rock. Normally, the crustal residence age of a sediment is greater than the stratigraphic age (Figure 6.5).

The interpretation of whole-rock ages

Whole-rock systems are more likely than mineral ages to determine the timing of crystallization of an igneous rock or the timing of the metamorphic peak during metamorphism. Sometimes it may be possible to determine the time of crustal formation.

(a) The Rb–Sr system The rubidium–strontium isotopic system is still one of most widely used isotopic whole-rock methods, for most crustal rocks contain sufficient Rb and Sr (anywhere between 10 and 1000 ppm) to make the chemical separation of the elements and the mass spectrometry relatively straightforward. It is a versatile method and can be applied to a range of whole-rock compositions as well as individual minerals. On the other hand, the results of Rb–Sr geochronology are not

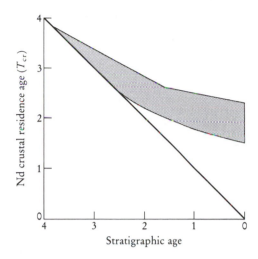

Figure 6.5   A plot of stratigraphic age vs neodymium crustal residence age (T_{cr}) for fine-grained clastic sediments between 4.0 Ga and the present day (shaded region). The departure from the 1:1 line at about 2.5 Ga shows that whilst sediments in the Archaean contain a mainly juvenile component, Phanerozoic sediments contain an old crustal component (after Miller and O'Nions, 1985).

always easy to interpret, for both Rb and Sr are relatively mobile elements so that the isotopic system may readily be disturbed either by the influx of fluids or by a later thermal event. Thus Rb–Sr isochrons are rarely useful in constraining crust formation ages. Nevertheless, if an Rb–Sr isochron is obtained it can usually be attributed to a definite event such as the age of metamorphism or alteration, or the age of diagenesis in sedimentary rocks, even if the primary age of the rock cannot be determined.

(b) Pb isotopes   Pb–Pb isochrons can be plotted for a wide range of rock types from granites to basalts and can be used to give the depositional age of some sediments. The method yields reliable ages in rocks with a simple crystallization history, although, because of differences in the mobility and the geochemistry of uranium and lead, there are severe problems of interpretation in systems where lead isotopes have been reset. Thus the Pb–Pb isochron method is rarely able to give crust formation ages although it can be used to give the age of metamorphism.

(c) The Sm–Nd system   The elements Sm and Nd are much less mobile than Rb, Sr, Th, U and Pb and may be used to 'see through' younger events in rocks whose Rb–Sr and Pb isotopic chemistry has been disturbed. Thus the Sm–Nd technique seems to be the most likely whole-rock method from which crust formation ages may be determined. The chief limitations on the method are the long half-life, so that it is only really applicable to old rocks, and the relatively small variations in Sm/Nd ratio found in most cogenetic rock suites. This latter problem has led some workers to combine a wide range of lithologies on the same isochron in order to obtain a spread of Sm/Nd ratios. However this resulted in samples extracted from different sources and with very different histories being plotted on the same isochron and has given rise to some spurious results.

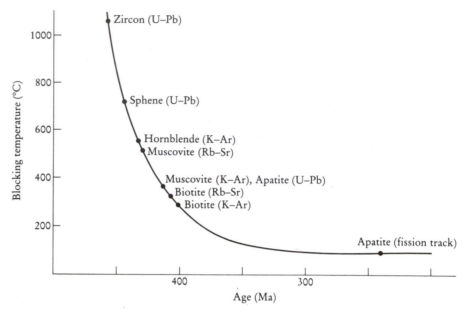

Figure 6.6 A plot of mineral age vs blocking temperature for the Glen Dessary syenite from Scotland (after van Breemen *et al.*, 1979; Cliff, 1985). The mineral ages define a cooling curve for the pluton. The mineral on which each age determination was made and the method used are indicated.

The interpreta-tion of mineral ages Isotopic measurements on minerals may be used to obtain a mineral isochron in which minerals from the same rock are used to define an isochron. This, of course, is only valid if all the minerals have the same blocking temperature. A second way in which minerals are used is in the calculation of mineral ages. In most cases these will be cooling ages, although highly refractory minerals such as zircon may be used to give the magmatic age of some igneous rocks, and may on some occasions see through a thermal event to the time of crust formation. Blocking temperatures for isotopic systems in minerals are illustrated in Figure 6.6.

(a) Rb–Sr mineral ages The minerals biotite and muscovite are commonly dated by the Rb–Sr method. Ages are calculated using a two-point isochron method in which a mineral with a low concentration of Rb such as plagioclase feldspar, or better (to avoid two different blocking temperatures) the whole rock, is used as control on the initial ratio. Blocking temperatures for biotite are in the range 300 ± 50 °C and are about 40 °C higher than the blocking temperatures for Ar in biotite (Harrison and McDougall, 1980). Muscovite blocking temperatures are 550 ± 50 °C, although Cliff (1985) points out that in coarse-grained muscovite from pegmatites blocking temperatures may reach 600–650 °C and are close to the crystallization temperature.

(b) Argon methods Argon isotopes are studied most commonly in the minerals biotite, muscovite, hornblende and alkali feldspar. The blocking temperatures for argon diffusion, applicable to the interpretation of K–Ar and ^{39}Ar/^{40}Ar ages are

more difficult to apply than those for Sr diffusion. For example, biotite blocking temperatures determined from hydrothermal diffusion experiments vary from *ca* 295 °C to 410 ± 50 °C according to the Fe/Mg ratio of the biotite (see compilation by Blanckenburg *et al.*, 1989). Muscovite has a blocking temperature of 350 ± 50 °C although very fine-grained white micas formed at low grades of metamorphism may yield crystallization ages. Reuter and Dallmeyer (1989) dated the timing of cleavage formation in low-grade pelites using this method. Hornblende has a blocking temperature of 530 ± 30 °C, although values up to 650 °C have also been recorded.

(c) Sm–Nd mineral ages   Garnet-bearing mineral assemblages offer the potential for precise Sm–Nd mineral isochron ages, for garnets have high Sm/Nd ratios which allows the slope of a mineral isochron to be accurately determined with errors of only 5–10 Ma on lower Palaeozoic ages (Cliff, 1985). Garnet ages are determined on two-point isochrons between garnet and one or more of the minerals plagioclase, orthopyroxene or clinopyroxene and hornblende. Mineral isochrons which do not involve garnet show a smaller range of Sm/Nd ratio but still yield reasonably precise ages with errors of ± 20 Ma. Blocking temperatures for Nd diffusion in garnet are *ca* 700 °C (Cohen *et al.*, 1988; Windrim and McCulloch, 1986).

(d) U–Pb dating of zircon   A plot of $^{206}Pb/^{238}U$ vs $^{207}Pb/^{235}U$ concentrations in a population of zircons tends to define a linear array with an upper and lower time intercept on a concordia–discordia diagram (Figure 6.7). The concordia curve is the curve which defines the locus of concordant ages for both ^{238}U and ^{235}U decay — see Box 6.1. A general interpretation of the linear array (the discordia) is one of lead loss (or uranium gain) and both upper and lower intercepts with the concordia have petrological significance. Magmatic zircons in felsic rocks can be used to date the time of crystallization from the upper intercept of a discordia curve. Granitic rocks which are melts of old crust may contain inherited zircons. In this case the lower intercept may give the age of new zircon crystallization (magmatic crystallization) and the upper intercept the age of the crustal source (Liew and McCulloch, 1985). Detrital zircons in sediments may date the provenance of the sediments, and zircons in metamorphic rocks may date the protolith or, in the case of meta-sediments, the provenance of the protolith.

Recently, the application of the ion microprobe to the U–Pb isotopic analysis of single zircon crystals (Compston *et al.*, 1982, 1984) has revolutionized zircon geochronology, allowing a number of analyses to be made on a single zircon crystal. Hence it is possible to unravel the geological history of complex zircon grains (see, for example, Kroner *et al.*, 1987 and Figure 6.7).

The U–Pb method is also applicable to the minerals sphene, monazite and epidote. Sphene has a blocking temperature of greater than 550 °C. The blocking temperature of epidote is less well known but could also be as high as sphene. Monazite also has a high blocking temperature (*ca* 600 °C; Gebauer and Grunenfelder, 1979) but it is not precisely known.

The interpreta-
tion of model
ages

A model age is an estimate of the time at which a sample separated from its mantle source region. Thus for igneous rocks and meta-igneous rocks this is a good estimate of the crust formation age. In detail, of course, this assumes (1) knowledge of the isotopic composition of mantle source, (2) absence of parent/daughter isotope

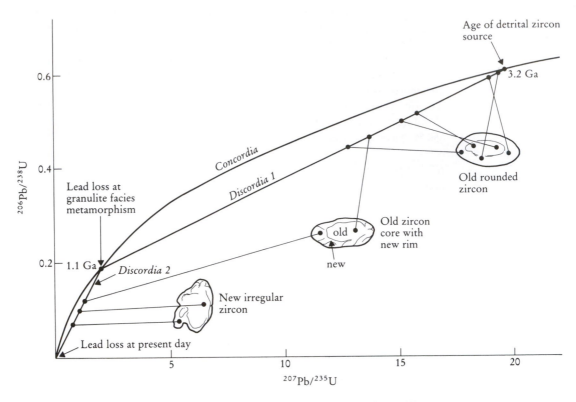

Figure 6.7   Concordia–discordia diagram showing $^{206}Pb/^{238}U$ and $^{207}Pb/^{235}U$ ratios in zircon grains in granulite facies metasediments from Sri Lanka (after Kroner *et al.*, 1987). The concordia curve is the locus of points for which the $^{206}Pb/^{238}U$ age equals the $^{207}Pb/^{235}U$ age. It is curved due to the different half-lives of ^{238}U and ^{235}U and it may be plotted using the values in Box 6.1. The discordia curve is a linear array of points which intersect the concordia curve at two points. The upper intersection is interpreted as the primary age of the sample whereas the lower intersection defines a period of lead loss and may have a number of possible geological meanings. The sketches of individual zircon grains show the spots analysed using the ion microprobe and their compositions are plotted on the discordia–concordia diagram. Old zircon grains and the cores of composite grains define a discordia curve (discordia 1) with an upper intercept at 3.2 Ga (the age of the source of the detrital zircons) and a lower intercept at 1.1 Ga (the time of the granulite facies metamorphism). New zircon grains (recognized on the basis of their irregular morphology) and new zircon rims on composite grains define a second discordia (discordia 2) with an upper intercept at 1.1 Ga (the granulite facies metamorphism) and a lower intercept at the present day. The new zircon growth took place during the granulite facies event.

fractionation after extraction from the mantle source and (3) the immobility of parent and daughter isotopes. In the case of Sr isotopes none of these criteria is usually fulfilled with any certainty either for igneous rocks or for sediments (Goldstein, 1988) and so at the present time model ages are not frequently calculated for the Rb–Sr system. Nd isotopes, however, are more useful and meaningful model ages can be calculated, although the reference reservoir (CHUR or depleted mantle) must be specified and in the case of depleted mantle model ages the depleted mantle model evolution curve must also be specified.

Model ages of granitoids may be used to estimate the age of their source. In the case of mantle-derived granites, the model age gives the time of mantle fractionation

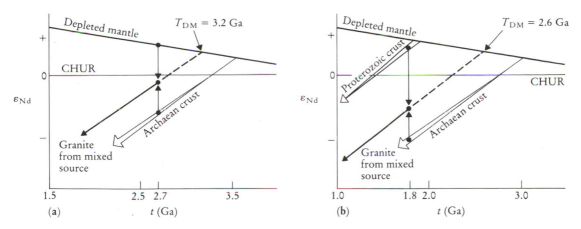

Figure 6.8 The evolution of ^{143}Nd/^{144}Nd (expressed as epsilon units — see Section 6.3.4) with time, for depleted mantle and continental crust. Relative to CHUR the depleted mantle shows increasing ^{143}Nd/^{144}Nd (increasingly positive ε-values) with time, whereas the continental crust shows retarded ^{143}Nd/^{144}Nd evolution (increasingly negative ε-values). (a) The depleted mantle model age for a granite formed at 2.7 Ga from a mixture of a juvenile component derived from the depleted mantle and from Archaean crust formed at 3.5 Ga will be 3.2 Ga. (b) The depleted mantle model age of a granite formed by the mixture of Proterozoic (1.9 Ga) and Archaean (3.0 Ga) crustal sources at 1.8 Ga is 2.6 Ga. The model ages have little real meaning for they neither reflect the crystallization age of the granite nor the age of the crustal source (Arndt and Goldstein, 1987).

of the basaltic precursor to the granite. This is usually assumed to be very close in time to the crystallization age of the granite. Granitoids which are derived by the melting of older continental crust will give model ages which are indicative of the age of the crustal source. This is possible because the intracrustal fractionation process does not greatly disturb the Sm/Nd ratio of the source. Often, however, granites are a mixture of crustal and mantle sources. When this is the case, calculated model ages may give very misleading results (Arndt and Goldstein, 1987, and Figure 6.8).

Model ages for clastic sedimentary rocks are thought to provide an estimate of their crustal residence age (O'Nions *et al.*, 1983), for minimal fractionation of Sm/Nd accompanies their generation. Since, however, many continental sediments are a mixture of materials from different sources, model Nd ages are the average model age of the sediment and provide a minimum estimate of the crustal residence age or an average crustal residence age.

6.3 Radiogenic isotopes in petrogenesis

Isotope ratios in a magma are characteristic of the source region from which the magma was extracted and the ratios remain unchanged during subsequent fractionation events. This is because the mass difference between any pair of radiogenic isotopes used in geochemistry (with the exception of helium) is so small that the isotope-pair cannot be fractionated by processes controlled by crystal–liquid equilibria. Thus during partial melting a magma will have the same isotopic character as its source region.

This simple observation has led to two important developments in isotope geochemistry. Firstly, distinct source regions can be recognized with their own unique isotopic character and, secondly, mixing can be recognized between isotopically distinct sources. Thus one of the main quests of isotope geology is to identify the different isotopic reservoirs in the crust and mantle and to characterize them for as many isotopic systems as is possible. An attempt to do this for crust and oceanic basalt sources is presented in Table 6.5. The larger and more difficult question of how the reservoirs acquired their own identities leads into the area of crust and mantle geodynamics, a topic treated briefly at the end of this chapter (see Section 6.3.6).

In detail a wealth of mixing and contamination processes can be recognized in and between sources, although these can normally be reduced to two essentially different types of process (Hawkesworth and van Calsteren, 1984). Firstly, radiogenic isotopes may be used to identify components from different sources which have contributed to a particular magmatic suite — for example, the contamination of continental flood basalt with old continental crust. The time-scale of these processes is short and equivalent to the time-scale of most familiar geological processes. The second application is to constrain models of the evolution of source regions of magmatic rocks. For example, we will show below that there are several sources for oceanic basalts in the upper mantle. It is of some importance to discover how and when these separate sources acquired their separate identities. The time-scale of these processes is long and is of the order of billions (Ga) rather than millions (Ma) of years.

6.3.1 The role of different isotopic systems in identifying reservoirs and processes

The different elements used in radiogenic isotope studies in geology vary greatly in their chemical and physical properties, so much so that different isotope systems vary in their sensitivity to particular petrological processes. This variability may show itself in two ways. Firstly, the parent and the daughter elements may under certain circumstances behave in different ways so that the two become fractionated. The order of incompatibility for elements of interest is

$$Rb > Th > U > Pb > (Nd, Hf) > (Sr, Sm, Lu)$$

These values indicate the extent to which an element is fractionated into the crust relative to the depleted mantle. For example, Rb is the element most concentrated in the crust relative to the depleted mantle whereas Sr, Sm and Lu are the least concentrated. Alternatively, a parent–daughter element-pair may behave coherently and not be fractionated and yet behave in a very different manner from the parent–daughter pair of another isotopic system. A good example is the contrast between the Sm–Nd system, in which both elements share very similar chemical and physical characteristics, and the Rb–Sr system, in which the elements are

Table 6.5   The isotopic character of crust and mantle reservoirs (present-day isotope ratios shown in parentheses)

	^{87}Rb–^{86}Sr	^{147}Sm–^{143}Nd	^{238}U–^{206}Pb	^{235}U–^{207}Pb	^{232}Th–^{208}Pb
Continental crustal sources					
Upper crust	High Rb/Sr; high ^{87}Sr/^{86}Sr	low Sm/Nd; low ^{143}Nd/^{144}Nd (Negative epsilon)	High U/Pb; high^{206}Pb/^{204}Pb	High U/Pb; high ^{207}Pb/^{204}Pb	High Th/Pb; high ^{208}Pb/^{204}Pb
Middle crust	Mod. high Rb/Sr(0.2–0.4); ^{87}Sr/^{86}Sr = (0.72–0.74)	Retarded Nd evolution in the crust	U-depleted; low ^{206}Pb/^{204}Pb	U-depleted; low ^{207}Pb/^{204}Pb	Mod. high Th; mod high ^{208}Pb/^{204}Pb
Lower crust	Rb depletion; Rb/Sr <*ca*0.04; low ^{87}Sr/^{86}Sr (0.702–0.705)	relative to chondritic source	Severe U depletion; very low ^{206}Pb/^{204}Pb (*ca* 14.0)	Severe U depletion; very low ^{207}Pb/^{204}Pb (*ca* 14.7)	Severe Th depletion; very low ^{208}Pb/^{204}Pb
Subcontinental lithosphere					
Archaean	Low Rb/Sr	Low Sm/Nd			
Proterozoic to Recent	High Rb/Sr	Low Sm/Nd			
Oceanic basalt sources (Zindler and Hart, 1986)					
Depleted mantle	Low Rb/Sr; low ^{87}Sr/^{86}Sr	High Sm/Nd; high ^{143}Nd/^{144}Nd (positive epsilon)	Low U/Pb; low ^{206}Pb/^{204}Pb (*ca* 17.2–17.7)	Low U/Pb low ^{207}Pb/^{204}Pb (*ca* 15.4)	Th/U=2.4±0.4; low ^{208}Pb/^{204}Pb (*ca* 37.2–37.4)
HIMU	Low Rb/Sr; low ^{87}Sr/^{86}Sr (=0.7029)	Intermediate Sm/Nd (<0.51282)*	High U/Pb; high ^{206}Pb/^{204}Pb (>20.8)	High U/Pb; high ^{207}Pb/^{204}Pb	High Th/Pb
Enriched mantle					
EM I	Low Rb/Sr; ^{87}Sr/^{86}Sr = ±0.705	Low Sm/Nd; ^{143}Nd/^{144}Nd < 0.5112*	Low U/Pb; ^{206}Pb/^{204}Pb = (17.6–17.7)	Low U/Pb; ^{207}Pb/^{204}Pb = (15.46–15.49)	Low Th/Pb; ^{208}Pb/^{204}Pb = (38.0–38.2)
EM II	High Rb/Sr; ^{87}Sr/^{86}Sr > 0.722	Low Sm/Nd; ^{143}Nd/^{144}Nd = (0.511–0.5121)*	High ^{207}Pb/^{204}Pb and ^{208}Pb/^{204}Pb at a given ^{206}Pb/^{204}Pb		
PREMA	^{87}Sr/^{86}Sr = 0.7033	^{143}Nd/^{144}Nd = 0.5130*	^{206}Pb/^{204}Pb = (18.2–18.5)		
Bulk Earth	^{87}Sr/^{86}Sr = 0.7052	^{143}Nd/^{144}Nd = 0.51264* (= chondrite)	^{206}Pb/^{204}Pb = 18.4 ± 0.3	^{207}Pb/^{204}Pb = 15.58 ± 0.08	Th/U = 4.2; ^{208}Pb/^{204}Pb = 38.9 ± 0.3

* Normalized to ^{146}Nd/^{144}Nd = 0.7219.

strongly fractionated from one another. Below we briefly review the properties of the five isotopic systems summarized in Table 6.5.

Sm and Nd isotopes are not significantly fractionated within the continental crust by metamorphic or sedimentary processes and thus preserve the parent/daughter ratio of their source region. In this respect Sm–Nd are very similar to

Lu–Hf but differ markedly from Rb–Sr, U–Pb and Th–Pb. Sm–Nd are immobile under hydrothermal conditions and so their isotopic composition reflects the actual proportions of rock or magma involved in specific petrological processes. The Sm–Nd system, however, has the disadvantage that small amounts of recycled crust mixed with a large proportion of a mantle component become isotopically invisible.

Lead isotopes are more complex because of the three different decay schemes employed (Box 6.1) and because they do not define linear trends on lead isotope evolution diagrams. In general, uranium and lead are relatively mobile in crustal processes, particularly in magmatic/hydrothermal situations, whereas thorium is highly insoluble. Both U and Pb are incompatible elements in silicates although U enters a melt more readily than Pb. In detail, the two isotopes of lead produced from uranium, ^{206}Pb and ^{207}Pb, show contrasting behaviour as a consequence of their differing radioactive decay rates. Early in the history of the Earth ^{235}U decayed rapidly relative to ^{238}U so that ^{207}Pb evolved rapidly with time. ^{207}Pb abundances therefore are an extremely sensitive indicator of an old source. Today, however, ^{235}U is largely extinct so that in the recent history of the earth ^{238}U decay is more prominent and consequently ^{206}Pb abundances show a greater spread than ^{207}Pb (Figure 6.18a and Table 6.6). The difference in behaviour between the different isotopes of lead allows the identification of several isotopic reservoirs (Table 6.5). The crustal reservoirs are best sampled by studying the isotopic composition of a mineral such as feldspar with a low U/Pb or low Th/Pb ratio and which preserves the 'initial' Pb isotopic composition of the source. This approach was developed by Doe and Zartman (1979) and is discussed in the section on 'plumbotectonics' (Section 6.3.6).

Strontium is relatively immobile under hydrothermal conditions, although Rb is more mobile. Sr therefore reflects fairly closely the original bulk composition of a suite of rocks, and Rb less so. In addition the Rb–Sr system shows the most extreme differences in incompatability between the parent and daughter elements. Rb and Sr are easily separated, so that there is extreme fractionation between crust and mantle leading to the accelerated strontium isotope evolution of the continental crust relative to the mantle (see Figure 6.15). Within the continental crust Rb and Sr are further separated by remelting, metamorphism, and sedimentation, for Sr is partitioned into and retained by plagioclase whereas Rb is preferentially partitioned into the melt or fluid phase.

6.3.2   Recognizing isotopic reservoirs

Taylor *et al.* (1984) recognize three isotopic reservoirs in the continental crust which they characterize with respect to Nd, Sr and Pb isotopes. Zindler and Hart (1986) have delineated five end-member compositions in the mantle which by a variety of mixing processes can explain all the observations on mid-ocean ridge and ocean-island basalts. The composition of each of these sources is summarized in Table 6.5 and plotted on a series of generalized isotope correlation diagrams (Figures 6.9 to 6.12). In addition Table 6.6 summarizes the present-day compositional ranges of Sr, Nd and Pb isotopes in oceanic and crustal rock types. In the following section each of the important mantle and crustal reservoirs is described and its particular isotopic character highlighted.

Table 6.6 Sr, Nd and Pb isotopic compositional ranges in common rock types

Rock type	$^{87}Sr/^{86}Sr$	$^{143}Nd/^{144}Nd$	$^{206}Pb/^{204}Pb$	$^{207}Pb/^{204}Pb$	$^{208}Pb/^{204}Pb$	Dominant mantle component	Ref.*
N-type MORB							
Atlantic	0.70229–0.70316	0.5130–0.5132	18.28–18.5	15.45–15.53	37.2–38.0	DM	(1)
Pacific	0.70240–0.70256	0.5130–0.5133	17.98–18.5	15.44–15.51	37.6–38.0	DM	(1)
Indian	0.70274–0.70311	0.5130–0.5131	17.31–18.5	15.43–15.56	37.1–38.7	DM	(1)
E-type MORB	0.70280–0.70334	0.51299–0.5130	18.50–19.69	15.50–15.60	38.0–39.3	DM + PREMA	(1)
Ocean-island basalts							
Mangaia	0.702720	0.512850	21.69	15.84	40.69	HIMU	(1)
Upolu–Samoa	0.705560	0.512650	18.59	15.62	38.78	EMII	(1)
Samoan Is.	0.704410–0.70651	0.512669–0.512935					(7)
Walvis Ridge	0.705070	0.512312	17.54	15.47	38.14	EMI	(1)
St Helena	0.702818–0.70309	0.512824–0.512970	20.40–20.89	15.71–15.81	39.74–40.17	HIMU	(3)
Cape Verde	0.702919–0.703875	0.512606–0.513095	18.88–20.30	15.52–15.64	38.71–39.45	HIMU/EM	(4)
Tr. da Cunha	0.704400–0.70505	0.512520–0.51267	18.60–18.76	15.52–15.59	38.93–39.24	EM (DUPAL)	(6)
Kerguelen	0.703880–0.70598	0.512498–0.513062	17.99–18.31	15.48–15.59	38.29–38.88	EMI (DUPAL)	(7,8)
Hawaii	0.703170–0.70412	0.512698–0.513060	17.83–18.20	15.44–15.48	37.69–37.86		(9)
Continental flood basalts							
Western USA	0.70351–0.70689	0.51224–0.512925					(2)
Parana	0.70468–0.71391	0.51221–0.51278					(5)
Mantle xenoliths							
Subcontinental lithosphere							
Scotland	0.703200–0.71410	0.510967–0.512798				EMI, EMII	(10)
E. China	0.702215–0.704300	0.512491–0.513585				DM, PREMA	(39)
Mass. Central	0.702440–0.70459	0.512368–0.513203					(40)
Sub–oceanic lithosphere							
Hawaii	0.703188–0.704207	0.512924–0.513100					(41)
Canary Is.	0.702967–0.703286	0.512856–0.513017					(41)
Kergeulen	0.704221–0.705025	0.512647–0.512816					(41)
Kimberlite / lamproite							
W. Australia	0.71066–0.72008	0.51104–0.51144					(11)
S. Africa	0.70280–0.70691	0.51274–0.51302					(37)
Subduction-related volcanic rocks							
Young volcanic-arc							
Philippines	0.70356–0.70476		18.27–18.47	15.49–15.64	38.32–38.83		(12)
Marianas	0.70332–0.70378	0.512966–0.513032	18.70–18.78	15.49–15.57	38.14–38.43		(12,26)
Java	0.70504–0.70576		18.70–18.72	15.63–15.65	38.91–38.96		(12)
Stromboli	0.70603–0.70750		18.93–19.10	15.64–15.97	39.01–39.08		(12)
Lr Antilles	0.70359–0.70897	0.512120–0.512978	19.17–19.93	15.67–15.85	38.85–39.75		(26,27)
Andesite							
Andes	0.70566–0.70951	0.512223–0.512556					(28)
Western USA	0.70386–0.70500	0.512660–0.512836	18.82–18.91	15.57–15.62	38.45–38.65		(29)
Upper Crust							
South Britain	0.71463–0.78662	0.511843–0.512261					(31)
Young G–toids	0.70400–0.82131	0.511700–0.51279					(33)
PreC G–toids	0.70330–0.8405	0.510660–0.51210					(33)
Archean	0.73307–1.54807	0.510236–0.510943	15.64–33.96	14.56–18.89	34.76–53.00		(38)
Hercynian granite (feldspars)			17.60–19.79	15.48–15.72	38.00–39.14		(36)

Table 6.6   Continued

Rock type	$^{87}Sr/^{86}Sr$	$^{143}Nd/^{144}Nd$	$^{206}Pb/^{204}Pb$	$^{207}Pb/^{204}Pb$	$^{208}Pb/^{204}Pb$	Dominant mantle component	Ref.*
S–type granitoids							
Australia	0.70940–0.87933	0.510791–0.511325					(14)
Malaysia	0.73709–0.81187	0.511480–0.51163					(35)
I–type granitoids							
Australia	0.70453–0.80803	0.510842–0.511657					(14)
Malaysia	0.70676–0.73006	0.511390–0.51164					(35)
Modern pelagic sediments							
Pacific	0.706900–0.72253		16.72–19.17	15.57–15.75	38.43–39.19		(12)
Atlantic	0.709288–0.723619	0.511646–0.512065	18.61–19.01	15.68–15.74	38.93–39.19		(13)
Terrigenous sediments							
Amazonian basin	0.714675–0.722524	0.512033–0.512266					(15)
S. Britain	0.711440–0.78919	0.511816–0.512259					(31)
Mesozoic — North Sea		0.511435–0.511954					(32)
Phanerozoic French shales		0.511851–0.512627					(17)
Archean metasediments		0.510418–0.512214					(16)
Chemical sediments							
Limestone	0.70821–0.72982	0.512012–0.512050					(31)
Archaean BIF		0.511179–0.512355					(34)
Lower crustal granulites							
Archean							
Lewisian	0.70320–0.7668	0.509818–0.513518	13.52–20.68	14.43–15.67	33.19–57.36		(18,19,20)
Enderby Land	0.70780–0.8160		15.68–27.05	15.61–19.52	35.50–126.6		(22)
South India	0.70210–0.72580	0.510377–0.511432	13.52–27.71	14.54–17.47	33.61–44.32		(25)
Proterozoic							
Arunta Block	0.70195–3.61759	0.510481–0.517585					(23)
Phanerozoic							
Beni Bousera	0.71958–0.72468	0.51198–0.51206					(24)
Ivrea Zone	0.71014–0.73911	0.51226–0.51237					(24)
Xenoliths							
Mass. Central	0.70469–0.71876	0.512027–0.512651	18.19–18.70	15.65–15.72	38.49–39.35		(21,36)
Lesotho	0.70372–0.70590	0.511764–0.512951					(30)

References: (1) Saunders *et al.* (1988) — compilation; (2) Fitton *et al.* (1988); (3) Chaffey *et al.* (1989); (4) Gerlach *et al.* (1988); (5) Piccirillo *et al.* (1989); (6) Cliff *et al.* (1991); (7) White and Hofmann (1982); (8) Storey *et al.* (1988); (9) Stille *et al.* (1983); (10) Menzies and Halliday (1988); (11) McCulloch *et al.* (1983); (12) McDermott and Hawkesworth (1991); (13) Hoernle *et al.* (1991); (14) McCulloch and Chappell (1982); (15) Basu *et al.* (1990); (16) Maas and McCulloch (1991); (17) Michard *et al.* (1985); (18) Whitehouse (1989a,b); (19) Moorbath *et al.* (1975); (20) Hamilton *et al.* (1979a); (21) Downes and Leyreloup (1986); (22) DePaolo (1982); (23) Windrim and McCulloch (1986); (24) Ben Othman *et al.* (1984) — compilation; (25) Peucat *et al.* (1989); (26) White and Patchett (1984); (27) Davidson (1983); (28) Hawkesworth *et al.* (1982); (29) Norman and Leeman (1990); (30) Rogers and Hawkesworth (1982); (31) Davies *et al.* (1985); (32) Mearns *et al.* (1989); (33) Allegre and Ben Othman (1980); (34) Miller and O'Nions (1985); (35) Liew and McCulloch (1985); (36) Vitrac *et al.* (1981); (37) Kramers *et al.* (1981); (38) Bickle *et al.* (1989); (39) Song and Frey (1989); (40) Downes and Dupuy (1987); (41) Vance *et al.* (1989).

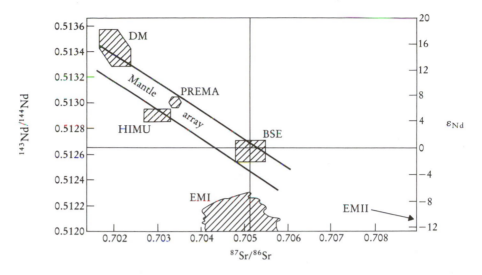

Figure 6.9 $^{143}Nd/^{144}Nd$ vs $^{87}Sr/^{86}Sr$ isotope correlation diagram, showing the main oceanic mantle reservoirs of Zindler and Hart (1986). DM, depleted mantle; BE, bulk silicate Earth; EMI and EMII, enriched mantle; HIMU, mantle with high U/Pb ratio; PREMA, frequently observed PREvalent MAntle composition. The Mantle array is defined by many oceanic basalts and a bulk Earth value for $^{87}Sr/^{86}Sr$ can be obtained from this trend.

Oceanic mantle sources

Young magmatic rocks record the isotopic composition of their source directly. This is because there is insufficient time for the parent isotope present in a newly formed magma to decay and produce additional daughter isotopes to add to those inherited from the source. Thus the present-day isotopic compositions of recent oceanic basalts were used by Zindler and Hart (1986) to identify five possible end-member mantle reservoirs (Figures 6.9 to 6.12). Their possible location in the mantle is given in the cartoon in Figure 6.27. The end-member compositions are as follows.

(a) Depleted mantle (DM) Depleted mantle is characterized by high $^{143}Nd/^{144}Nd$, low $^{87}Sr/^{86}Sr$ and low $^{206}Pb/^{204}Pb$. It is the dominant component in the source of many MORBs (see Figures 6.9, 6.11, 6.12). An extreme example of depleted mantle is recorded in mantle xenoliths from eastern China (Table 6.6) described by Song and Frey (1989).

(b) HIMU mantle The μ-value in lead isotope geochemistry is the ratio $^{238}U/^{204}Pb$. The very high $^{206}Pb/^{204}Pb$ and $^{208}Pb/^{104}Pb$ ratios observed in some ocean islands (Table 6.6) coupled with low $^{87}Sr/^{86}Sr$ (*ca* 0.7030) and intermediate $^{143}Nd/^{144}Nd$ suggest a mantle source that is enriched in U and Th relative to Pb without an associated increase in Rb/Sr. The enrichment is thought to have taken place between 1.5 and 2.0 Ga. A number of models have been proposed to explain the origin of this mantle reservoir — the mixing into the mantle of altered oceanic crust (possibly contaminated with seawater), the loss of lead from part of the mantle into the Earth's core and the removal of lead (and Rb) by metasomatic fluids in the mantle.

(c) Enriched mantle Enriched mantle has variable $^{87}Sr/^{86}Sr$, low $^{143}Nd/^{144}Nd$ and

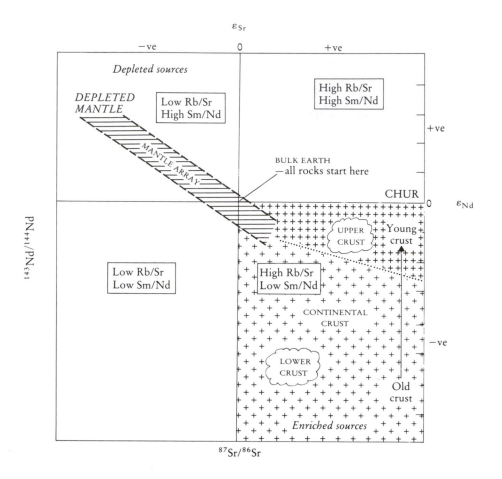

Figure 6.10 ^{143}Nd/^{144}Nd vs ^{87}Sr/^{86}Sr (ε_{Nd} vs ε_{Sr}) isotope correlation diagram showing the relative positions of depleted and enriched mantle sources. Most non-enriched mantle reservoirs plot in the upper left 'depleted' quadrant (cf. Figure 6.9) whereas most crustal rocks plot in the lower right 'enriched' quadrant. Upper and lower crust tend to plot in different positions in the crustal quadrant (from DePaolo and Wasserburg, 1979)

high ^{207}Pb/^{204}Pb and ^{208}Pb/^{204}Pb at a given value of ^{206}Pb/^{204}Pb. Zindler and Hart (1986) differentiate between enriched mantle type-I (EMI) with low ^{87}Sr/^{86}Sr (an even more extreme example is given by Richardson et al., 1984) and enriched mantle type-II (EMII) with high ^{87}Sr/^{86}Sr. One striking and large-scale example of enriched mantle (EMII) has been identified in the southern hemisphere by Hart (1984) and is known as the DUPAL anomaly (named after the authors DUPre and ALlegre (1983) who first identified the isotopic anomaly). The enriched mantle was identified with respect to a Northern Hemisphere Reference Line (NHRL) defined by the linear arrays on ^{207}Pb/^{204}Pb vs ^{206}Pb/^{204}Pb and ^{208}Pb/^{204}Pb vs ^{206}Pb/^{204}Pb plots for mid-ocean ridge basalts and ocean islands (Figure 6.11). The equations are

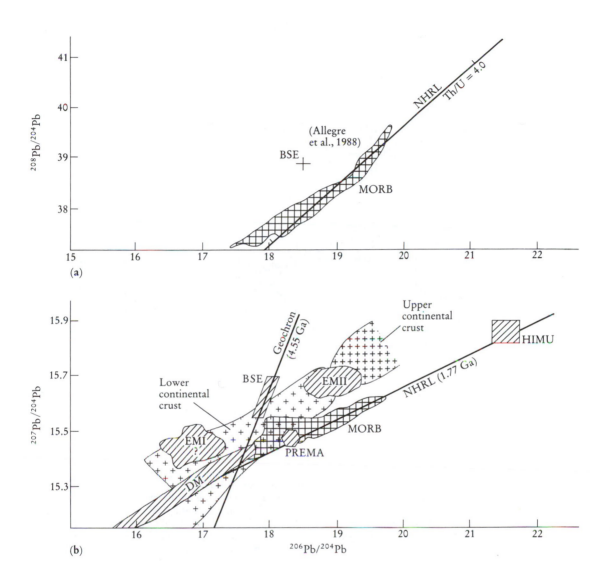

Figure 6.11   (a) $^{208}Pb/^{204}Pb$ vs $^{206}Pb/^{204}Pb$ isotope correlation diagram showing the position of the northern hemisphere reference line (NHRL) with Th/U = 4.0. The bulk silicate Earth value (BSE) is from Allegre *et al.* (1988). The field of MORB is shown with a cross-hatched pattern. (b) $^{207}Pb/^{204}Pb$ vs $^{206}Pb/^{204}Pb$ isotope correlation diagram showing the position of the northern hemisphere reference line (NHRL), the slope of which has an age significance of 1.77 Ga, and the geochron. Volcanic rocks which plot above the NHRL are said to have a DUPAL signature. The mantle reservoirs of Zindler and Hart are plotted as follows: DM, depleted mantle; BSE, bulk silicate Earth; EMI and EMII, enriched mantle; HIMU, mantle with high U/Pb ratio; PREMA, frequently observed PREvalent MAntle composition. EMII also coincides with the field of oceanic pelagic sediment. The fields of the upper and lower continental crust are shown with crosses, and the field of MORB with cross-hatching.

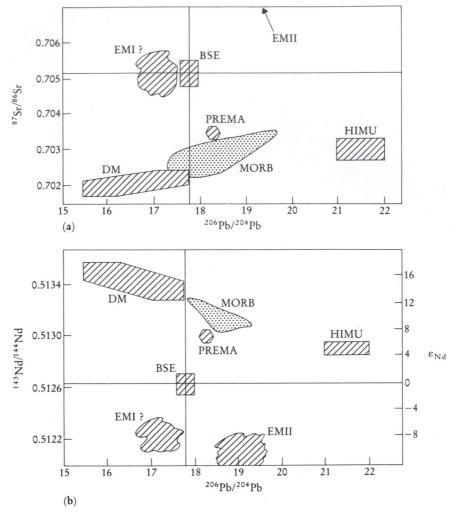

Figure 6.12 (a) $^{87}Sr/^{86}Sr$ vs $^{206}Pb/^{204}Pb$ isotope correlation diagram; (b) $^{143}Nd/^{144}Nd$ vs $^{206}Pb/^{204}Pb$ isotope correlation diagram. Both diagrams show the positions of the mantle reservoirs identified by Zindler and Hart (1986): DM, depleted mantle; BSE, bulk silicate Earth; EMI and EMII, enriched mantle; HIMU, mantle with high U/Pb ratio; PREMA, frequently observed PREvalent MAntle composition. The stippled field is mid-ocean ridge basalts (MORB). The $^{206}Pb/^{204}Pb$ value of the bulk Earth differs from the value in Table 6.5, which is taken from Allegre *et al.* (1988).

$$^{207}Pb/^{204}Pb = 0.1084\ (^{206}Pb/^{204}Pb) + 13.491 \qquad [6.6]$$

$$^{208}Pb/^{204}Pb = 1.209\ \ (^{206}Pb/^{204}Pb) + 15.627 \qquad [6.7]$$

The isotopic anomaly was expressed by Hart (1984) in terms of $\Delta\,7/4$ and $\Delta\,8/4$, defined as the vertical deviation from the reference line so that for a given data-set (DS):

$$\Delta\,7/4 = [(^{207}Pb/^{204}Pb)_{DS} - (^{207}Pb/^{204}Pb)_{NHRL}] \times 100 \qquad [6.8]$$
$$\Delta\,8/4 = [(^{208}Pb/^{204}Pb)_{DS} - (^{208}Pb/^{204}Pb)_{NHRL}] \times 100 \qquad [6.9]$$

A similar notation was used for $^{87}Sr/^{86}Sr$:

$$\Delta\,Sr = [(^{87}Sr/^{86}Sr)_{DS} - 0.7] \times 10\,000 \qquad [6.10]$$

There are a number of models which explain the origin of enriched mantle. In general terms enrichment is likely to be related to subduction, whereby crustal material is injected into the mantle (Figure 6.27). EMII has affinities with the upper continental crust and may represent the recycling of continentally derived sediment, continental crust, altered oceanic crust or ocean–island crust. An alternative model is based upon the similarity between enriched mantle and the subcontinental lithosphere, and suggests that the enrichment is due to the mixing of the subcontinental lithosphere into the mantle. EMI has affinities with the lower crust and may represent recycled lower crustal material, but an alternative hypothesis suggests that it is enriched by mantle metasomatism. Weaver (1991) proposed that EMI and EMII are produced by mixing between HIMU mantle and subducted oceanic sediment.

(d) PREMA The great frequency of basalts from ocean islands, intra-oceanic island arcs and continental basalt suites with $^{143}Nd/^{144}Nd = 0.5130$ and $^{87}Sr/^{86}Sr = 0.7033$ suggests that there is an identifiable mantle component with this isotopic character. Zindler and Hart (1986) refer to this as the PREvalent MAntle reservoir. It has $^{206}Pb/^{204}Pb = 18.2$ to 18.5.

(e) Bulk Earth (Bulk Silicate Earth — BSE) = Primary Uniform Reservoir It can be argued that there is a mantle component which has the chemistry of the bulk silicate Earth (the Earth without the core). This composition is equivalent to that of a homogeneous primitive mantle which formed during the degassing of the planet and during core formation, prior to the formation of the continents. Some oceanic basalts have isotopic compositions which closely approximate to the composition of the bulk Earth although at present there are no geochemical data which require that such a mantle reservoir still survives.

(f) The origin of oceanic basalts The recognition of a variety of possible mantle reservoirs for the source of oceanic basalts has led to an appreciation of the complexity of mantle processes, particularly beneath ocean islands. For example, a study of the island Gran Canaria in the Canary Islands by Hoernle *et al.* (1991) showed that at least four different mantle components (HIMU, DM, EMI and EMII) have contributed to the production of these volcanic rocks since the Miocene. Furthermore, the isotopic evidence indicates that the different reservoirs have made their contributions to the lava pile at different times in the history of the volcano.

(g) Trace elements and mantle end-member compositions Following the work of Zindler and Hart (1986) a number of workers have sought to characterize further the proposed mantle end-member compositions in terms of their trace element concentrations. Ranges of values currently in use for highly incompatible trace element ratios are given in Table 6.7.

Continental crustal sources The isotopic composition of rocks from the continental crust is extremely variable and isotope ratios are only strictly comparable if the samples are all of the same age. Thus compositional differences are best considered relative to a normalizing parameter which takes into account the age of the sample. One such reference point is the chondritic model of Earth composition (CHUR) and its evolution through time. Thus crustal samples may be plotted on a graph of isotopic composition vs time relative to a reference line such as the CHUR reservoir (see Section 6.3.3). For

Table 6.7   Incompatible trace element ratios in crust and mantle reservoirs (from compilations by Saunders *et al.*, 1988; Weaver, 1991)

	Zr/Nb	La/Nb	Ba/Nb	Ba/Th	Rb/Nb	K/Nb	Th/Nb	Th/La	Ba/La
Primitive mantle	14.8	0.94	9.0	77	0.91	323	0.117	0.125	9.6
N-MORB	30	1.07	1.7–8.0	60	0.36	210–350	0.025–0.071	0.067	4.0
E-MORB			4.9–8.5			205–230	0.06–0.08		
Continental crust	16.2	2.2	54	124	4.7	1341	0.44	0.204	25
HIMU OIB	3.2–5.0	0.66–0.77	4.9–6.9	49–77	0.35–0.38	77–179	0.078–0.101	0.107–0.133	6.8–8.7
EMI OIB	4.2–11.5	0.86–1.19	11.4–17.8	103–154	0.88–1.17	213–432	0.105–0.122	0.107–0.128	13.2–16.9
EMII OIB	4.5–7.3	0.89–1.09	7.3–13.3	67–84	0.59–0.85	248–378	0.111–0.157	0.122–0.163	8.3–11.5

Nd isotopes this may be quantified in terms of the epsilon notation (see Section 6.3.4). The general isotopic characteristics of crustal reservoirs are given in Table 6.5 and typical isotopic compositions of crustal rocks are presented in Table 6.6.

(a) Upper continental crust   The upper continental crust is characterized by high Rb/Sr and consequently has high ^{87}Sr/^{86}Sr ratios. Neodymium isotope ratios, on the other hand, are low relative to mantle values as a consequence of the light rare earth element enrichment and the low Sm/Nd ratios which characterize the continental crust. U and Th are enriched in the upper continental crust and give rise to high ^{206}Pb, ^{207}Pb and ^{208}Pb isotope ratios.

(b) Middle continental crust   The mid-continental crust here describes extensive areas of amphibolite facies gneisses found in granite–gneiss terrains. These rocks have retarded ^{143}Nd/^{144}Nd as discussed above and ^{87}Sr/^{86}Sr ratios lower than in the upper crust. Uranium, however, is depleted and ^{206}Pb/^{204}Pb and ^{207}Pb/^{204}Pb ratios may be lower than in the mantle. Th levels are lower than those in the upper crust but not as depleted as uranium.

(c) Lower continental crust   The lower continental crust is characterized by granulite facies metamorphism and is often strongly Rb depleted. Thus it has low ^{87}Sr/^{86}Sr ratios which are not greatly different from modern mantle values. This means that a modern granite derived from the lower crust and one derived from the mantle will have very similar ^{87}Sr/^{86}Sr initial ratios. U/Pb and Th/Pb ratios in the lower crust are lower than modern mantle values so that ^{206}Pb, ^{207}Pb and ^{208}Pb isotope ratios are all very low and may be used to distinguish between lower crust and mantle reservoirs (see Figure 6.26).

(d) Subcontinental lithosphere   The subcontinental lithosphere is not so easily characterized as other domains, for it is extremely variable in isotopic composition. For example, Menzies and Halliday (1988) show that the subcontinental lithosphere beneath Scotland shows extreme Nd–Sr isotopic heterogeneity which encompasses both of the enriched mantle domains EMI and EMII described above. Some of the variability can be correlated with the age of the subcontinental lithosphere.

Archaean subcontinental lithosphere is normally underlain by enriched lithosphere (EMI of Zindler and Hart (1986) — low Rb/Sr, low Sm/Nd), but depleted examples are also known. The lithosphere beneath Proterozoic mobile belts, however, more closely resembles the depleted mantle found beneath older ocean basins (Menzies, 1989). Proterozic to Phanerozoic subcontinental lithosphere is characterized by enrichment in Rb and the light REE resulting in radiogenic Sr and non-radiogenic Nd isotopes. This is similar to enriched mantle EMII described above and in Table 6.5.

Seawater   The isotopic composition of seawater can be understood in terms of the **residence time** of an element in seawater. This is the average time an element remains in seawater and is defined by the expresion

$$t_A = A/(dA/dt)$$ [6.11]

where t_A is the residence time, A is the total amount of the element disolved in the oceans and dA/dt is the amount of the element introduced into or removed from the ocean per unit time, assuming a steady state. It is estimated that the oceans mix in the order of 1000 years so that an element with a residence time longer than this will be well mixed and homogeneous whereas elements which have a shorter residence time will reflect regional variations in isotopic composition (Hawkeworth and van Calsteren, 1984).

The present-day strontium isotope composition of seawater is bracketed by the values 0.709241 ± 32 and 0.709211 ± 37 (Elderfield, 1986). This is the result of the complex interplay of processes which include continental run-off, mid-ocean ridge hydrothermal activity, input from old sediments and extraction of Sr by new sediments (Veizer, 1989). The residence time of Sr in seawater is of the order of 4 Ma and so the oceans are regarded as a well mixed reservoir with uniform composition. However, the strontium isotopic composition of seawater has changed with time. Veizer (1989), in an excellent review, shows how the change may be resolved on a range of time-scales. For instance, over the total history of the Earth (Figure 6.13) Sr isotopic compositions remained close to mantle values until about 2.5 Ga and then increased with time. In the Phanerozoic $^{87}Sr/^{86}Sr$ ratios have oscillated in the order of 100 Ma but since the Cretaceous they have increased gradually from about 0.7077 to the present value and this curve is known with great precision (Elderfield, 1986). The seawater curve for the Phanerozoic and particularly the high-precision curve for the last 70 Ma has been used to date Tertiary sediments, to determine the timing of diagenetic processes and to date the timing of formation of gangue minerals in mineral deposits.

Nd and Pb, by way of contrast to Sr, have short residence times, Nd 100 to 1000 years (Hooker *et al.*, 1981), Pb *ca* 50 years (Taylor and McLennan, 1985). The Nd isotopic composition of seawater has been estimated by direct analysis of seawater and from the composition of manganese nodules precipitated from it. Different ocean basins have slightly different $^{143}Nd/^{144}Nd$ isotopic compositions: the Atlantic is 0.5119–0.5122, the Pacific is 0.5120–0.5122 and the Indian Ocean is 0.5124–0.5126 (Hooker *et al.*, 1981). The values vary because the Nd budget of an ocean is dominated by the continent-derived Nd, a principle which Hooker *et al.* (1981) used to deduce the nature of the land adjacent to the lower Palaeozoic Iapetus Ocean at 490 Ma.

The difference in solubility between U (very soluble) and Pb (relatively

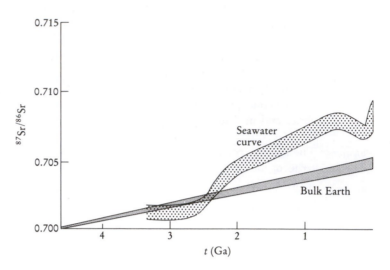

Figure 6.13   The variation of $^{87}Sr/^{86}Sr$ in seawater through time relative to the bulk Earth (from Veizer, 1989).

insoluble) means that the U/Pb ratio (μ value) of modern seawater is very high (*ca* 95 000; Taylor and McLennan, 1985), but the short residence time means that significant quantities of radiogenic lead will not develop in the oceans. Flegal *et al.* (1986) measured the lead isotopic composition of the east Pacific and found that the $^{206}Pb/^{204}Pb$ value varies with depth (17.39–19.05) but attributed much of the effect to contamination from modern industrial lead.

6.3.3   The evolution of mantle reservoirs with time — mantle evolution diagrams

The isotopic compositions of mantle source regions change with time according to the parent/daughter element ratio. These changes may be plotted on isotopic evolution diagrams to show the change in isotopic composition of the reservoir with time. Such diagrams can be constructed for each isotopic system by measuring the isotopic composition of the mantle at the present day from modern lavas and by calculating the initial ratio from isochron diagrams of ancient mantle melts. Diagrams of this type are particularly useful for depicting the isotopic evolution of crustal rocks of differing ages and with different parent/daughter isotope ratios. It should be noted, however, that as yet there are insufficient data to draw detailed mantle evolution diagrams for all the postulated mantle reservoirs of Zindler and Hart (1986).

The evolution of Sr isotopes with time   The starting point for the evolution of strontium isotopes is the $^{87}Sr/^{86}Sr$ ratio at the formation of the Earth. This is taken to be the isotopic composition of basaltic achondrite meteorites, thought to have a composition approximating to that of the solar nebula at the time of planetary formation. It is usually referred to as BABI (Basaltic Achondrite Best Initial) and the measured value is 0.69897 ± 0.000003.

Table 6.8   Parameters used in the calculation of model ages and epsilon values for strontium isotopes

	Bulk Earth values*	Depleted mantle values*
$\left[\dfrac{^{87}Sr}{^{86}Sr}\right]_{4.6\ Ga}$	0.69897 ± 0.00003 (1)	
$\left[\dfrac{^{87}Rb}{^{86}Sr}\right]_{today}$	0.0847 (1) 0.085 (2) 0.0827 (4) 0.0892 (7)	0.052 (1) 0.046 (2) 0.024–0.0007 (7)
Rb/Sr	0.03 (3) 0.032 (5)	
$\left[\dfrac{^{87}Sr}{^{86}Sr}\right]_{today}$	0.7047 (2) 0.7045 (4) 0.7052 (5) 0.7047–0.7050 (7)	0.7026 (1,2) 0.70321 ± 0.00015 (6) 0.70265–0.70216 (7)

* References: (1) McCulloch and Black (1984); (2) Taylor and McLennan (1985); (3) O'Nions *et al.* (1977); (4) DePaolo (1988); (5) Zindler *et al.* (1982); (6) Bell *et al.* (1982); (7) Allegre *et al.* (1983a).

Estimates of $^{87}Sr/^{86}Sr$ for the bulk Earth today vary (see Table 6.8) but are between 0.7045 and 0.7052 indicating Rb/Sr ratios between 0.03 and 0.032. This gives a broad band for the $^{87}Sr/^{86}Sr$ evolution curve for the bulk Earth (Figure 6.14). Bell *et al.* (1982) estimated the composition of a depleted mantle source beneath the Superior Province, Canada, which appears to have evolved from the primitive mantle at around 2.8 Ga (Figure 6.14, curve 1) although this differs from the depleted mantle evolution curve of Ben Othman *et al.* (1984) (Figure 6.14, curve 2) and the present-day composition of the depleted mantle of Zindler and Hart (1986). Estimates for enriched mantle (Zindler and Hart, 1986) suggest present-day $^{87}Sr/^{86}Sr$ ratios of about 0.705 and about 0.722. Measured values in present-day basalts from Samoa (White and Hofmann, 1982) scatter about the lower of these two values.

Jahn *et al.* (1980) show that $^{87}Sr/^{86}Sr$ ratios for Archaean mafic and felsic igneous rocks scatter above and below the bulk Earth curve indicating that there may have been a variety of mantle sources with different compositions even in the Archaean. However mantle $^{87}Sr/^{86}Sr$ ratios need to be interpreted with great caution, especially in old rocks, because the high ratios can result from metasomatic processes. For this reason studies of mantle evolution are better based on other isotopic systems where the parent and daughter isotopes are less mobile; this is why the study of Nd isotopes has become so popular over the last ten years.

A number of authors (e.g. Faure, 1977) have proposed on the basis of measured initial ratios that the growth of $^{87}Sr/^{86}Sr$ in the mantle with time defines a curvilinear path (cf. Figure 6.14, curves 1 and 2) and that this reflects the irreversible loss of Rb from the mantle into the crust during the formation of the continental crust. The loss of Rb from the mantle and its enrichment in the continental crust lead to very different patterns of strontium isotope evolution in the two reservoirs as a consequence of their different Rb/Sr ratios (Figure 6.15). The high Rb/Sr ratios found in the continental crust give rise to an accelerated

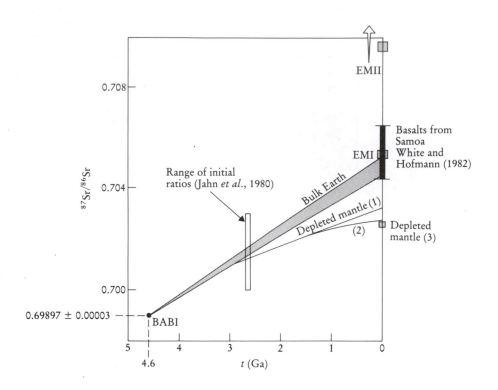

Figure 6.14   The evolution of strontium isotopes with time. The shaded field indicates the evolution of the bulk earth from 4.6 Ga to the present-day value of between 0.7045 and 0.7052, reflecting Rb/Sr ratios of between 0.03 and 0.032. The depleted mantle curve (1) is from Bell *et al.* (1982) and indicates the separation of a depleted mantle reservoir from the primitive mantle at *ca* 2.8 Ga beneath the Superior Province, Canada. Curve (2) is from Ben Othman *et al.* (1984) and is based upon the data from MORB, ophiolites, komatiites and meteorites. It represents the equation $(^{87}Sr/^{86}Sr) = At^2 + Bt + C$ where $A = -1.54985776 \times 10^{-4}$, $B = -1.6007234$ and $C = 0.70273029$, where *t* is time from the present in Ga; (3) is the present-day depleted mantle value of McCulloch and Black (1984) . The present-day enriched mantle compositions (EMI and EMII) are from Zindler and Hart (1986) and the composition of enriched basalts from Samoa from White and Hofmann (1982). The spread of initial ratios from 2.7 to 2.6 Ga is for mafic and felsic igneous rocks, from Jahn *et al.* (1980).

increase in $^{87}Sr/^{86}Sr$ with time whereas the low Rb/Sr of depeleted mantle gives rise to only a small increase in $^{87}Sr/^{86}Sr$ since the formation of the Earth.

　　There are many examples in the literature of authors who have used mantle evolution diagrams of the type illustrated in Figure 6.15 to plot the initial strontium isotope ratios of measured samples relative to mantle and crustal evolution curves, in order to determine their likely source region. For example, it is easy to see from Figure 6.15 that a suite of rocks produced by partial melting of the mantle at 1.0 Ga will have a very different initial ratio $(^{87}Sr/^{86}Sr = 0.7034)$ from rocks produced by partial melting of the crust at this time $(^{87}Sr/^{86}Sr = 0.7140)$. It is this principle which may be used to identify the source of magmatic rocks of known age.

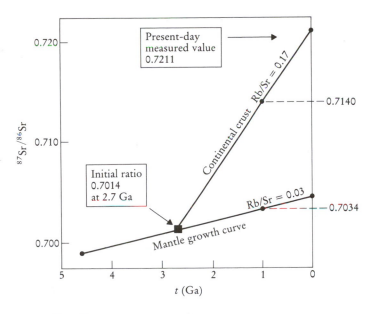

Figure 6.15   The evolution of $^{87}Sr/^{86}Sr$ with time in the continental crust and mantle. At 2.7 Ga mantle differentiation led to the formation of new continental crust. The new crust inherited an initial $^{87}Sr/^{86}Sr$ ratio of 0.7014 from its parent mantle but acquired a substantially different Rb/Sr ratio (0.17) compared with 0.03 in the mantle. The higher Rb/Sr ratio in the continental crust led to an accelerated growth of $^{87}Sr/^{86}Sr$ with time in the continental crust relative to that in the mantle so that the present-day measured value in the crust is 0.7211 compared with 0.7045 in the mantle. The $^{87}Sr/^{86}Sr$ ratios shown on the right-hand side of the diagram indicate the initial ratios in melts formed from continental crust at 1.0 Ga (0.7140) and from the mantle at 1.0 Ga (0.7034).

The evolution of   The $^{143}Nd/^{144}Nd$ isotopic composition of the bulk Earth is thought to be
Nd isotopes with   approximated by the composition of chondritic meteorites (CHUR). The
time   compositions of CHUR for the present day and at 4.6 Ga are given in Table 6.3
and the linear evolution of CHUR with time is shown in Figure 6.16. Studies of
modern oceanic basalts and both ancient and modern granitic rocks indicate that
many igneous rocks are derived from a mantle composition which has a higher Sm/
Nd than CHUR, is enriched in $^{143}Nd/^{144}Nd$ relative to CHUR, but which is
known as depleted mantle. Currently there are several models for this depleted
mantle source and these are portrayed relative to CHUR (the epsilon notation) in
Figure 6.17. The main differences between the different models of the depleted
mantle are (1) the time at which the depleted mantle differentiated from the bulk
Earth and (2) whether the depletion of the mantle was linear or whether it varied
with time. Liew and McCulloch (1985) proposed that the light REE depletion from
the chondritic reservoir took place between 2.5 and 3.0 Ga at the time of a major
episode of crust generation and they propose a linear evolution of $^{143}Nd/^{144}Nd$ in
the depleted mantle between this time and the present. Goldstein *et al.* (1984)
preferred a linear evolution between the present day and the formation of the Earth.
DePaolo (1981a) and Nelson and DePaolo (1984) plot curves which decrease to zero
at 4.6 Ga, whilst the curve of Allegre and Rousseau (1984), also given by Ben

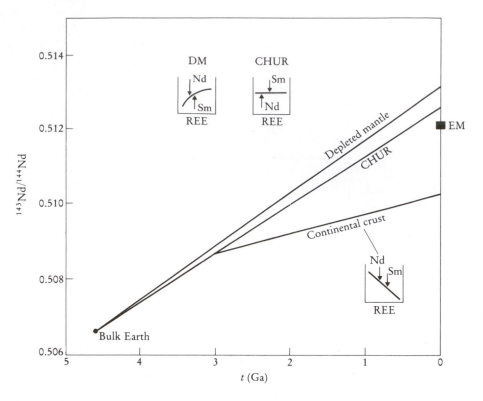

Figure 6.16   The evolution of $^{143}Nd/^{144}Nd$ isotopes with time in the mantle, the continental crust and the bulk earth. Relative to the bulk Earth (CHUR) in which the fractionation of Sm/Nd is normalized to unity, the depleted mantle (DM) has a high Sm/Nd ratio and shows higher $^{143}Nd/^{144}Nd$. The continental crust has lower Sm/Nd and shows a retarded $^{143}Nd/^{144}Nd$ evolution with time. Enriched mantle (EM) shows some affinity with the continental crust inasmuch as it also has a retarded $^{143}Nd/^{144}Nd$ evolution. Note that in the Sm–Nd system the crust shows retarded isotopic evolution whilst in the Rb–Sr system it shows accelerated evolution relative to the bulk Earth and mantle (cf. Figure 6.15).

Othman *et al.* (1984), reduces to an epsilon value of +2.5 at 4.0 Ga. Some support for depleted mantle very early in the history of the earth comes from the work of Collerson *et al.* (1991).

The Sm/Nd ratio of the continental crust is highly fractionated (<1.0) relative to CHUR and shows retarded $^{143}Nd/^{144}Nd$ evolution with time (Figure 6.16). This is the opposite of the Rb–Sr system where the strontium-isotopic evolution of the crust shows accelerated evolution with time relative to the mantle.

The evolution of
Pb isotopes with
time

The isotopic evolution of lead isotopes with time is illustrated in Figure 6.18. Because there are so many variables in lead isotopic systems (since we are dealing with three different decay schemes) it is not convenient to display the data on the same type of time vs isotope ratio diagram as is used for Sr and Nd isotopes, and the concordia diagram is a better choice. On this diagram families of concordia (growth) curves can be drawn for different $^{238}U/^{204}Pb$ ratios (μ values) (Figure 6.18b). Each curve can be calibrated for time and equal time points can be joined to give isochrons (Figure 6.18b and c).

Brevart *et al.* (1986) use Pb isotope initial ratios to demonstrate mantle heterogeneity at 2.7 Ga and have constructed a depleted mantle curve from 2.7 Ga to the present day from the initial ratios of komatiites and basalts on a ^{206}Pb–^{207}Pb isotope evolution diagram.

6.3.4 The epsilon notation

An alternative way of expressing isotope ratios which allows greater flexibility in the way in which isotopic data are presented is the epsilon notation (DePaolo and Wasserburg, 1976). Isotope ratios are only strictly comparable if the samples plotted are of the same age. The epsilon value is a measure of the deviation of a sample or sample suite from the expected value in a uniform reservoir and may be used as a normalizing parameter for samples of different age. It is normally calculated for Nd, although some authors also calculate epsilon parameters for Sr isotopes.

Calculating epsilon values

In the study of neodymium isotopes the epsilon parameter is a measure of the difference between the $^{143}Nd/^{144}Nd$ ratio of a sample or suite of samples and a reference value, which in this case is CHUR. Epsilon values are used in three slightly different ways (see the worked examples in Box 6.2) which are described below.

Box 6.2

Worked example of ε_{Nd} calculations

(1) ε_{Nd} values for an isochron

Using the data from Hamilton *et al.* (1979b) in Table 6.2, presented in Figure 6.2:

initial ratio = 0.50809
isochron age = 3.54 Ga
decay constant $\lambda = 6.54 \times 10^{-12}$ yr^{-1}

$$\varepsilon_{Nd}^{isochron,t} = \left[\frac{^{143}Nd/^{144}Nd_{initial}}{^{143}Nd/^{144}Nd_{CHUR,t}} - 1 \right] \times 10^4 \qquad [6.12]$$

First calculate the value of CHUR at time t from the equation using the CHUR values from Table 6.3 (normalized to $^{146}Nd/^{144}Nd = 0.7219$), Sm/Nd = 0.325.

$$\left[\frac{^{143}Nd}{^{144}Nd} \right]_{CHUR,t} = \left[\frac{^{143}Nd}{^{144}Nd} \right]_{CHUR,today} - \left[\frac{^{147}Sm}{^{144}Nd} \right]_{CHUR,today} \times (e^{\lambda t} - 1) \qquad [6.13]$$

$$CHUR,t = 0.512638 - (0.1967 \times [e^{(6.54 \times 10^{-12}) \times (3.54 \times 10^9)} - 1])$$
$$= 0.512638 - 0.004607$$
$$= 0.508031$$

Then using Eqn [6.12],

$$\varepsilon_{Nd}^{isochron,t} = \left[\frac{0.50809}{0.508031} - 1 \right] \times 10^4$$
$$= 1.16$$

(2) ε^{Nd} values for individual rocks at their time of formation

Taking a single rock from Table 6.2 — sample HSS-74 (sodic porphyry), $^{147}Sm/^{144}Nd = 0.1030$, $^{143}Nd/^{144}Nd = 0.51048\prime$ — and using the equation

Box 6.2
(continued)

$$\varepsilon_{Nd}^{t} = \left[\frac{^{143}Nd/^{144}Nd_{rock,t}}{^{143}Nd/^{144}Nd_{CHUR,t}} - 1 \right] \times 10^4 \qquad [6.14]$$

First calculate $^{143}Nd/^{144}Nd$ rock at 3.54 Ga from

$$\left[\frac{^{143}Nd}{^{144}Nd} \right]_{rock,t} = \left[\frac{^{143}Nd}{^{144}Nd} \right]_{rock,today} - \left[\frac{^{147}Sm}{^{144}Nd} \right]_{rock,today} \times [e^{\lambda t} - 1] \qquad [6.15]$$

$$\left[\frac{^{143}Nd}{^{144}Nd} \right]_{rock,t} = 0.510487 - (0.1030 \times [e^{(6.54 \times 10^{-12}) \times (3.54 \times 10^9)} - 1])$$

$$= 0.510487 - 0.002412$$

$$= 0.508075$$

Then from Eqn [6.14]

$$\varepsilon_{Nd}^{t} = \left[\frac{0.508075}{0.508031} - 1 \right] \times 10^4$$

$$= 0.87$$

(3) ε_{Nd} values for individual rocks at the present day

The present-day deviation from CHUR may be easily calculated from the equation below.

$$\varepsilon_{Nd}^{today} = \left[\frac{(^{143}Nd/^{144}Nd)_{rock,today}}{(^{143}Nd/^{144}Nd)_{CHUR,today}} - 1 \right] \times 10^4 \qquad [6.16]$$

$$\varepsilon_{Nd}^{today} = \left[\frac{0.510487}{0.512638} - 1 \right] \times 10^4$$

$$= -41.96$$

Summary

$$\varepsilon_{Nd,3.54\ Ga}^{isochron} = 1.16$$

$$\varepsilon_{Nd,3.54\ Ga}^{sample\ HSS-74} = 0.87$$

$$\varepsilon_{Nd,today}^{sample\ HSS-74} = -41.96$$

(a) Epsilon values calculated for an isochron Epsilon values are frequently quoted for Sm–Nd isochrons and are a measure of the difference between the initial ratio of the sample suite and CHUR at the time of formation, expressed in parts per 10^4 (see Figure 6.2). They are calculated from the expression

$$\varepsilon_{Nd}^{isochron,t} = \left[\frac{(^{143}Nd/^{144}Nd)_{initial}}{(^{143}Nd/^{144}Nd)_{CHUR,t}} - 1 \right] \times 10^4 \qquad [6.12]$$

where $(^{143}Nd/^{144}Nd)_{initial}$ is the experimentally determined initial ratio of the sample suite at the time of its formation (t) and is calculated from the isochron, $(^{143}Nd/^{144}Nd)_{CHUR,t}$ is the isotope ratio of CHUR at time t and is given by the expression

$$\left[\frac{^{143}Nd}{^{144}Nd} \right]_{CHUR,t} = \left[\frac{^{143}Nd}{^{144}Nd} \right]_{CHUR,today} - \left[\frac{^{147}Sm}{^{144}Nd} \right]_{CHUR,today} \times (e^{\lambda t} - 1) \qquad [6.13]$$

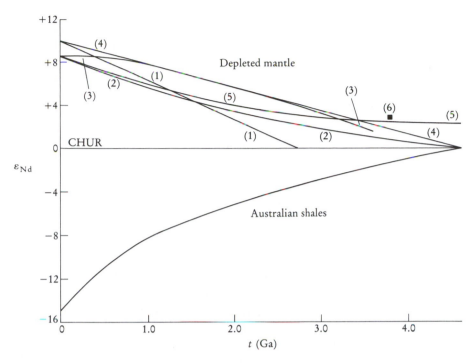

Figure 6.17   The contrasting evolution of ε_{Nd} in the continental crust and mantle. CHUR is the chondritic uniform reservoir; it is equivalent to the bulk Earth value and is used as the reference against which other measurements are made. Hence ε_{Nd} for CHUR is zero. Five curves are shown for the evolution of the depleted mantle. Curve (1), after Liew and McCulloch (1985), assumes the derivation of the depleted mantle from CHUR at 2.7 Ga and linear evolution until the present-day value of +10. Curve (2), after DePaolo (1981a,b), may be plotted from the expression $0.25t^2 - 3t + 8.5$, where t is the time before the present in Ga. Curve (3) is taken from Nelson and DePaolo (1984 — Figure 2). Curve (4) is from Goldstein *et al.* (1984) and assumes the linear evolution of the depleted mantle between a value of +10 at the present day and 0 at 4.5 Ga. Curve (5) is based upon data from MORB, ophiolites, komatiites and meteorites and is calculated from the expression given both by Ben Othman *et al.* (1984) and Allegre and Rousseau (1984): $(^{143}Nd/^{144}Nd)_{DM} = At^2 + Bt + C$, where $A = 1.53077 \times 10^{-5}$, $B = -0.22073 \times 6.54 \times 10^{-3}$ and $C = 0.513078$, ($^{146}Nd/^{144}Nd$ is normalized to 0.7219 and t is time before the present, in Ga). Point (6) is for samples of 3.8 Ga residual mantle from Collerson *et al.* (1991). It should be noted that both curves (3) and (5) use the data of Claoue-Long *et al.* (1984) for the Kambalda volcanics which are now known to be incorrect. The lower curve is for Australian shales from Allegre and Rousseau (1984) and is thought to approximate to the evolution of average continental crust through time.

Values for CHUR are given in Table 6.3. It should be noted that different laboratories use different normalizing values for the $^{143}Nd/^{144}Nd$ ratio and the correct value of CHUR must be selected. However, if this is done, then the interlaboratory differences in $^{143}Nd/^{144}Nd$ virtually disappear, making the epsilon notation a very useful way of displaying data from different laboratories.

(b) Epsilon values for individual rocks at their time of formation   Epsilon values calculated at time t, the time of formation, are frequently tabulated along with Sm–Nd isotopic data. They are calculated from the expression

Figure 6.18

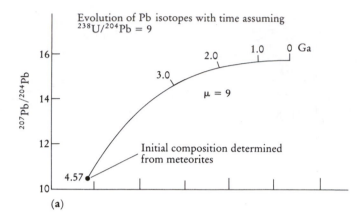

(a)

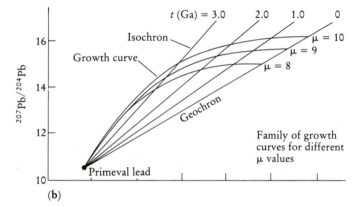

(b)

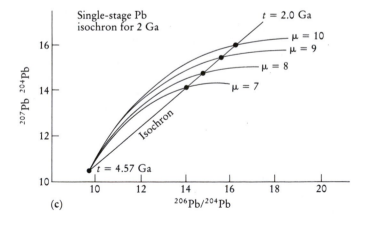

(c)

$$\varepsilon_{Nd}^{t} = \left[\frac{^{143}Nd/^{144}Nd_{rock,t}}{^{143}Nd/^{144}Nd_{CHUR,t}} - 1 \right] \times 10^{4} \qquad [6.14]$$

where the composition of CHUR at time t is calculated from Eqn [6.13] and the composition of the sample at time t is calculated using the expression

$$\left[\frac{^{143}Nd}{^{144}Nd} \right]_{rock,t} = \left[\frac{^{143}Nd}{^{144}Nd} \right]_{rock,today} - \left[\frac{^{147}Sm}{^{144}Nd} \right]_{rock,today} \times [e^{\lambda t} - 1] \qquad [6.15]$$

In calculating ε_{Nd}^{t} for individual samples, the value for t may be taken either from the isochron to which the samples belong or, where the Sm–Nd isochron may not record the true age of the samples, from an independent estimate of the age from some other geochronological technique.

(c) Epsilon values for individual rocks at the present day Epsilon values are also calculated for individual samples at the present day from the expression

$$\varepsilon_{Nd}^{today} = \left[\frac{(^{143}Nd/^{144}Nd)_{rock,today}}{(^{143}Nd/^{144}Nd)_{CHUR,today}} - 1 \right] \times 10^{4} \qquad [6.16]$$

The magnitude of the present-day ε_{Nd} values reflects the degree of time integrated depletion in Nd relative to CHUR as expressed by the factor $f^{Sm/Nd}$ (see below).

Epsilon values Epsilon values may be calculated from $^{87}Sr/^{86}Sr$ ratios in exactly the same way
for Sr isotopes as for neodymium isotopes. There is however a problem with calculating ε_{Sr} values, for as can be seen from the data in Table 6.8 and in Figure 6.14 the bulk Earth values for $^{87}Sr/^{86}Sr$ are not well known. For this reason when ε_{Sr} values are plotted

---

Figure 6.18 The evolution of lead isotopes with time. (a) The growth curve for $^{207}Pb/^{204}Pb$ and $^{206}Pb/^{204}Pb$ through time. The locus of points is a curve because the two isotope ratios are produced at different rates. The starting composition at the formation of the Earth (4.57 Ga) is taken from the isotopic composition of troilite in the Canyon Diablo meteorite (Box 6.1). This growth curve assumes a $^{238}U/^{204}Pb$ ratio (μ value) of 9.0. (b) A family of growth curves for lead isotopes identical to that depicted in (a), showing lead isotopic growth from the primeval composition at 4.57 Ga until the present for $\mu = 8$, 9 and 10. Isochrons for 0, 1.0, 2.0 and 3.0 Ga are the lines which join the 0, 1.0, 2.0 and 3.0 Ga points on each growth curve. The 0 Ga isochron is known as the geochron. (c) A single-stage lead isochron for 2.0 Ga. The isochron line joins the 2.0 Ga points on the growth curves for four different μ values and passes through the primeval lead composition at 4.57 Ga; it represents a suite of samples which evolved from 4.57 Ga until 2.0 Ga, when they were isolated from uranium. These values have remained frozen in the rocks since 2.0 Ga. The growth curves in these diagrams are based upon the assumption that the isotopes of lead have evolved uninterrupted since the formation of the Earth. This is thought to be a simplification and normally a more complex model of lead isotope evolution is adopted (Stacey and Kramers, 1975; Cumming and Richards, 1975).

many authors show both the initial ratio and the epsilon value. Because of the uncertainty over bulk Earth values and to save unnecessary confusion and ambiguity, it is wise at the present time to plot only the $^{87}Sr/^{86}Sr$ ratio, applying an age correction if necessary.

Calculating the uncertainties in epsilon values when they are determined for isochron diagrams

Points on an Sm–Nd isochron typically plot far away from the origin. In order to obtain a greater degree of accuracy in determining the initial ratio, some geochronologists have plotted felsic rocks (with small Sm/Nd ratios) on the same isochron as basic igneous rocks with larger Sm/Nd ratios. However, this stretches the assumption that rocks plotted on a common isochron are cogenetic. An alternative method of reducing the error in calculating initial ratios and hence epsilon values was proposed by Fletcher and Rosman (1982), who suggested that the origin of the $^{147}Sm/^{144}Nd$ axis be translated from zero to 0.1967, the present-day composition of CHUR (Table 6.3), prior to performing the line-fitting operation. This is achieved simply by removing 0.1967 from all the $^{147}Sm/^{144}Nd$ coordinates. All other data parameters are unchanged. With this translation the $^{147}Sm/^{144}Nd$ origin usually falls within the compositional range of the data-set resulting in a more precise determination of the intercept. The intercept value determined in this way is the present-day composition of $^{143}Nd/^{144}Nd$ in a hypothetical sample which has the same $^{147}Sm/^{144}Nd$ as CHUR. This value may then be compared with CHUR to calculate the epsilon value using the equation

$$\varepsilon^i_{Nd} = \left[\frac{(^{143}Nd/^{144}Nd)_{sample\ at\ 0.1967,today}}{(^{143}Nd/^{144}Nd)_{CHUR,today}} - 1 \right] \times 10^4 \qquad [6.17]$$

The definition of ε^i_{Nd} differs from ε^t_{Nd} (Eqn [6.12], above) but the difference between the calculated values of the two is less than 1 %, i.e. less than 0.1 epsilon unit in all terrestrial rocks.

The meaning of epsilon values

The symbol ε^t_{Nd} represents the 'initial' value of ε_{Nd} in the rock at the time of its crystallization and is useful because it provides information about the magma source. For example, if in an isochron calculation $\varepsilon^t_{Nd} = 0$, then the epsilon value indicates that the magma was derived from a mantle reservoir which has had a chondritic Sm/Nd from the origin of the earth until time t. A positive value of epsilon for igneous rocks implies a magma derived from a source with a greater Sm/Nd than CHUR, i.e. a depleted mantle source region, whereas a negative epsilon value implies a source with a lower Sm/Nd than CHUR, i.e. an enriched mantle source or a crustal source. Epsilon values for individual samples may be calculated as a test of cogenicity or of contamination in samples which have been used in an isochron plot.

Epsilon values (today) for individual samples are a measure of the extent to which the individual samples are fractionated relative to CHUR as defined by the fractionation factor $f^{Sm/Nd}$ (see below and Figure 6.19).

Epsilon values vary with time and the following approximation may be used to calculate epsilon (t) from epsilon (today)

$$\varepsilon^t_{Nd} = \varepsilon^{today}_{Nd} - Q_{Nd}f^{Sm/Nd} \times t \qquad [6.18]$$

where $f^{Sm/Nd}$ is given by Eqn [6.19] (below) and Q is 25.13 Ga^{-1} when $^{146}Nd/^{142}Nd$ is normalized to 0.63151 (DePaolo, 1988) and when Q is 25.09 Ga^{-1} when $^{146}Nd/^{144}Nd$ is normalized to 0.7219.

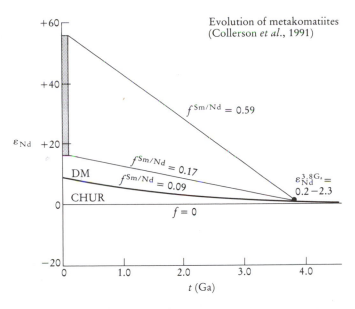

Figure 6.19 Time (t)–epsilon evolution diagram for 3.8 Ga metakomatiites from Labrador (adapted from Collerson *et al.*, 1991). Komatiites were produced from depleted mantle at 3.8 Ga with individual samples ranging in composition from $\varepsilon = +0.2$ to $+2.3$, and with variable Sm/Nd ratios ($f^{Sm/Nd} = 0.17$ to 0.59). Individual samples have evolved to present-day epsilon values between $+16$ and $+56$. The metakomatiites show extreme depletion relative to the depleted mantle curve ($f^{Sm/Nd} = 0.09$) of DePaolo (1981a).

The fractionation factor $f^{Sm/Nd}$

The parameter f is a fractionation factor (often written as $f^{Sm/Nd}$) and is the $^{147}Sm/^{144}Nd$ analogue of epsilon, for it is a measure of Sm/Nd enrichment in a given reservoir relative to CHUR. It is calculated from the expression

$$f = [(^{147}Sm/^{144}Nd)_{today} / (^{147}Sm/^{144}Nd)_{CHUR}) - 1] \qquad [6.19]$$

Thus an $f^{Sm/Nd}$ value of $+0.09$, as is adopted for one particular model of the depleted mantle, has a time-averaged Sm/Nd which is 9 % higher than CHUR. Over time this leads to positive epsilon values for the depleted mantle. Metakomatiites of 3.8 Ga from Labrador have values between $+0.16$ to $+0.59$ and indicate a depleted mantle source with a substantially higher Sm/Nd ratio (Collerson *et al.*, 1991). This is illustrated in Figure 6.19 and shows how samples with high positive f values evolve to very high present-day epsilon values.

Epsilon–Nd time plots

A convenient way of plotting calculated epsilon values is on an epsilon–time diagram analogous to a mantle evolution diagram described in Section 6.3.3. A range of depleted mantle evolution curves are presented in this form in Figure 6.17, for the ε_{Nd} notation is a convenient way of eliminating interlaboratory differences in $^{143}Nd/^{144}Nd$ ratio. Many workers plot ε values relative to CHUR and a model-depleted mantle as a means of determining the likely mantle source of the sample (Figure 6.8). Strictly this should be done only for an ε value calculated from an isochron and the method of Fletcher and Rosman (1982) should be employed to calculate two sigma error polygons for each data point (Figure 6.20); ε values

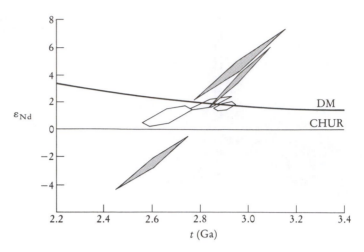

Figure 6.20   Diagram of ε_{Nd} vs time showing 2 σ error polygons calculated using the method of Fletcher and Rosman (1982). The data are for the Lewisian complex of north-west Scotland (from Whitehouse, 1989a) and are plotted relative to the depleted mantle curve (DM) of DePaolo (1981a) and CHUR. The error polygons show the uncertainty in the age and epsilon values determined from isochrons for quartzofeldspathic gneisses (shaded) and for layered mafic–ultramafic intrusions (unshaded) from three different regions of the Lewisian complex. The mafic–ultramafic intrusions (unshaded) lie close to the depleted mantle curve and may be derived from a depleted mantle source. The large error and high positive ε_{Nd} values for the quartzofeldspathic gneisses make them more difficult to interpret.

calculated for individual samples may also be plotted, provided the age of a sample is known with certainty. In addition, individual sample evolution curves may be drawn for different values of $f^{Sm/Nd}$ (Figure 6.19).

In contrast to the depleted mantle, the continental crust shows retarded ^{143}Nd/^{144}Nd evolution relative to CHUR with time (Figure 6.16) and generally has negative ε values. Figure 6.17 shows the evolution of Australian shales with time from Allegre and Rousseau (1984), for these are thought to approximate to the evolution of average continental crust through time.

6.3.5   Isotope correlation diagrams

In the light of the contrasting behaviour of the different isotopic systems outlined in Section 6.3.1, it is instructive to investigate more than one isotope system in a given suite of rocks. This allows us to identify correlations between pairs of isotope ratios which, in turn, lead to a better understanding of the petrogenesis. Most simply the relationships are investigated on an isotope correlation diagram — an x–y graph on which a pair of isotope ratios is plotted. Isotope correlation diagrams were used to display the contrasting compositions of crust and mantle reservoirs (Figures 6.9 to 6.12) and are a means of investigating mixing processes between sources of contrasting composition.

It is unlikely, however, that petrogenetic processes can be satisfactorily reduced to two dimensions, and this must be borne in mind when using two-dimensional projections (Zindler and Hart, 1986). In fact Zindler *et al.* (1982) proposed a three-dimensional Sr–Pb–Nd plot for oceanic basalts and showed that averaged data

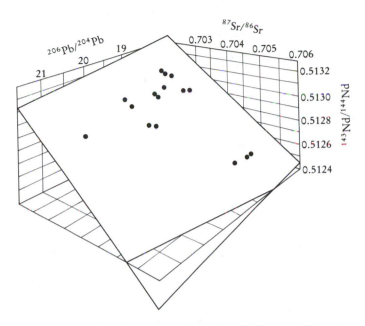

Figure 6.21   Three-dimensional plot of average ^{206}Pb/^{204}Pb, ^{143}Nd/^{144}Nd and ^{87}Sr/^{86}Sr for oceanic basalts, indicating coherence of isotopic ratios (after Zindler *et al.*, 1982).

define a planar surface, implying coherence between Sr–Pb–Nd in the source of oceanic basalts (Figure 6.21). This too, however, is an oversimplification, as is apparent from the Pb–Sr–Nd–Hf study of Stille *et al.* (1983), who show that data from Hawaii do not plot in the mantle plane of Zindler *et al.* (1982) and suggest that U–Pb fractionation may be decoupled from Rb–Sr, Sm–Nd and Lu–Hf fractionation.

Using isotope correlation diagrams and epsilon plots to recognize processes

Trends on isotope correlation diagrams are most commonly interpreted as mixing lines. The mixing may be in the source region, in a magma chamber or between a melt and a 'contaminant', such as the wallrock into which a magma is emplaced or through which it has travelled. Langmuir *et al.* (1978) describe a general mixing equation which can be applied to isotope ratios. Compositions resulting from mixing normally lie on a hyperbolic curve (see also Section 4.9.3), but if the ratios have a common denominator (as in the case of ^{87}Sr/^{86}Sr vs ^{87}Rb/^{86}Sr and ^{207}Pb/^{204}Pb vs ^{206}Pb/^{204}Pb), then mixing will produce a linear trend.

(a) Mixing between sources   For a number of years, students of basalt chemistry have regarded the isotopic compositions of oceanic basalts as the result of mixing of a variety of mantle sources. This is borne out by linear and curvilinear arrays on isotope correlation diagrams such as the Nd–Sr mantle array (Figures 6.9 and 6.10) and the Pb–Pb isotope NHRL (Figure 6.11). Similar principles apply to mixing between crust and mantle sources.

Many authors portray mixing on an Nd–Sr isotope correlation diagram following the principles outlined by DePaolo and Wasserburg (1979). For instance, McCulloch *et al.* (1983) use this diagram (Figure 6.22) to show how the

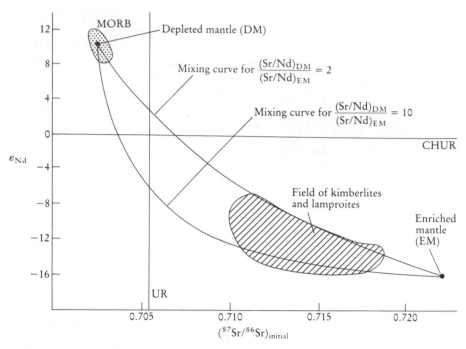

Figure 6.22 ^{143}Nd/^{144}Nd vs ^{87}Sr/^{86}Sr correlation diagram showing two hyperbolic mixing curves for different ratios of (Sr/Nd)$_{EM}$/(Sr/Nd)$_{DM}$ between depleted mantle (DM) and enriched mantle (EM). The mixing can account for the range of ^{143}Nd/^{144}Nd and ^{87}Sr/^{86}Sr values observed in kimberlites and lamproites from western Australia (after McCulloch *et al.*, 1983).

composition of kimberlites and lamproites may be derived by the mixing of MORB-type depleted mantle and an enriched mantle similar to EMII. McCulloch and Chappell (1982) also used the Nd–Sr diagram to explain the contrasting isotope chemistry of S- and I-type granites in terms of mixing between depleted mantle and a sedimentary crustal component. More complex examples of mixing between three end-members are discussed by Shirey *et al.* (1987) and Ellam and Hawkesworth (1988) in studies of oceanic basalts and subduction-related magmas respectively.

The identification of mixing as an important process both in the mantle and between the crust and mantle inevitably leads to more profound questions about mechanisms of mixing and how they relate to major plate tectonic processes. This takes us into the field of geodynamics — a topic briefly discussed at the end of this chapter.

(b) Mixing in a magma chamber   Sharpe (1985) showed that Sr isotopes can be used to monitor magma mixing and multiplicity of magmas in a layered intrusion. He documented the strontium isotopic change with stratigraphic height within the Bushveld layered intrusion and showed that there is a gradual increase in initial ^{87}Sr/^{86}Sr with height up the intrusion and that there is a marked increase in initial ratio in the central part of the layered intrusion — the Main Zone (Figure 6.23). The gradual change in ^{87}Sr/^{86}Sr from 0.7065 at the base of the intrusion to 0.7073 at the top is interpreted as magma addition and mixing. The high initial ratio of the Main Zone (0.7085) represents a different liquid layer (probably contaminated with

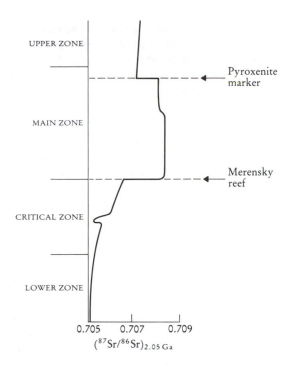

Figure 6.23   Change in ^{87}Sr/^{86}Sr with stratigraphic height in the Bushveld intrusion, South Africa. The
gradual increase in ^{87}Sr/^{86}Sr with stratigraphic height reflects the addition and mixing of a
magma with a higher ^{87}Sr/^{86}Sr ratio. The marked increase in isotopic composition between
the pyroxenite marker and the Merensky Reef marks the influx of a new pulse of magma
(after Sharpe, 1985).

shales) intruded into the density-stratified magma pile near the base of the magma
chamber.

Applications to   **(a) Contamination of magmas by the continental crust**   The expression **crustal**
contamination   **contamination** can have a number of different meanings. Most normally it means
'the contamination of mantle derived melts by continental crust after they have left
the source region' (Hawkesworth and van Calsteren, 1984). However, it can also be
used in the sense of a mantle source region from which magmas are derived which
was contaminated by crustal material at some time in the past by, for example, the
incorporation of subducted sediment into the mantle. One of the problems for
geochemists is that it is not always easy to discriminate between the two processes
on the basis of rock chemistry.

   Crustal contamination may arise in a variety of ways. Mechanisms include the
bulk assimilation of crustal material, the assimilation of a partial melt derived from
crustal materials and the selective exchange of specific elements aided by the
transfer of fluids from crust to melt. The more evolved members of an igneous suite
are more likely to show evidence of contamination, for they have spent the longest
time in the continental crust.

   Stable isotopes, particularly oxygen isotopes are probably the most sensitive of
all isotopic systems to the process of crustal contamination (Hawkesworth and van

Calsteren, 1984) and this topic will be treated in the next chapter (Section 7.2.3). Of the radiogenic isotopes Nd isotopes are the least sensitive to this process whereas Sr and particularly Pb isotopes are of great value. The ability of lead isotopes to detect the contamination of mantle-derived melts by old continental crust has been elegantly demonstrated by Taylor *et al.* (1980). They show that the late Archaean (2.85 Ga) Nuk gneisses of west Greenland were contaminated to a varying extent by lead derived from early Archaean (3.7 Ga) Amitsoq gneisses. The essence of their argument, summarized in Figure 6.24, is that the isotopic composition of lead in the old crust evolved very slowly from 3.7 Ga to 2.85 Ga compared with that in the mantle so that at 2.85 Ga the two had distinctly different isotopic compositions, so much so that the degree of mixing between the two sources of lead could be determined. Sr isotopic compositions in these same rocks are low, even in the most contaminated rocks, indicating that the contamination was selective and leading the authors to suggest that the contamination process was probably related to a fluid phase.

Gray *et al.* (1981) use a Nd–Sr isotope correlation diagram to show the effects of contamination of the Proterozoic Kalka layered basic intrusion with the quartzofeldspathic granulite country rock (Figure 6.25a). Their samples lie on a mixing line between the composition of the granulites and a basaltic source within the mantle array. It should be noted, however, that the isotopic compositions of Jurassic dolerites in Tasmania plot in a similar position on a Nd–Sr isotope correlation diagram (Hergt *et al.*, 1989), but in this case there is a very different interpretation. Hergt *et al.* (1989) believe that the Nd–Sr isotopic character of the melts was inherited from their mantle source region; that a small amount of sediment was introduced into the source region by subduction and that the isotopic signature of this component overprinted that of the mantle (Figure 6.25b). Differentiating between contamination in the source region and contamination during transport through the continental crust can be difficult using isotope correlation diagrams alone and additional information, such as that from stable isotope studies and from radiogenic isotope–trace element plots is necessary.

(b) Crustal contamination and AFC processes   DePaolo (1981b) has shown that Assimilation and Fractional Crystallization (AFC), a popular mechanism of contamination, may shift compositions on an isotope correlation diagram far from what might be expected on the basis of simple mixing. On an Nd–Sr isotope correlation diagram simple mixtures will define a straight line trend between the magma and contaminant, whereas in AFC processes when the bulk solid/liquid distribution coefficients for Nd and Sr differ markedly then there is a significant departure from the simple mixing curve. Powell (1984) inverted the equations of DePaolo (1981b) in an attempt to calculate the composition of the contaminant from an AFC trend on an isotope correlation diagram.

(c) Contamination with seawater   The high $^{87}Sr/^{86}Sr$ ratio of seawater relative to mantle values as seen in oceanic basalts (Figure 6.13) means that the exchange of Sr between seawater and ocean crust in a mid-ocean ridge system has capacity to produce relatively radiogenic $^{87}Sr/^{86}Sr$ ratios even in young MORB and associated sulphides (Spooner *et al.* 1977; Vidal and Clauer, 1981). Nd isotopes are in contrast relatively insensitive to this type of contamination and a plot on an Nd–Sr diagram shows enhanced $^{87}Sr/^{86}Sr$ at constant $^{143}Nd/^{144}Nd$. The length of the vector is

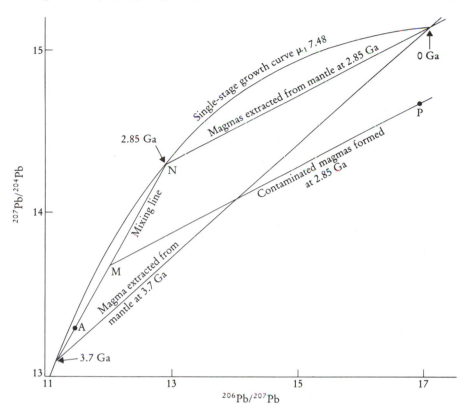

Figure 6.24   The isotopic evolution of Pb isotopes on a $^{207}Pb/^{204}Pb$ vs $^{206}Pb/^{204}Pb$ diagram of the type described in Figure 6.18, showing the effect of crustal contamination on a mantle-derived melt. The single stage growth curve is for $\mu_1 = 7.48$. At 3.7 Ga gneisses separated from a mantle source and evolved along a series of growth curves with average $\mu_2 = 1.16$ indicating extreme depletion in U. The present-day composition of the samples plots at the lower end of the curve joining 3.7 Ga with 0 Ga. At 2.85 Ga the average composition of these gneisses is at A. The mantle source with $\mu_1 = 7.48$ has a composition N at 2.85 Ga and magmas extracted from a mantle source at this time will lie along the line between N and 0 Ga. Magmas extracted from the mantle at 2.85 Ga and contaminated by 3.7 Ga crust at 2.85 Ga will lie on the mixing line AN at a point such as M. Rocks crystallizing from this melt M will evolve along line MP (after Taylor *et al.*, 1980).

proportional to the effective water/rock mass ratio. Pb isotopes are less predictable. Vidal and Clauer (1981), in a study of recent suphides from the East Pacific Rise, found that the Sr isotopes showed evidence of seawater contamination whereas the Pb isotopes did not. Spooner and Gale (1982), however, in a similar study of larger Cretaceous mid-ocean ridge sulphide deposits, did find evidence of contamination from seawater Pb. Chivas *et al.* (1982) used Sr isotopes to calculate water/rock ratios in diorites and granodiorites altered by seawater using an equation analogous to that used by Taylor (1974) for oxygen isotopes (Eqn [7.14]).

Isotope vs trace (and major) element plots   Since an isotope ratio cannot be fractionated by crystal–liquid equilibria and therefore is indicative of the magmatic source, the correlation between an isotope ratio and a major or trace element can be used as a guide to the major element or trace element composition of the source. Typically, ratios of highly incompatible trace elements (Section 4.7.1 and Table 6.7) are most useful in characterizing the

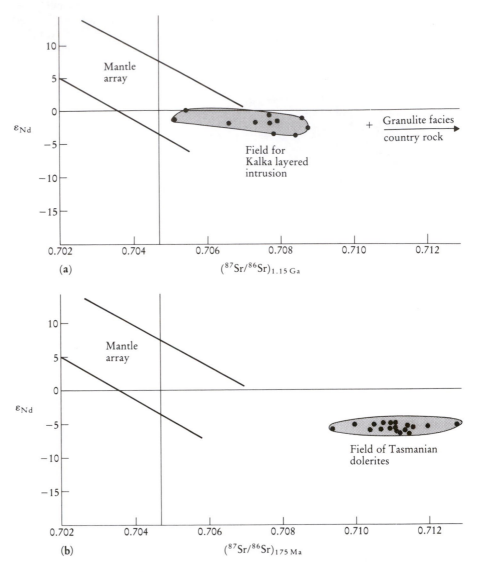

Figure 6.25   (a) ^{143}Nd/^{144}Nd vs ^{87}Sr/^{86}Sr isotope correlation diagram showing the compositional variation within the Proterozoic Kalka layered basic intrusion (shaded) and the composition of the granulite country rock (+ and higher values of ^{87}Sr/^{86}Sr) relative to the mantle array. The high ^{87}Sr/^{86}Sr ratios in the layered intrusion are interpreted by Gray *et al.* (1981) as contamination from the granulite facies country rock. (b) ^{143}Nd/^{144}Nd vs ^{87}Sr/^{86}Sr isotope correlation diagram showing the compositional variation within the Tasmanian dolerites. The high ^{87}Sr/^{86}Sr ratios in the dolerites are interpreted by Hergt *et al.* (1989) as inherited from the mantle source region of the magma, which may be contaminated with subducted sediment.

elemental composition of the source region. Thus, correlations between isotope ratios and ratios of highly incompatible trace elements are likely to indicate mixing between compositionally distinct sources.

The equations of Langmuir *et al.* (1978) predict that data plotted on an isotope ratio vs trace element diagram with a common denominator (e.g. $^{87}Sr/^{86}Sr$ vs $1/^{86}Sr$) will define a straight line. Mixing diagrams of this type, however, can give rise to spurious correlations because of the common denominator effect (Dodson, 1982; and see discussion in Section 2.5.3). An additional problem with the $^{87}Sr/^{86}Sr$ ratio vs $1/Sr$ diagram, highlighted by Mensing *et al.* (1984), is that, in the case of basaltic lavas undergoing fractional crystallization and assimilating crustal material (AFC), the composition of the contaminant does not lie on the magma evolution trajectory. Thus the linear mixing trend cannot be used to determine the composition of the contaminant.

6.3.6   Mantle–crust geodynamics

The ultimate result of defining crust and mantle isotopic reservoirs must be a model which explains how the reservoirs interact and how they obtained their present-day compositions. One of the major successes of modern isotope geology is that such unifying models now exist and are able to link reservoir compositions to the processes that govern plate tectonics. In detail there are two types of approach. Models based upon Pb isotopes and which chiefly emphasize crustal reservoirs have contributed greatly to our understanding of the evolution of the continental crust. This type of modelling is known as plumbotectonics. Models which emphasize mixing processes in the mantle, and which are based upon a number of different isotope systems, are described here under the heading 'geodynamic models'.

Plumbotectonics   A model based on the discriminatatory power of Pb isotopes was developed by Doe and Zartman (1979), who showed that there are variations in initial lead isotopic composition which are related to tectonic setting. They showed that three reservoirs (the upper continental crust, the lower crust and the upper mantle) can be characterized by their concentrations of U, Th and Pb (see Table 6.5). U, Th and Pb are concentrated in the upper crust with U and Th enriched relative to Pb so that the upper crust evolves radiogenic lead. The lower crust is depleted in Th and U and evolves unradiogenic lead. The mantle has lower concentrations of U, Th and Pb than the continental crust but has U/Pb and Th/Pb ratios which lie between the two crustal reservoirs and so evolves Pb of an intermediate character. The mixing of Pb in its three reservoirs takes place in 'orogenies' (or orogenes) producing a fourth 'mixed' reservoir.

In the several versions of their models, Doe and Zartman (1979) and Zartman and Haines (1988) have plotted curves to show the Pb isotopic evolution of the four reservoirs with time. A plot of $^{207}Pb/^{204}Pb$ vs $^{206}Pb/^{204}Pb$ discriminates well between the upper crust and the lower crust/mantle and a plot of $^{208}Pb/^{204}Pb$ vs $^{206}Pb/^{204}Pb$ discriminates between the lower crust and the upper crust/mantle (Figure 6.26).

Geodynamics   Correlations between isotopic tracers has led to a search for an explanation of these phenomena and has given rise to a series of tectonic models for the chemical structure of the Earth. These models are constrained both by the isotopic data and our present understanding of plate tectonic processes.

Consider, for example, one of the early observations drawn from isotope correlation diagrams. In Figure 6.10 the Nd–Sr compositions of the continental

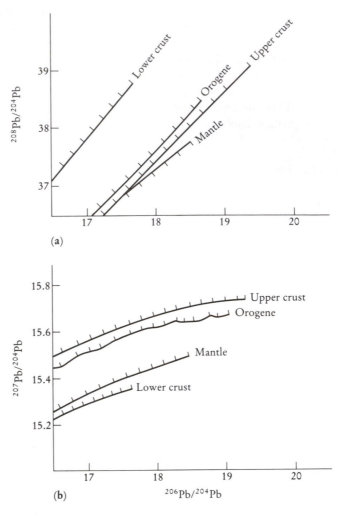

Figure 6.26   Lead isotope evolution curves for the lower crust, upper crust, mantle and orogene plotted for (a) $^{208}Pb/^{204}Pb$ vs $^{206}Pb/^{204}Pb$ and (b) $^{207}Pb/^{204}Pb$ vs $^{206}Pb/^{204}Pb$ for plumbotectonics version IV (from Zartman and Haines, 1988). The ticks on each curve are at 100 Ma intervals.

crust and oceanic basalts are plotted relative to the composition of the bulk Earth. Oceanic basalts are enriched in Nd and depleted in Sr relative to the bulk Earth whilst the continental crust shows the opposite relationship. This suggests that the continental crust and the mantle source of oceanic basalts are complementary reservoirs of Nd and Sr and that the continental crust has been extracted from the Earth's mantle leaving a reservoir enriched in Nd and depleted in Sr.

In a similar way the isotopic composition of crust and mantle reservoirs have been interpreted by means of a series of mass balance equations between crust, mantle and the original composition of the bulk silicate Earth. This has allowed us to explore the interrelationships between different reservoirs, set limits on the proportion of mantle involved in the formation of the continental crust and offer insights into the nature of mantle convection (Allegre *et al.*, 1983a,b; Allegre, 1987;

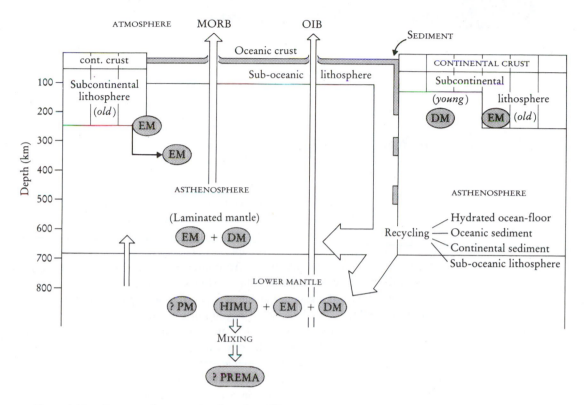

Figure 6.27   Cartoon diagram showing the different crust and mantle reservoirs and the possible relationships between them, based upon the observations of isotope geochemistry. The mantle reservoirs are identified using the nomenclature of Zindler and Hart (1986): EM, enriched mantle; DM, depleted mantle; HIMU, mantle with high U/Pb ratio; PREMA, prevalent mantle; PM, primitive mantle.

Galer and O'Nions, 1985; Zindler and Hart, 1986). A cartoon illustrating a possible way in which the different reservoirs interact is given in Figure 6.27.

Chapter 7

Using stable isotope data

7.1 Introduction

Most naturally occurring elements consist of more than one stable isotope. In elements with an atomic mass of less than 40 (i.e. lighter than Ca) it is possible for the isotopes to be fractionated through physical processes as a consequence of the mass difference between the isotopes. The degree of mass fractionation is proportional to the mass difference. At atomic masses higher than 40 the relative mass differences are too small to allow isotopes to become physically separated.

In geochemistry the study of stable isotopes is a powerful means of studying the light elements H, C, N, O and S — a group of elements not easily studied in other ways. These elements are often the main constituents of geologically important fluids, thus affording a means of directly studying both the fluids and the effects of fluid–rock interaction. In addition, stable isotopes are used as tracers to determine the source of an element, as palaeothermometers and as a means of studying diffusion and reaction mechanisms in geological processes. In this chapter we will survey the use of the stable isotopes of hydrogen, oxygen, carbon and sulphur and illustrate their role in elucidating geochemical processes in igneous, metamorphic and sedimentary rocks. More detailed treatments are given by Hoefs (1987) and Valley *et al.* (1986).

Conventionally stable isotopes are converted into a gas (usually H_2, CO_2 or SO_2) for the purposes of isotopic analysis and the mass differences are measured in a mass spectrometer. With such commonly occurring elements as O, H, C and S, contamination during sample preparation and analysis is a particular problem and great care must be taken to ensure clean sample handling. Increasingly, however, the ion probe and laser microprobe are being used in stable isotope analysis and a number of recent studies have illustrated how a much finer spatial resolution of isotopic compositions is possible using these techniques.

7.1.1 Notation

Stable isotope ratios are measured relative to a standard and are expressed in parts per thousand, i.e. parts per mil ($^0/_{00}$). The isotope ratio is expressed as a δ value, or 'del value' as it is sometimes called. Using oxygen isotopes as an example, the δ value is calculated as follows:

$$\delta^{18}O\,^0/_{00} = \left[\frac{^{18}O/^{16}O \text{ (sample)} - {}^{18}O/^{16}O \text{ (standard)}}{^{18}O/^{16}O \text{ (standard)}} \right] \times 1000 \qquad [7.1]$$

Thus a $\delta^{18}O$ value of $+10.0$ means that the sample is enriched in ^{18}O relative to the standard by 10 parts in a thousand and a value of -10.0 means that the sample is depleted in ^{18}O relative to the standard by 10 parts in a thousand.

7.1.2 Isotope fractionation

The chief purpose of studying stable isotopes is as a means of investigating the processes which in nature separate isotopes on the basis of their mass rather than on the basis of their chemistry. This is known as isotopic fractionation and takes place in nature in three different ways:

(1) Isotopic exchange reactions. Isotope fractionation may take place in a conventional exchange reaction in which, for example, oxygen is exchanged between quartz and magnetite

$$2Si^{16}O_2 + Fe_3\,^{18}O_4 = 2Si\,^{18}O_2 + Fe_3\,^{16}O_4$$

The isotopic fractionation is controlled by bond-strength and follows the general rule that the lighter isotope forms a weaker bond than the heavier isotope.
(2) Kinetic processes. Kinetically controlled stable isotope fractionation reflects the readiness of a particular isotope to react. Kinetic effects are only observed when a reaction does not go to completion.
(3) Physico-chemical processes such as evaporation and condensation, melting and crystallization and diffusion.

The fractionation of an isotope between two substances A and B can be defined by the fractionation factor α:

$$\alpha_{A-B} = \frac{\text{ratio in A}}{\text{ratio in B}} \qquad [7.2]$$

For example, in the reaction in which ^{18}O and ^{16}O are exchanged between magnetite and quartz, the fractionation of $^{18}O/^{16}O$ between quartz and magnetite is expressed as

$$\alpha_{\text{quartz–magnetite}} = \frac{(^{18}O/^{16}O)\ \text{in quartz}}{(^{18}O/^{16}O)\ \text{in magnetite}}$$

where '$^{18}O/^{16}O$ in quartz' and '$^{18}O/^{16}O$ in magnetite' are the measured isotopic ratios in coexisting quartz and magnetite. If the isotopes are randomly distributed over all the possible atomic positions in the compounds measured, then α is related to an equilibrium constant K such that

$$\alpha = K^{1/n} \qquad [7.3]$$

where n is the number of atoms exchanged. Normally, exchange reactions are written so that only one atom is exchanged, in which case $\alpha = K$, and the equilibrium constant is equivalent to the fractionation factor.

Values for α are very close to unity and typically vary in the third decimal place. Most values therefore are of the form 1.00X. For example, the fractionation factor for ^{18}O between quartz and magnetite at 500 °C is 1.009 (Javoy, 1977). This may be expressed as the third decimal place value — the per mil value — such that the quartz–magnetite fractionation factor is 9 (or 9.0 per mil). A useful mathematical approximation for the fractionation factor a stems from the relationship

$$1000\ln(1.00X) \sim X \tag{7.4}$$

In the case cited above where $\alpha = 1.009$, $1000\ln\alpha = 9.0$. This relationship has the added value that experimental studies have shown that $1000\ln\alpha$ is a smooth and often linear function of $1/T^2$ for mineral–mineral and mineral–fluid pairs. This gives rise to the general relationship for the fractionation factor:

$$1000\ln\alpha_{mineral1-mineral2} = A(10^6/T^2) + B \tag{7.5}$$

where T is in kelvin and A and B are constants, normally determined by experiment. In the case of the quartz–magnetite pair, the values for A and B are 6.29 and zero respectively (Chiba *et al.*, 1989), giving the expression

$$1000\ln\alpha_{quartz-magnetite} = 6.29 \times 10^6/T^2$$

A further useful approximation is the relationship between $1000\ln\alpha$ and measured isotope ratios expressed as δ values. The difference between the δ values for two minerals is expressed as Δ and this approximates to $1000\ln\alpha$, when the δ values are less than 10. In the case of oxygen isotopic exchange between quartz and magnetite,

$$\Delta_{qz-mgt} = \delta_{qz} - \delta_{mgt} \sim 1000\ln\alpha_{qz-mgt} \tag{7.6}$$

When δ values are larger than 10 the expression given as Eqn [7.11] in Box 7.1 should be used.

7.1.3   Physical and chemical controls on stable isotope fractionation

Understanding the physical and chemical controls on stable isotope fractionation is vital to a correct interpretation of measured stable isotope ratios, for when the fractionation processes are fully understood, then the measured δ–values can be used to identify correctly the source of the element in question and the geological processes involved. In this section we examine briefly the chief controls on stable isotope fractionation.

There is an important **temperature** control on isotopic fractionation. This has already been described in Eqn [7.5] and has an obvious application in isotopic thermometry (see Sections 7.2.2., 7.4.6 and 7.5.2).

Relative volume changes in isotopic exchange reactions, on the other hand, are very small except for hydrogen isotopes and therefore there is a minimal **pressure** effect. Clayton (1981) showed that at pressures of less than 20 kb the effect of pressure on oxygen isotope fractionation is less than 0.1 %. and lies within the measured analytical uncertainties. The absence of a significant pressure effect on stable isotope fractionation means that isotopic exchange reactions can be investigated at high pressures where reaction rates are fast, and the results extrapolated to lower pressures.

Some isotope fractionations, notably those in biological systems, are primarily controlled by **kinetic effects**. For example, the bacterial reduction of seawater sulphate to sulphide proceeds 2.2 % faster for the light isotope ^{32}S than for ^{34}S. For the reactions

$$^{34}SO_4^{2-} \xrightarrow{k_2} H_2^{34}S \tag{7.7}$$

$$^{32}SO_4^{2-} \xrightarrow{k_1} H_2^{32}S \tag{7.8}$$

the rate constant k_1 is greater than the rate constant k_2 and the ratio $k_1/k_2 = 1.022$. The effects of this fractionation in a closed system may be modelled using the Rayleigh fractionation equation (Section 4.2.2).

When isotopic fractionation takes place as a result of **diffusion** the light isotope is enriched relative to the heavy one in the direction of transport. Diffusion-controlled isotopic fractionation can be important when interpreting the results of oxygen isotopes as thermometers (Section 7.2.2). A related process to that of diffusion is the microfiltration effect in which isotopes are fractionated by adsorption onto clay minerals in sediments. It is thought that isotopically lighter hydrogen, oxygen and sulphur may be preferentially adsorbed onto clay leading to isotopic enrichments in formation waters (Ohmoto and Rye, 1979).

During **distillation** the light isotopic species is preferentially enriched in the vapour phase according to the Rayleigh fractionation law (Section 4.2.2). This process applies to the evaporation and condensation of meteoric water and accounts for the marked fractionation of $\delta^{18}O$ and δD in rainwater and in ice.

Generally **major element chemistry** has a very small effect on stable isotope fractionation. However, Taylor and Epstein (1962) observed that the oxygen isotope fractionation in silicate minerals can be accounted for in terms of the average bonding arrangement of the mineral structure, and heavy isotopes associated with elements with a high ionic potential. This is seen in the fractionation of ^{18}O between quartz and magnetite, for quartz containing the small, highly charged Si^{4+} is enriched mineral in ^{18}O, whereas magnetite with the large Fe^{2+} ion is ^{18}O-deficient. In the case of feldspars, the oxygen isotope composition plagioclase is a function of An content.

Heavy isotopes are concentrated in more closely packed **crystal structures**. The fractionation of carbon isotopes between diamond and graphite is well known and there are smaller changes between calcite and aragonite. There is a also a small change in ^{18}O between α- and β-quartz.

Recently it was discovered that there is a **crystallographic control** on the fractionation of oxygen and carbon isotopes in calcite. Dickson (1991) found that, in a single crystal of calcite, crystal faces from different crystallographic forms have different isotopic compositions. This observation indicates that different surfaces in the same crystal have slightly different bonding characteristics which are sufficient to fractionate the isotopes of oxygen and carbon.

7.2   Using oxygen isotopes

There are three stable isotopes of oxygen which have the following abundances:

^{16}O = 99.763 %
^{17}O = 0.0375 %
^{18}O = 0.1995 %

The isotope ratio $^{18}O/^{16}O$ is the ratio which is normally determined in oxygen isotope studies and δ values are calculated from Eqn [7.1]. There are currently two isotopic standards in use for oxygen isotope measurements. Low-temperature geothermometry measurements (Section 7.2.2) are made relative to PDB (a belemnite from the Cretaceous Peedee formation of South Carolina, a standard otherwise used for carbon isotope measurements) whilst all other measurements are calculated relative to concentrations in Standard Mean Ocean Water (SMOW). SMOW was originally a hypothetical water sample with oxygen and hydrogen isotope ratios similar to those of standard ocean water. Currently a water standard distibuted by the Atomic Energy Agency in Vienna, known as Vienna-SMOW or V-SMOW is used. This has an $^{18}O/^{16}O$ ratio identical to SMOW and a D/H ratio which is within error of the original definition of SMOW (Gonfiantini, 1978). V-SMOW and PDB $\delta^{18}O$ values are related by the expressions

$$\delta^{18}O_{V-SMOW} = 1.03091 \, \delta^{18}O_{PDB} + 30.01 \qquad [7.9]$$

and

$$\delta^{18}O_{PDB} = 0.97002 \, \delta^{18}O_{V-SMOW} - 29.98 \qquad [7.10]$$

(Coplen *et al.*, 1983). In addition to SMOW the standard SLAP (Standard Light Antarctic Precipication) is sometimes used. This has a $\delta^{18}O$ value of -55.5 %. relative to SMOW (Gonfiantini, 1978).

Oxygen is liberated from silicates and oxides through fluorination with F_2 or BrF_5 and then reduced to CO_2 at high temperature for measurement in a mass spectrometer. In carbonates carbon dioxide is liberated with >103 % phosphoric acid. When oxygen isotope ratios are determined in water the sample is equilibrated with a small amount of CO_2 and the oxygen isotope ratio in the CO_2 is measured. From the known water–CO_2 fractionation factor, the $^{18}O/^{16}O$ ratio in the water is calculated. The precision of $\delta^{18}O$ values is of the order of 0.1–0.2 $^o/_{oo}$.

In this section we consider first the distribution of oxygen isotopes in nature and then the use of oxygen isotopes in thermometry. This is followed by a discussion of correlation diagrams which combine both oxygen isotopes and radiogenic isotopes and the way in which such correlations may be used to infer geological processes.

7.2.1   Variations of $\delta^{18}O$ in nature

$\delta^{18}O$ values vary in nature by about 100 $^o/_{oo}$, about half of this range occurring in meteoric water (Figure 7.1). Chondritic meteorites have a very resticted range of $\delta^{18}O$ values and the mantle has a $\delta^{18}O$ value of 5.7 $\pm$ 0.3$^o/_{oo}$ and this seems to have been constant through time for the Earth and the Moon (Taylor, 1980). However, Kyser *et al.* (1982) found that the alkali basalts in Hawaii are enriched in $\delta^{18}O$ by values of 0.5 to 1.0 $^o/_{oo}$ over tholeiites and suggested on the basis of measured

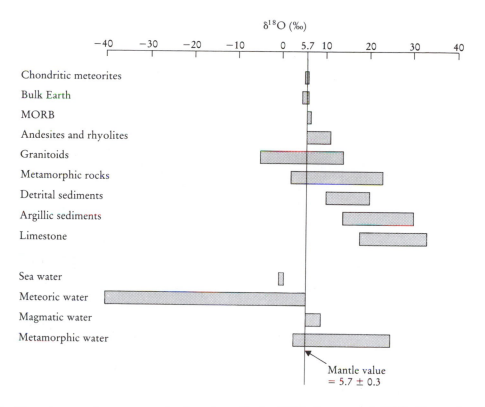

Figure 7.1   Natural oxygen isotope reservoirs. Data from: Taylor (1974), Onuma *et al.* (1972), Sheppard (1977), Graham and Harmon (1983) and Hoefs (1987).

diffusion rates (Graham and Harmon, 1983) that the two had distinct mantle sources. Thus there is some evidence for small isotopic heterogeneities in the mantle.

Most granites, metamorphic rocks and sediments are enriched in $\delta^{18}O$ relative to the mantle value, whereas seawater and meteoric waters are depleted, thus forming complementary $\delta^{18}O$ reservoirs.

7.2.2   Oxygen isotope thermometry

One of the first applications of the study of oxygen isotopes to geological problems was to geothermometry. Urey (1947) suggested that the enrichment of ^{18}O in calcium carbonate relative to seawater was temperature-dependent and could be used to determine the temperature of ancient ocean waters. The idea was quickly adopted and palaeotemperatures calculated for the Upper Cretaceous seas of the northern hemisphere. Subsequently, a methodology was developed for application to higher-temperature systems based upon the distribution of ^{18}O between mineral-pairs. An excellent review of the methods and applications of oxygen isotope thermometry is given by Clayton (1981).

The expression summarizing the temperature dependence of oxygen isotopic exchange between a mineral-pair is given in Eqn [7.5]. Often the *B* term is zero and the fractionation factor is simply a function of $1/T^2$. Empirical observations indicate that a graph of ln α vs $1/T^2$ is linear over a temperature range of several hundred degrees (Figure 7.2) and a plot of this type for a pair of anhydrous phases should

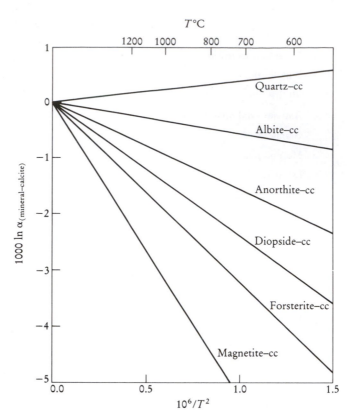

Figure 7.2   1000 ln α vs $10^6/T^2$ calibrations of mineral–calcite pairs using the data of Chiba *et al.* (1989).

also pass through the origin. Isotopic fractionations decrease with increasing temperature and so oxygen isotope thermometers might be expected to be less sensitive at high temperatures. Experimental studies, however, are most precise at high temperatures (see for example Clayton *et al.*, 1989), and reliable thermometers have been calibrated for use with igneous and metamorphic rocks. Oxygen isotope thermometry has a number of advantages over conventional cation-exchange thermometry; for example oxygen isotopic exchange can be measured between many mineral-pairs in a single rock. In addition, minerals with low oxygen diffusivities such as garnet and pyroxene are capable of recording peak temperature conditions.

Calibration of oxygen isotope thermometers

There are a large number of different calibrations of oxygen isotope exchange reactions (Table 7.1), some of which give conflicting results. This has given rise to much confusion over which calibrations can be used as a basis for reliable thermometry. In brief, there are three different approaches to the calibration of oxygen isotope exchange thermometers — the theoretical approach, experimental methods and empirical methods.

Theoretical calculations of oxygen isotope fractionations are based upon studies of lattice dynamics. Recent results of this type have been found to agree with new

Table 7.1   Oxygen isotopic thermometer calibrations

Experimentally determined coefficients A for the fractionation of oxygen isotopes between anhydrous mineral-pairs according to $1000 \ln \alpha = A \times 10^6/T^2$. The data are arranged so that all coefficients are positive.

(a) Experimentally determined — equilibration with calcite (Chiba et al., 1989)

	Cc	Ab	An	Di	Fo	Mt
Q	0.38	0.94	1.99	2.75	3.67	6.29
Cc		0.56	1.61	2.37	3.29	5.91
Ab			1.05	1.81	2.73	5.35
An				0.76	1.68	4.30
Di					0.92	3.54
Fo						2.62

(b) Experimentally determined — equilibration with water (Matthews et al., 1983)

	Ab	Jd	An	Di	Wo	Mt
Q	0.50	1.09	1.59	2.08	2.20	6.11
Ab		0.57	1.09	1.58	1.70	5.61
Jd			0.50	0.99	1.11	5.02
An				0.49	0.61	4.52
Di					0.12	4.03
Wo						3.91

(c) Empirically determined (Bottinga and Javoy, 1975; Javoy, 1977)

	Ab	Pl	Px	Ol	Gt	Mt	Il
Q	0.97	1.59	2.75	3.91	2.88	5.57	5.29
Pl			1.08	2.32	1.29	3.98	3.70
Px				1.24	—	0.21	—

(d) Polynomial functions for individual minerals used for calculating oxygen isotopic fractionation between mineral-pairs (Clayton, 1991) [$x = 10^6/T^2$; T in kelvin units]

Mineral	Function
Calcite	$f_{Cc} = 11.781x - 0.420x^2 + 0.0158x^3$
Quartz	$f_{Q} = 12.116x - 0.370x^2 + 0.0123x^3$
Albite	$f_{Ab} = 11.134x - 0.326x^2 + 0.0104x^3$
Anorthite	$f_{An} = 9.993x - 0.271x^2 + 0.0082x^3$
Diopside	$f_{Di} = 9.237x - 0.199x^2 + 0.0053x^3$
Forsterite	$f_{Fo} = 8.326x - 0.142x^2 + 0.0032x^3$
Magnetite	$f_{Mt} = 5.674x - 0.038x^2 + 0.0003x^2$

Key: Ab, albite; An, anorthite; Cc, calcite; Di, diopside; Fo, forsterite; Il, ilmenite; Jd, jadeiite; Mt, magnetite; Ol, olivine; Pl, plagioclase (An_{60}); Px, pyroxene; Q, quartz; Wo, wollastonite.

experimental studies by Clayton *et al.* (1989) and used to extrapolate experimental results outside their temperature range (Clayton and Kieffer, 1991).

Experimental studies based upon mineral–water isotopic exchange have been used to calibrate oxygen isotope thermometers, although the more reliable exchange reaction with calcite is now preferred (Table 7.1).

Calcite–mineral oxygen isotope exchange is stable to relatively high temperatures and the results can be extrapolated outside the experimental range. Mineral–calcite pairs are combined to give mineral–mineral oxygen isotope fractionation equations. The best thermometers are between mineral–calcite pairs which show the greatest divergence on a ln α vs $1/T^2$ diagram (Figure 7.2). Thus the mineral-pair quartz–diopside is a sensitive thermometer, for there is significant fractionation of oxygen isotopes between the two minerals, whereas a pair such as quartz–albite is not sufficiently sensitive. Quartz–magnetite fractionation is not widely used because the high diffusivity of oxygen in magnetite means that it cannot record peak temperatures.

Empirical calibrations of oxygen-isotope thermometers are based upon experimental data which is then applied to a natural assemblage (Bottinga and Javoy, 1973, 1975; see Table 7.1). Given that all the minerals in a rock are in isotopic equilibrium, thermometers can then be calibrated for mineral-pairs which have not been experimentally studied. However, the underlying assumption of isotopic equilibrium is rarely fulfilled, making the application of this method questionable (Clayton, 1981).

Currently the most reliable calibration of oxygen isotope thermometers is based upon a combination of experimental and theoretical studies. Clayton (1991) combined calcite–mineral experimental data with theoretical studies and calculated polynomial expressions for a number of common rock-forming minerals (Table 7.1). These expressions may be combined to give mineral-pair thermometers.

Tests of isotopic equilibrium

One of the merits of oxygen isotope thermometry is that there are potentially a large number of thermometers available in a single rock, a situation which rarely arises with cation-exchange thermometers. This allows the assumption of isotopic equilibrium to be tested.

Javoy *et al.* (1970) proposed the use of an isotherm plot to measure isotopic equilibrium. This diagram plots the A parameter of the oxygen isotope thermometer equation (Eqn [7.5] and Table 7.1) and the term $[1000 \ln \alpha_{(quartz–mineral)} - B]$ as the axes of a bivariate plot. Appropriate isotherms for the system under investigation are calculated and the values for sets of mineral-pairs are plotted. Minerals from an individual rock which are in isotopic equilibrium should show a smooth trend parallel to the isotherms. Isotherm diagrams of this type are used by Huebner *et al.* (1986) to demonstrate isotopic disequilibrium between mineral pairs in granulite facies metapelites (Figure 7.3).

More recently it has been argued that in slowly cooled rocks isotopic equilibrium should not be expected. Giletti (1986) showed that in a slowly cooled rock individual minerals cease to exchange oxygen as they pass through their own blocking temperatures and it is only the last mineral-pair to exchange oxygen that will be in equilibrium. Diffusion data summarized by Farver (1989) indicate that oxygen isotope closure in a cooling rock is in the order diopside (first to close), hornblende, magnetite, quartz and anorthite (Figure 7.4). The blocking temperature

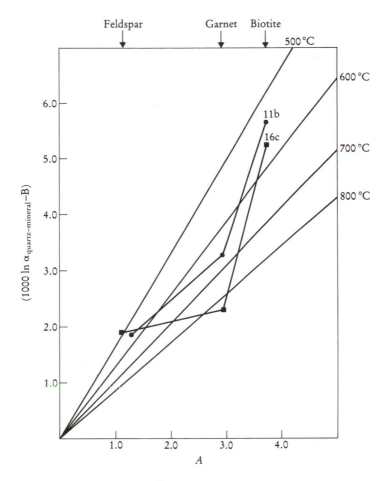

Figure 7.3   Isotherm diagram of Javoy *et al.* (1970) used to assess the degree of relative isotopic equilibrium amongst minerals in the same sample. The diagram is based upon the oxygen isotope thermometer equation (Eqn [7.5])

$$1000 \ln \alpha_{\text{mineral1–mineral2}} = A(10^6/T^2) + B$$

Variable A is plotted along the x-axis and $[1000 \ln \alpha_{\text{(quartz–mineral)}} - B]$ along the y-axis. Isotherms from 500 °C to 800 °C are constructed from the thermometer equation. The $1000 \ln \alpha$ values are plotted for the mineral pairs quartz–feldspar, quartz–garnet and quartz–biotite for samples 11b and 16c from Huebner *et al.* (1986). They are determined from the $\delta^{18}O$ values using Eqn [7.6]. The plotted compositions show that the isotopic composition of garnet is out of equilibrium with biotite and feldspar. Note that the feldpars in the two samples have slightly different compositions.

approach of Giletti forms the basis of a computer model for estimating isotopic equilibrium by Jenkin *et al.* (1991). The authors give the source code in FORTRAN–77.

It should be noted, however, that the difference in temperatures of equilibration is only one reason for isotopic disequilibrium and isotopic exchange with an externally derived fluid is an equally possible explanation. This may be assessed on a plot of δ-values for two coexisting minerals. Equilibrium is measured relative to a straight line with a slope of 45 ° which represents a Δ ^{18}O value for the mineral-pair

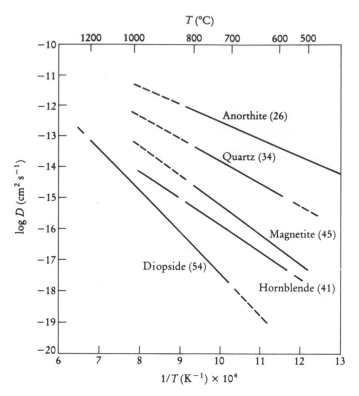

Figure 7.4   Log diffusion coefficient vs $1/T$ plot (Arrhenius plot) for the diffusion of oxygen in selected minerals. The numbers in parentheses are the activation energies in kcal/g-atom O (after Farver, 1989).

and which is an isotherm. Any departure of trends from parallelism to the 45 ° line is indicative of isotopic disequilibrium (Figure 7.5; Gregory *et al.*, 1989).

Applications   There are both low-temperature and high-temperature applications of oxygen isotope thermometry.

(a) Low-temperature thermometry   The earliest application of oxygen isotopes to geological thermometry was in the determination of ocean palaeotemperatures. The method assumes isotopic equilibrium between the carbonate shells of marine organisms and ocean water and uses the equation of Epstein *et al.* (1953) which is still applicable despite some proposed revisions (Friedman and O'Neil, 1977):

$$T\,°C = 16.5 - 4.3(\delta_c - \delta_w) + 0.14(\delta_c - \delta_w)^2 \qquad [7.12]$$

where δ_c and δ_w are respectively the $\delta^{18}O$ of CO_2 obtained from $CaCO_3$ by reaction with H_3PO_4 at 25 °C and the $\delta^{18}O$ of CO_2 in equilibrium with the seawater at 25 °C.

The method assumes that the oxygen isotopic composition of seawater was the same in the past as today, an assumption which has frequently been challenged and which does not hold for at least parts of the Pleistocene when glaciation removed ^{18}O-depleted water from the oceans. This has the effect of amplifying the temperature variations (Clayton, 1981). The method also assumes that the isotopic

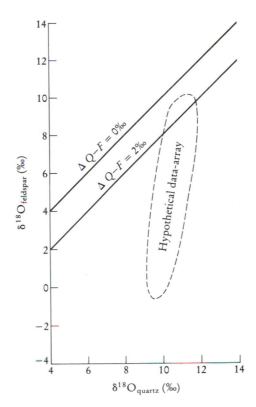

Figure 7.5 $\delta^{18}O_{quartz}$ vs $\delta^{18}O_{feldspar}$ diagram. Two 45° equilibrium lines are shown for quartz–feldspar Δ values of 0 ‰ and 2 ‰. A hypothetical disequilibrium array is shown with a steep positive slope.

composition of oxygen in the carbonate is primary and that the carbonate precipitation was an equilibrium process. Both these assumptions should also be carefully examined. Because the temperatures of ocean bottom water vary as a function of depth it is also possible to use oxygen isotope thermometry to estimate the depth at which certain benthic marine fauna lived — palaeobathymetry.

Low-temperature isotopic thermometry is also applicable to ascertaining the temperatures of diagenesis and low-grade metamorphism, and estimating the temperatures of active geothermal systems, both in the continental crust and on the ocean floor.

(b) High-temperature thermometry Stable isotope systems are frequently out of equilibrium in rocks which formed at high temperatures as a result of equilibration with a fluid phase following crystallization. This fact can be used to make inferences about the nature of rock–water interaction (see Section 7.3.4) but does not help establish solidus or peak-metamorphic temperatures in igneous and metamorphic rocks. In systems where there is minimal water present, such as on the Moon, oxygen isotope thermometers yield meaningful temperatures. Kyser *et al.* (1981) have obtained high-temperature results on terrestrial lavas and mantle nodules. In metamorphic rocks, in which there has been minimal fluid interaction, the newly calibrated thermometers of Chiba *et al.* (1989) and Clayton *et al.* (1989) hold

promise for mineral-pairs with slow diffusion rates such as garnet–quartz and pyroxene–quartz (see Connolly and Muehlenbachs, 1988). An example of a thermometric calculation using oxygen isotope exchange between quartz and pyroxene in a metapelite is given in Box 7.1.

Box 7.1

Example of an oxygen-isotope thermometer calculation

Data

Coexisting quartz–orthopyroxene pair from a granulite facies metapelite (from Huebner *et al.* 1986, sample Bb25c).

$\delta^{18}O_{quartz} = 10.2\ ‰$
$\delta^{18}O_{orthopyroxene} = 7.9\ ‰$

Theory

The temperature dependence of the fractionation of ^{18}O between quartz and pyroxene is given by Chiba *et al.* (1989) and Javoy (1977) as

$$1000 \ln \alpha_{qz-px} = 2.75 \times 10^6/T^2$$

Calculation

In this case the approximation given in Eqn [7.6] is inappropriate, since the δ-value for quartz is > 10.0, and so we use

$$\alpha_{A-B} = \frac{1000 + \delta_A}{1000 + \delta_B} \qquad [7.11]$$

Thus: $\alpha_{quartz-px} = 1010.2/1007.9 = 1.00228$
$1000 \ln \alpha = 2.279$

From the quartz–pyroxene thermometer:

$$2.279 = 2\,750\,000/T^2$$

$$T = 1098\ K = 825\ °C$$

7.2.3 Oxygen isotope — radiogenic isotope correlation diagrams

Correlations between radiogenic and oxygen isotopes are of particular importance because variations in the two types of isotope come about through totally different mechanisms. This means that correlation diagrams of this type convey information of quite a different type from either stable isotope correlation diagrams (Sections 7.3.3, 7.4.2) or radiogenic isotope correlation diagrams (Section 6.3.5).

Recognizing crust and mantle reservoirs

Oxygen isotopes provide a very effective way of distinguishing between rocks which formed in equilibrium with the mantle and those which formed from the continental crust. In general the continental crust is enriched in $\delta^{18}O$ relative to the Earth's mantle (Figure 7.1). This has come about largely as a consequence of the long interaction between the continental crust and the hydrosphere and the partitioning of ^{18}O into crustal minerals during low-temperature geological processes. Oxygen isotopes, therefore, are a valuable indicator of surface processes and a useful tracer of rocks which at some time have had contact with the Earth's surface.

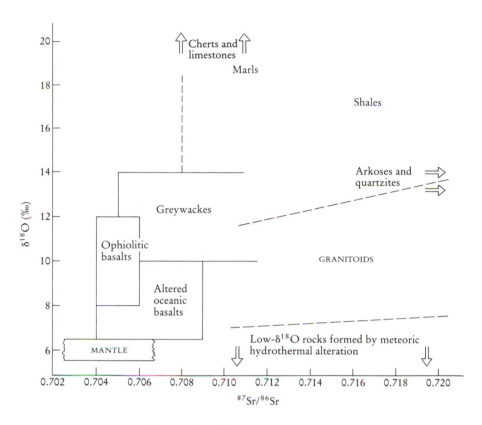

Figure 7.6 Approximate ranges of $\delta^{18}O$ and $^{87}Sr/^{86}Sr$ in common igneous and sedimentary rocks (adapted from Magaritz *et al.*, 1978).

Radiogenic isotopes, on the other hand, show differences between crust and mantle reservoirs which are a function of long-lived differences in parent/daughter element ratios and indicate the isolation of the reservoirs from one another for long periods of Earth history. This gives rise to crustal reservoirs which generally are enriched in $^{87}Sr/^{86}Sr$ and in radiogenic lead isotopes but depleted in $^{143}Nd/^{144}Nd$ relative to the mantle.

The discussion below is restricted to the isotopes of strontium and oxygen although the principles enunciated can apply equally to other radiogenic isotope–oxygen isotope pairs. The range of combined oxygen and strontium isotopic compositions in common rock types is shown in Figure 7.6.

Recognizing crustal contamination in igneous rocks Crustal rocks are enriched in both strontium and oxygen isotopes relative to the mantle (Figure 7.6); thus a bivariate Sr–O isotope correlation diagram is a powerful means of recognizing crustal contamination in mantle-derived rocks. There are two contamination mechanisms — the contamination of the source region and contamination of a magma during its ascent through the continental crust. These two types of contamination may be distinguished from one another from the shape of the mixing curve on an Sr–O isotope plot (James, 1981). In the case when the contaminant in a source region is enriched in Sr relative to the mantle and forms a

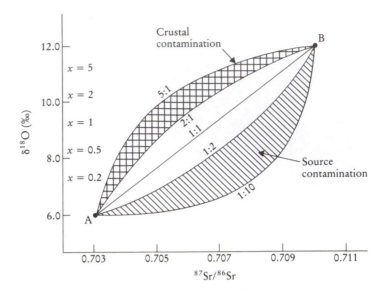

Figure 7.7   Hypothetical mixing diagram (James, 1981) illustrating the effects of source contamination and crustal contamination on Sr and O isotope concentrations in a mantle melt. The mantle source (A) has $^{87}Sr/^{86}Sr = 0.703$ and $\delta^{18}O = 6.0\,^o/_{oo}$; the contaminant (B) has $^{87}Sr/^{86}Sr = 0.710$ and $\delta^{18}O = 12\,^o/_{oo}$. The values on the curves show the relative proportions of (Sr in mantle):(Sr in contaminant). In the case of source contamination (diagonal ruling) the ratio (Sr in mantle):(Sr in contaminant) < 1.0 giving rise to a convex-down mixing curve in which a large change in $^{87}Sr/^{86}Sr$ is produced by a small amount of contaminant. Where there is crustal contamination (cross-hatching) the (Sr in melt):(Sr in crust) ratio may be >1.0 and the mixing curve is convex-upward. x indicates the proportion of contaminant (B) to mantle (A).

relatively small proportion of the whole, then contamination on an $^{87}Sr/^{86}Sr$ vs $\delta^{18}O$ mixing diagram is characterized by the convex-downward curvature of the mixing line (Figure 7.7). This arises because crustal materials are not only enriched in Sr relative to the mantle but also their $^{87}Sr/^{86}Sr$ ratio is greater than that of the mantle and thus dominates any mixture of the two. Oxygen concentrations, however, are broadly similar in all rocks so that there is no massive increase in the oxygen isotope ratio of the derivative melt. The small increase in $\delta^{18}O$ is a simple linear function of the bulk proportion of crustal to mantle materials. This pattern is seen in the data of Magaritz *et al.* (1978) for the lavas of the Banda Arc (Figure 7.8).

In the case when the melt is enriched in Sr relative to the contaminant and the relative proportion of the contaminant is high, then compositions on an Sr–O isotope plot will define a mixing curve with a convex-up curvature (James, 1981). This could be the case for a mantle-derived melt passing through the continental crust. In addition, crustal contamination requires significant changes in major and trace element chemistry and these are expected to correlate with the oxygen and radiogenic isotope ratios.

However, crustal contamination is rarely a simple mixing process and frequently it involves three components — a melt, a precipitating cumulate phase and a

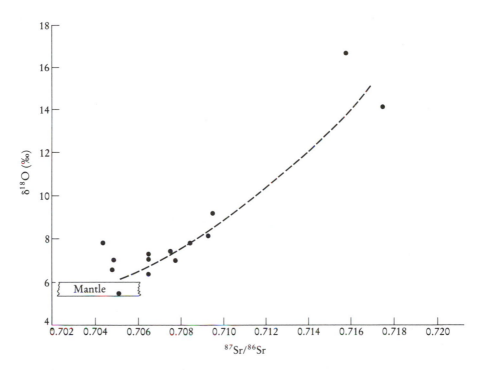

Figure 7.8 Plot of $\delta^{18}O$ vs $^{87}Sr/^{86}Sr$ for whole-rock samples of lavas from the Banda Arc, Indonesia. The range of possible mantle compositions is also shown. The data define a concave-up curve (in part controlled by only two data points) which is suggestive of source contamination (cf. Figure 7.7). Data from Magaritz *et al.* (1978).

contaminant (Taylor, 1980; James, 1981). This is the familiar AFC process first proposed by Bowen (1928). AFC processes can be recognized on a radiogenic isotope–oxygen isotope correlation diagram (Figure 7.9) by a sigmoidal mixing curve which does not extrapolate back to the position of either the source or the contaminant. Details of the modelling are given by Taylor (1980) and an example of the sigmoidal distribution is given by Mauche *et al.* (1989).

Recognizing An igneous system which has not suffered crustal contamination will exhibit the
simple crystal radiogenic isotope characteristics of the source, for radiogenic isotope ratios are not
fractionation in altered by crystal–liquid equilibria such as crystal fractionation. Oxygen isotopes,
igneous rocks on the other hand, do show small changes in isotope ratio with crystal fractionation, although extreme fractionation is required to produce small changes in $\delta^{18}O$. This has been documented by Chivas *et al.* (1982) in a study of a highly fractionated oceanic-arc plutonic suite. In these rocks oxygen isotope ratios range from $\delta^{18}O = 5.4\,°/_{00}$ in gabbros to $\delta^{18}O = 7.2\,°/_{00}$ in an aplite dyke. The increase in $\delta^{18}O$ correlates with an increase in SiO_2 content — a measure of the degree of differentiation. $^{87}Sr/^{86}Sr$ ratios, on the other hand, do not correlate with SiO_2 or with $\delta^{18}O$ values and remain constant within the limits of error of their determination (Figure 7.10).

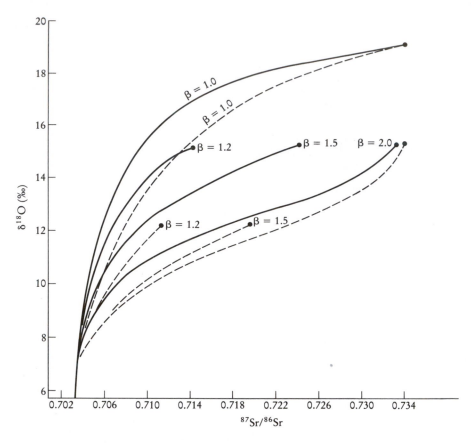

Figure 7.9   Plot of $^{87}Sr/^{86}Sr$ ratio vs $\delta^{18}O$ showing the trajectories that would be followed during assimilation and fractional crystallization in a magma chamber. The initial composition of the melt is $^{87}Sr/^{86}Sr = 0.703$, $\delta^{18}O = 5.7$. The two sets of curves represent two different ratios of Sr in the melt to Sr in the assimilated country rock. The solid lines represent a melt:country rock ratio of 10:1 and the dashed lines a ratio of 5:1. The ratio of cumulates to assimilated country rock is 5:1. Higher values produce curves with lower $\delta^{18}O$ values. β is the ratio of Sr in the cumulates to Sr in the melt, i.e. the bulk partition coefficient of the cumulates. The end-points are placed at 99 % fractional crystallization. (Adapted from Taylor, 1980.)

7.3   Fingerprinting hydrothermal solutions using oxygen and hydrogen isotopes — water–rock interaction

The study of oxygen isotopes in conjunction with the isotopic study of hydrogen has proved to be a very powerful tool in investigating geological processes involving water. When plotted on a bivariate δD vs $\delta^{18}O$ graph, waters from different geological environments are found to have very different isotopic signatures (Figure 7.11).

Hydrogen is a minor component of most rocks and so, excepting when the fluid/rock ratio is very low, the hydrogen isotope composition of rocks and minerals is very sensitive to the hydrogen isotope composition of interacting fluids. Oxygen, on the other hand, comprises 50 % by weight (and in some cases more than 90 % by volume) of common rocks and minerals and so is less sensitive to the oxygen isotope ratio of interacting fluids, except at very high fluid/rock ratios (see Figure 7.12).

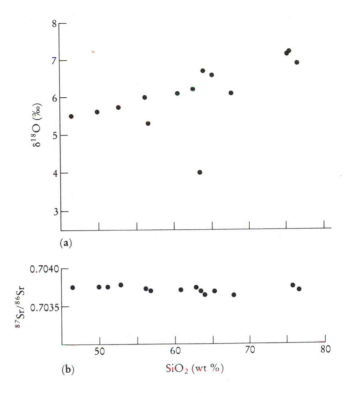

Figure 7.10   (a) Plot showing a positive correlation between $\delta^{18}O$ and SiO_2 for a fractional crystallization-related calc–alkaline gabbro–diorite–tonalite–trondhjemite suite from an oceanic-arc plutonic complex in the Solomon Islands. (b) The same suite of rocks shows no correlation between $^{87}Sr/^{86}Sr$ and SiO_2. (Data from Chivas *et al.*, 1982.)

7.3.1   Hydrogen isotopes

There are two naturally occurring stable isotopes of hydrogen which occur in the following proportions:

$^1H = 99.9844\,\%$
$^2D = 0.0156\,\%$ (deuterium)

Hydrogen isotopes show the largest relative mass difference between two stable isotopes, with the result that there are huge variations in measured hydrogen isotope ratios in naturally occurring materials. In addition, hydrogen isotopes are ubiquitous in nature in the forms H_2O, OH^- and H_2 and as hydrocarbons.

Hydrogen isotopes are measured in parts per thousand relative to the SMOW standard and are calculated in an analogous manner to that for oxygen isotopes (see Eqn [7.1]) and expressed as $\delta D\,^o/_{oo}$. Precision is between 1 and $2\,^o/_{oo}$. δD values for the SLAP standard relative to SMOW are $-428\,^o/_{oo}$. D/H ratios are usually measured on H_2 gas which is produced from the reduction of water at high temperatures.

A summary of δD values for common rock types and waters is given in Figure 7.12. Mantle values are normally in the range -40 to -80 although Deloule *et al.*

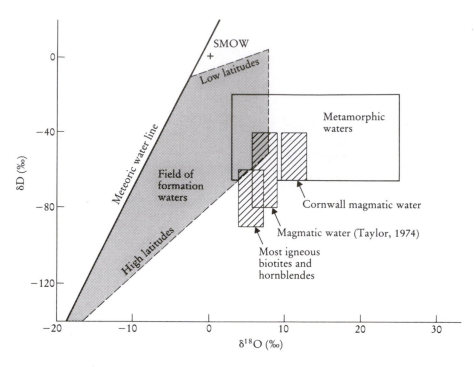

Figure 7.11   Plot of δD vs δ¹⁸O diagram for different water types. The fields of magmatic water and
formation waters are taken from Taylor (1974). The field for igneous hornblendes and
biotites from Taylor (1974) and that of magmatic water from the granites of Cornwall from
Sheppard (1977). The meteoric water line is from Epstein *et al.* (1965) and Epstein (1970).
The metamorphic water field combines the values of Taylor (1974) and Sheppard (1981).

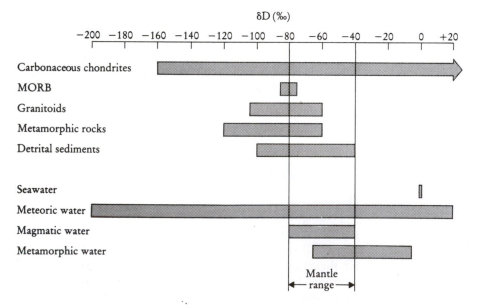

Figure 7.12   Natural hydrogen isotope reservoirs. Data from Taylor (1974), Graham and Harmon (1981),
Kyser and O'Neil (1984), Deloule *et al.* (1991) and Hoefs (1987).

(1991) reported values as low as −125 $^{o}/_{oo}$. The MORB reservoir is thought to have $\delta D = -80 \pm 5\ ^{o}/_{oo}$ (Kyser and O'Neil, 1984).

7.3.2 Calculating the isotopic composition of water from mineral compositions

The isotopic composition of waters from different geological settings can be measured directly as 'fossil' water preserved in fluid inclusions (Ohmoto and Rye, 1974; Richardson *et al.*, 1988). Most commonly, however, 'fossil' water is sampled indirectly and its isotopic composition determined from the isotopic composition of minerals which were in equilibrium with it.

If it can be assumed that there was a close approach to isotopic equilibrium between a given mineral and a hydrothermal solution, then laboratory calibrations of equlibria between rock-forming minerals and water can be used to calculate the isotopic composition of the hydrothermal solution. There are experimental calibrations for both oxygen and hydrogen isotopes (Tables 7.2 and 7.3) allowing the isotopic composition of water to be fully specified. The calculation requires a knowledge of the temperature of equilibration, which may have to be estimated or measured independently by a technique such as fluid inclusion thermometry. An excellent example of this approach is given by Hall *et al.* (1974 — Table 4) in a study of the origin of the water in the formation of the Climax molybdenum deposit, Colorado. In this study the temperature of the hydrothermal fluid was known from fluid inclusion thermometry and the $\delta^{18}O$ and δD composition of the water was calculated from the isotopic composition of muscovite and sericite using the experimental calibrations for muscovite–water. An example of the calculation is given in Box 7.2.

7.3.3 The isotopic composition of natural waters

The isotopic composition of natural waters may be obtained either by direct measurement or by calculation using the method outlined above. Taylor (1974) describes six types of naturally occurring water, the compositions of which are summarized on a δD vs $\delta^{18}O$ diagram (Figure 7.11). The isotopic character of the different types of water described here can be used to trace the origin of hydrothermal solutions.

(a) Meteoric water This shows the greatest variation of all natural waters. δD–$\delta^{18}O$ variations define a linear relationship, the meteoric water line, which may be represented by the expression:

$$\delta D\ ^{o}/_{oo} = 8\delta\ ^{18}O + 10 \tag{7.13}$$

(Taylor, 1979). The $\delta^{18}O$ and δD values for meteoric water vary according to latitude. Values are close to zero for meteoric waters on tropical oceanic islands whereas at high latitudes in continental areas $\delta^{18}O$ values are as low as −20 to −25

Box 7.2

Calculation of the isotopic composition of water in equilibrium with muscovite from the Climax molybdenum deposit, Colorado

Data

$\delta^{18}O_{muscovite}$ +7.4 ‰
$\delta D_{muscovite}$ = −91 ‰

Temperature = 500 °C (from fluid inclusion thermometry)
(Data from Hall *et al.*, 1974 — Table 4, sample CL31-70)

Calculation of oxygen isotope composition of the water

The equation for muscovite–water (O'Neil and Taylor, 1967) is:

$$1000 \ln \alpha = -3.89 + 2.38(10^6/T^2)$$

At 500 °C,

$$1000 \ln \alpha = 0.0931$$

Since $\Delta_{muscovite-water}$ = 1000 ln α

$$\delta^{18}O_{muscovite} - \delta^{18}O_{water} = 0.0931$$

$$+7.4 - \delta^{18}O_{water} = 0.0931$$

$$\delta^{18}O_{water} = 7.3 ‰$$

Calculation of hydrogen isotope composition of the water

The equation for muscovite–water (Suzuoki and Epstein, 1976) is:

$$1000 \ln \alpha = 19.1 - 22.1(10^6/T^2)$$

At 500 °C

$$1000 \ln \alpha -17.89$$

Since $\Delta_{muscovite-water}$ = 1000 ln α,
$$\delta D_{muscovite} - \delta D_{water} = -17.89$$
$$-91 - \delta D_{water} = -17.89$$
$$\delta D_{water} = -73.1 ‰$$

and δD values range between −150 to −250. Both the extreme variation and the linear relationship arise from the condensation of H_2O from the Earth's atmosphere. The extreme variation reflects the progressive lowering of ^{18}O in an air mass as it leaves the ocean and moves over a continent. The linearity of the relationship indicates that fractionation is an equilibrium process and that the fractionation of D/H is proportional to $^{18}O/^{16}O$.

(b) Ocean water Present-day ocean water is very uniform in composition with values of $\delta^{18}O = 0$ ‰ and $\delta D = 0$ ‰. The only exception to this is from areas where there have been high rates of evaporation, e.g. the Red Sea, where there are elevated values of $\delta^{18}O$ and δD, or from areas where seawater is diluted with fresh water. Muehlenbachs and Clayton (1976) suggested that the oxygen isotopic composition of ocean water is buffered by exchange with the ocean crust, a view which is strongly supported by the study of Gregory and Taylor (1981) on the distribution of oxygen isotopes in the Semail ophiolite, Oman (Figure 7.16).

Table 7.2   Constants for the fractionation of oxygen isotopes between minerals and water according to the equation $1000 \ln \alpha_{mineral-water} = A + B(10^6/T^2)$

Mineral	Experimental range (°C)	A	B	Reference
Baryte	100–350	−6.79	3.00	Friedman and O'Neil (1977)
Calcite	0–700	−3.39	2.78	O'Neil *et al.* (1969)
Dolomite	252–295	−3.24	3.06	Matthews and Katz (1977)
Quartz	200–500	−3.40	3.38	Clayton *et al.* (1972)
	500–750	−1.96	2.51	Clayton *et al.* (1972)
	250–500	−3.31	3.34	Matsuhisa *et al.* (1979)
	500–800	−1.14	2.05	Matsuhisa *et al.* (1979)
Alkali feldspar	350–800	−3.41	2.91	O'Neil and Taylor (1967)
	500–800	−3.70	3.13	Bottinga and Javoy (1973)
Albite	400–500	−2.51	2.39	Matsuhisa *et al.* (1979)
	500–800	−1.16	1.59	Matsuhisa *et al.* (1979)
Anorthite	350–800	−3.82	2.15	O'Neil and Taylor (1967)
	400–500	−2.81	1.49	Matsuhisa *et al.* (1979)
	500–800	−2.01	1.04	Matsuhisa *et al.* (1979)
Plagioclase	350–800	(−3.41–0.14An)	(2.91–0.76An)	O'Neil and Taylor (1967)
Feldspar	500–800	−3.70	(3.13–1.04An)	Bottinga and Javoy (1973)
	(where An is mole fraction of anorthite)			
Muscovite	400–650	−3.89	2.38	O'Neil and Taylor (1967)
	500–800	−3.10	1.90	Bottinga and Javoy (1973)
Rutile	575–775	1.46	4.10	Addy and Garlick (1974)
Magnetite	500–800	−3.70	−1.47	Bottinga and Javoy (1973)
Kaolinite	Uncertain	−2.87	2.5	Land and Dutton (1978)
Chlorite	66–175			Savin and Lee (1988)

For $[Mg_{2.5} \, Fe_{0.5} \, (OH)_6][Al_{1.5} \, Fe_{1.5}][Al,Si_3O_{10}][OH_2]$:

$$-11.97 + 2.67x + 2.93x^2 - 0.415x^3 + 0.037x^4$$

[where $x = 10^3/T$]

Table 7.3   Constants for the fractionation of hydrogen isotopes between minerals and water according to the equation $1000 \ln \alpha_{mineral-water} = A + B(10^6/T^2)$

Mineral	Experimental temperature range (°C)	A	B	Reference
Muscovite	450–800	19.1	−22.1	Suzuoki and Epstein (1976)
Biotite	450–800	−2.8	−21.3	Suzuoki and Epstein (1976)
Hornblende	450–800	7.9	−23.9	Suzuoki and Epstein (1976)
	350–850*	−23.1 ± 2.5		Graham *et al.* (1984)
	850–950*	1.1	−31.0	Graham *et al.* (1984)
Tremolite	350–650	−21.7		Graham *et al.* (1984)
	650–950	14.9	−31.0	Graham *et al.* (1984)
Actinolite	400	−29.0		Graham *et al.* (1984)
Arfvedsonite	Uncertain	−52.0		Graham *et al.* (1984)
Kaolinite/Dickite	100–250	0.972–0.985		Marumo *et al.* (1980)
Sericite	100–250	0.973–0.977		Marumo *et al.* (1980)
Chlorite	100–250	0.954–0.987		Marumo *et al.* (1980)
Zoisite	280–650	−27.73	−15.7	Graham *et al.* (1980)
Epidote	<300	−138.8	29.2	Graham *et al.* (1980)
	300–650	−35.9 ± 2.5		Graham *et al.* (1980) Suzuoki and Epstein (1976)
All minerals	$1000 \ln\alpha_{mineral-water} = 28.2 - 22.4(10^6/T^2) + (2X_{Al}-4X_{Mg}-68X_{Fe})$ where X_{Al} etc., is the mole fraction of Al in biotite, muscovite or hornblende.			

* Ferroan pargasitic hornblende.

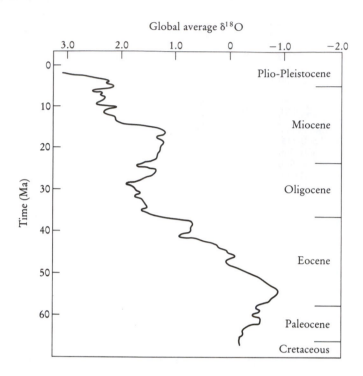

Figure 7.13   The change in the δ¹⁸O composition of the oceans throughout the Tertiary. The record is based upon the composition of benthic foraminifera (after Miller and Fairbanks, 1985).

Less certain is the isotopic composition of seawater in the past. There is evidence from the oxygen isotope composition of marine carbonates that there were global changes in the isotopic chemistry of the oceans during the Tertiary (Figure 7.13). These changes are thought to have been brought about by the storage of isotopically light oxygen in ice in the polar regions and are known in sufficient detail to allow the composition curve to be used as a stratigraphic tool. Similar changes are reported in the Ordovician from the oxygen isotope chemistry of unaltered parts of brachiopod shells (Marshall and Middleton, 1990), although one of the uncertainties in models of this kind, particularly prior to the Plio-Pleistocene, is the extent to which diagenetic change has influenced the calculated compositions of former ocean waters (Williams *et al.*, 1988).

(c) Geothermal water   Modern geothermal water is meteoric in origin but isotopic compositions are transposed to higher δ¹⁸O values through isotopic exchange with the country rocks. δD values are the same as in the parent meteoric water or slightly enriched. Similarly, ocean-floor geothermal systems have δ¹⁸O values of between +0.37 and +2.37, close to the value for unmodified seawater (Campbell *et al.*, 1988).

(d) Formation water   Formation waters from sedimentary basins show a wide range in δ¹⁸O and δD values (Figure 7.11). Individual basins have water compositions which define a linear trend representing mixing either between meteoric water and

water from another source such as trapped seawater, or between meteoric water and the country rock.

(e) Metamorphic water   Several attempts have been made to calculate the δD and δ^{18}O values of water in equilibrium with metamorphic minerals over a range of metamorphic grades (Taylor, 1974; Rye *et al.*, 1976; Sheppard, 1981). A combination of these values gives a metamorphic water 'box' with δ^{18}O values between +3 and +25 °/$_{\infty}$ and with δD values between −20 and −65 °/$_{\infty}$. (Figure 7.11).

(f) Magmatic water   Calculating the composition of magmatic water is difficult because many magmas interact with groundwater. However, primary magmatic waters calculated by Taylor (1974) define a region on δD vs δ^{18}O diagrams between δD values of −40 and −80 and δ^{18}O values of +5.5 and +9.0. Sheppard (1977), however, showed that the magmatic waters associated with the Permian granites of south-west England, produced by intracrustal melting, plot in a different field between δD values of −40 to −65, and δ^{18}O values of +9.5 to +13 (Figure 7.11).

7.3.4   Quantifying water/rock ratios

Water–rock interaction can vary between two extremes. When the water/rock ratio is small and the δ^{18}O in the rock dominates the system it is the fluid composition which is changed, as happens in geothermal systems. On the other hand, when the water/rock ratio is large and the δ^{18}O of the water dominates, the δ^{18}O value of the rock is modified. Taylor (1974, 1977) derived mass balance equations from which the water/rock ratio may be calculated from δ^{18}O values. For a closed system, from which none of the water is lost, the water/rock (W/R) ratio, integrated over the lifetime of the hydrothermal system, is:

$$\text{W/R}_{\text{closed}} = \frac{\delta^{18}O_{\text{rock}}^{\text{final}} - \delta^{18}O_{\text{rock}}^{\text{initial}}}{\delta^{18}O_{\text{fluid}}^{\text{initial}} - \delta^{18}O_{\text{fluid}}^{\text{final}}} \qquad [7.14]$$

This is the effective water/rock ratio, which can differ from the actual water/rock ratio depending upon the efficiency of the exchange reaction. The initial value for the rock is obtained from 'normal' values for the particular rock type (see Box 7.3), or from an unaltered sample of the rock suite being analysed. The final value for the rock is the measured value. The initial value for the fluid is assumed (for example, modern seawater) or in the case of meteoric water calculated from the D/H ratio of the alteration assemblage and the meteoric water equation. The composition of the final fluid can be calculated from the mineralogy of the altered rock. This is sometimes done by using the approximation that δ^{18}O for the rock equals δ^{18}O for plagioclase feldspar (An$_{30}$), for feldspar is generally an abundant mineral in most rocks and it exhibits the greatest rate of exchange of ^{18}O with an external fluid phase. Provided that the temperature can be independently estimated, then the feldspar–water fractionation equation can be used to calculate the water composition (see Box 7.3).

The equation for an open system through which the water makes only a single pass is given, Taylor (1977) by:

$$\text{W/R}_{\text{open}} = \ln(\text{W/R}_{\text{closed}} + 1) \qquad [7.15]$$

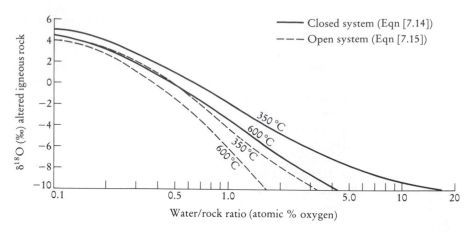

Figure 7.14   Plot of $\delta^{18}O$ values in a hydrothermal rock calculated from the open-system water/rock ratio equation (Eqn [7.15]) and the closed-system water/rock ratio equation (Eqn [7.14]) (Taylor, 1974). The model assumes an initial $\delta^{18}O$ value of +6.5 in the rock and an initial $\delta^{18}O$ value of −14 in the water; curves are shown for 350 °C and 600 °C.

It is likely that the behaviour of any given hydrothermal system will be somewhere between the two extremes. An example of how Eqns [7.14] and [7.15] might be used is given in Box 7.3 and depicted in Figure 7.14.

Box 7.3

Calculation of water/rock ratio from the equations of Taylor (1974, 1977)

Data

Initial rock composition: $\delta^{18}O = 6.5\,\%_{00}$
Final rock composition: $\delta^{18}O = -4.0\,\%_{00}$
Initial fluid composition: $\delta^{18}O = -14.0\,\%_{00}$

Calculation of final fluid composition

The equation for plagioclase (An_{30})–water exchange, from Table 7.2, is:

$$1000 \ln \alpha_{fsp-water} = (-3.41 - 0.14An) + 2.91 - 0.76An)(10^6/T^2)$$
$$= -3.52 + 2.682(10^6 T^2)$$

At 500 °C
$$\Delta_{fsp-water} = 0.97$$

Assuming $\delta^{18}O_{fsp} \sim \delta^{18}O_{whole\ rock}$ (final composition),

$$\delta^{18}O_{water} \sim -4.0 - 0.97 = -4.97$$

Final water composition: $\delta^{18}O = -4.97\,\%_{00}$

Water/rock ratio calculation

From the closed system equation (Eqn [7.14]):

$$W/R_{closed} = \frac{\delta^{18}O_{rock}^{final} - \delta^{18}O_{rock}^{initial}}{\delta^{18}O_{fluid}^{initial} - \delta^{18}O_{fluid}^{final}}$$

$$= \frac{-4 - 6.5}{-14 - (-4.97)} = 1.16$$

From the single-pass open system equation (Eqn [7.15]):
$$W/R_{open} = \ln(W/R_{closed} + 1) = 0.77$$

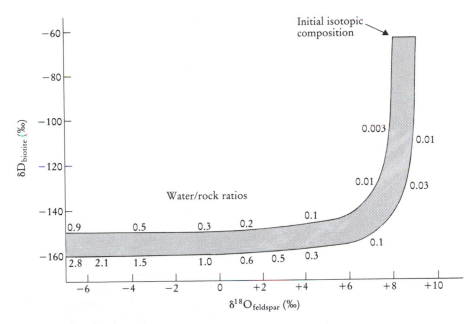

Figure 7.15 δD vs δ^{18}O diagram showing the isotopic change in D in biotite and ^{18}O in feldspar in a granodiorite undergoing isotopic exchange with groundwater. The plotted curve shows the range of δD and δ^{18}O values with changing water/rock ratios. The curve was calculated for an initial feldspar composition of δ^{18}O +8.0 to +9.0, biotite δD −65.0 and groundwater δD −120, δ^{18}O −16. The fractionation of D between biotite and water at 400–450 °C is given by $\Delta_{biotite-water}$ = −30 and −40; the fractionation of ^{18}O between feldspar and water at 400–450 °C is given by $\Delta_{feldspar-water}$ = 2.0. A small water/rock ratio has a dramatic effect on the isotopic composition of D in biotite whereas higher water/rock ratios affect both δD in biotite and δ^{18}O in feldspar. The upper boundary of the curve shows the water/rock ratio as weight units, the lower boundary as volume units. (Adapted from Taylor, 1977.)

The effects of differing water/rock ratios on the isotopic composition of a granodiorite are shown on a δD vs δ^{18}O plot in Figure 7.15, using the isotopic composition of biotite to monitor change in δD and the isotopic composition of feldspar to monitor change in δ^{18}O (Taylor, 1977). At small water/rock ratios (up to 0.1 rock volumes) oxygen isotopic compositions are virtually unchanged whilst δD values are reduced by about 100 $^o/_{oo}$. As the water/rock ratio increases, the δ^{18}O value decreases rapidly at almost constant δD values.

7.3.5 Examples of water–rock interaction

The studies of Taylor and coworkers into water–rock interaction have wide applicability in many fields of geochemistry. In general terms the sensitivity of oxygen isotopes to hydrothermal solutions means that they are an excellent tool for detecting hydrothermal alteration in otherwise fresh rocks (see for example Rautenschlein *et al.*, 1985). More specifically, the origin of the water may be identified and its volume relative to the country rock quantified. Some examples are briefly reviewed.

(a) Interaction between igneous intrusions and groundwater In a number of pioneering studies Taylor and his coworkers showed that high-level igneous intrusions are frequently associated with hydrothermal convective systems (see reviews by Taylor, 1977, 1978). They found that the country rocks surrounding such intrusions are massively depleted in ^{18}O and D relative to 'normal' values and that the minerals both in the intrusion and the country rock are isotopically out of equilibrium for magmatic values. They concluded that the isotopic effects were due to the interaction with meteoric water and proposed that the intrusion acted as a heat engine which initiated a hydrothermal convection cell in the groundwater of the enclosing country rocks. Water/rock ratios were found to vary from <<1.0 to about 7.0.

These studies offer an important insight into attempts to establish the original isotopic composition of igneous rocks, for clearly isotopically disturbed rocks must first be screened out. Similarly, caution needs to be exercised in interpreting the results of isotopic thermometry in igneous plutons emplaced at a high level in the crust. Further, the fact that some igneous rocks have totally re-equilibrated with groundwater injects a word of caution into radiogenic isotope studies.

(b) Interaction between ocean-floor basalt and seawater A large number of studies have shown that the rocks of the ocean-floor, now preserved as ophiolites, have undergone massive exchange with hydrothermal solutions, which were most probably seawater. Stakes and O'Neil (1982), for example, have calculated water/rock ratios in excess of 50 for recent basalts at the East Pacific rise. It is proposed that a hydrothermal convective system draws in cold seawater, which is heated adjacent to an active magma chamber. The heated seawater thus produced is thought to be the source of the hydrothermal solutions responsible for the formation of the massive sulphide deposits found in ophiolites. This hypothesis has been confirmed by the comparatively recent and exciting discovery of very high-temperature springs venting onto the ocean floor at mid-ocean ridges and giving rise to sulphide-rich deposits (Section 7.5.4).

Schiffman *et al.* (1987) showed that in the Troodos ophiolite in Cyprus, altered basalts have reacted with fluids which have $\delta^{18}O$ values of close to the seawater value of $0\,°/_{\infty}$ at temperatures of between 310 and 375 °C. Trace element concentrations were changed from the unaltered basalt values and fluid/rock ratios were calculated to be between 10 and 20. In a detailed study of the Semail ophiolite in Oman, Gregory and Taylor (1981) showed that upper layers were enriched in $\delta^{18}O$ whereas the lower gabbro layers were depleted in $\delta^{18}O$ relative to average ocean crust (Figure 7.16). This cross-over of values is due to the temperature-dependent partitioning of oxygen isotopes between silicate minerals and water. Gregory and Taylor (1981) also made the important observation that the net exchange of ^{18}O between seawater and the ocean crust in the Semail ophiolite was zero. This suggests that the $\delta^{18}O$ composition of seawater is buffered by the composition of the ocean floor, a view confirmed by the recent measurements of Campbell *et al.* (1988).

(c) Water–rock interaction in metamorphic rocks Stable isotope studies are used to determine two different features of water–rock interaction in metamorphic rocks. On the one hand, oxygen isotope studies can determine the pattern of fluid movement, whereas the combined study of hydrogen and oxygen isotopes can be

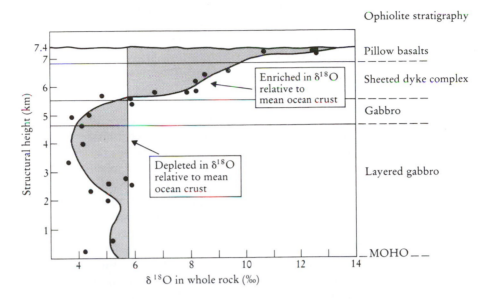

Figure 7.16   Whole-rock $\delta^{18}O$ profile through the Semail ophiolite, Oman, showing the relative enrichment and depletion in $\delta^{18}O$ as a function of the ophiolite stratigraphy. Individual measurements are shown as solid circles. Modified from Gregory and Taylor (1981).

used to determine the isotopic composition of the fluid originally in equilibrium with the metamorphic rock.

The presence or absence of water is an important control on isotopic equilibrium in metamorphic rocks and facilitates the equilibration of stable isotopes both between mineral pairs and between rock types. For example, the presence or absence of a fluid phase explains the apparent discrepancy between studies which show that interbedded sedimentary rocks at a range of metamorphic grades have not attained isotopic equilibrium during metamorphism (see for example Bottrell *et al.*, 1990; Valley and O'Neil, 1984), whereas in other metamorphic terrains the rock types show great uniformity in their oxygen isotopic compositions suggestive of isotopic equilibration (Chamberlain and Rumble, 1988; Rumble and Spear, 1983; Wickham and Taylor, 1985). Differences of this type can be further explained by considering the effects of differing water/rock ratios and patterns of fluid flow in the metamorphic terrain.

Water/rock ratios are very variable in metamorphic rocks and range between massive influxes of externally derived water (Wickham and Taylor, 1985) to very small values (water/rock ratios of between 0.1 and 0.0) in rocks at granulite grade (Valley and O'Neil, 1984). As a generalization, the water/rock ratio might be expected to be progressively reduced by dehydration reactions during prograde metamorphism. Fluid flow may be **pervasive**, in which the fluid moves independently of lithological or structural control, or **channelized**, in which water leaves the rock through cracks and fissures (Valley and O'Neil, 1984). Maps of fluid flow postulated on the basis of oxygen isotope studies are given by Chamberlain and Rumble (1985), Rumble and Spear (1983) and Wickham and Taylor (1985).

Oxygen isotopes alone, however, cannot specify the exact nature of an oxygen-rich fluid, for H_2O, CO_2 and a silicate melt are all oxygen-rich reservoirs. It is therefore necessary to study hydrogen isotopes also in order to specify the fluid as water and to identify its origin. This approach was used by Wickham and Taylor (1985), who showed from the isotopic composition of muscovites that the homogenization of pelites in the French Pyrenees had been achieved by the influx of a large amount of seawater which had exchanged its oxygen isotopes with the country rock and had a measured composition at about 450 °C close to that of basinal formation waters.

(d) Water–rock interaction during the formation of hydrothermal ore deposits The subject of 'fingerprinting' waters used in the formation of hydrothermal ore deposits is discussed at some length in *Economic Geology* (1974, No. 6), in which there is also an excellent review by Taylor (1974). As might be expected, there is a huge variation in the types of water involved in the formation of hydrothermal ore deposits. Porphyry copper deposits, for example, appear to have formed through the interaction of magmatic water and heated groundwater of meteoric origin, whereas massive sulphides of the Cyprus type are clearly seawater-related. Mississippi Valley-type Pb–Zn deposits have formed from heated formation waters and are identical in salinity and isotopic composition to oil-field brines. Kerrich and Fryer (1979), in a study of gold mineralization in the Archean Abitibi greenstone belt in Canada, concluded from the high calculated $\delta^{18}O$ of water in equilibrium with quartz that the water was originally metamorphic and that the source of the mineralizing fluids was dehydration during prograde metamorphism.

(e) Diagenesis of clastic sediments The fractionation of ^{18}O between diagenetic minerals and sedimentary pore waters can be used to calculate either water composition or, if the water composition is known, temperature of mineral growth. Ayalon and Longstaff (1988) show how a sequence of diagenetic minerals in an upper Cretaceous sandstone from Alberta was used to track the fluid evolution and thermal history of the sandstone from burial to uplift. They cite mineral–water equations for use with chlorite, illite, smectite, calcite, kaolinite and quartz (see also Table 7.3). A similar study by Girard *et al.* (1989) combines fluid inclusion temperature data with K/Ar age determinations and the calculated $\delta^{18}O$ values of the pore fluids to estimate the time–temperature–burial history of lower Cretaceous arkoses.

7.4   Using carbon isotopes

Carbon has two stable isotopes with the following abundances:

^{12}C = 98.89 %
^{13}C = 1.11 %

Measurements are made relative to a standard belemnite sample known as PDB — *belemnitella americana* from the Cretaceous Peedee formation, South Carolina. This standard is used because its ^{13}C and ^{18}O values are close to those of average marine limestone. The original material of the PDB standard is now exhausted and current standard materials are a carbonatite (NBS–18) and a marine limestone (NBS–19). The ratio $^{13}C/^{12}C$ is measured in parts per thousand in a similar way to that of oxygen isotopes:

$$\delta^{13}C \,^{0}/_{00} = \left[\frac{^{13}C/^{12}C \text{ (sample)} - {}^{13}C/^{12}C \text{ (standard)}}{^{13}C/^{12}C \text{ (standard)}} \right] \times 1000 \quad [7.16]$$

Carbon isotopes are measured as CO_2 gas and precision is normally better than 0.1 $^{0}/_{00}$. The CO_2 is liberated from carbonates with >103 % phosphoric acid or by thermal decomposition. Organic compounds are normally oxidized at very high temperatures in a stream of oxygen or with an oxidizing agent such as CuO.

7.4.1   The distribution of carbon isotopes in nature

Carbon occurs in nature in its oxidized form (CO_2, carbonates and bicarbonates), as reduced carbon (methane and organic carbon) and as diamond and graphite. The range of carbon isotope compositions in natural substances is summarized in Figure 7.17. Meteorites have a wide range of carbon isotope compositions. For example, carbonaceous chondrites have bulk $\delta^{13}C$ compositions in the range −25 to 0 (Kerridge, 1985). Mantle values, determined from the isotopic study of carbonatites, kimberlites and diamonds, are in the narrow range $\delta^{13}C$ = −3 to −8 $^{0}/_{00}$ and a mean value of about −6 $^{0}/_{00}$ is often used for the mantle. MORB has a mean value of $\delta^{13}C$ = −6.6 $^{0}/_{00}$ (Exley *et al.*, 1986). Seawater has by definition a $\delta^{13}C$ value of close to 0 $^{0}/_{00}$ and marine carbonate has a narrow range of values between −1 and +2 $^{0}/_{00}$ whilst marine bicarbonate values are between −2 and 1 $^{0}/_{00}$. Ancient seawater, however, has not had such a constant composition and excursions from the present-day value are discussed further below (Section 7.4.4).

Biologically derived (organic) carbon is isotopically light, i.e. depleted in $\delta^{13}C$. The conversion of inorganic carbon through a CO_2-fixing mechanism into living, organic carbon entails the preferential concentration of the light ^{12}C isotope in organic carbon. This process is chiefly controlled by reaction kinetics which favour the light isotope. The net effect of the fractionation is that relative to mantle-derived carbon ($\delta^{13}C \sim$ −6 $^{0}/_{00}$) there are two complementary reservoirs. Biological materials on the one hand are strongly depleted in $\delta^{13}C$ (−20 to −30 $^{0}/_{00}$; the mean value of the terrestrial biomass is −26 ± 7 $^{0}/_{00}$ according to Schidlowski, 1987) whereas seawater and marine carbonates ($\delta^{13}C\sim$ 0) are enriched. Methane is the most depleted of all carbon compounds and is commonly formed in nature either by the anaerobic fermentation of organic matter or by the thermal degradation of petroleum or kerogen at temperatures greater than 100 °C. Methane of biological origin has $\delta^{13}C$ values of about −80 $^{0}/_{00}$.

Controls on the
fractionation of
carbon isotopes

The fractionation of carbon isotopes is controlled by both equilibrium and kinetic processes. Equilibrium fractionation processes are illustrated in Figure 7.18, from which it can be seen that in many cases the fractionation of $\delta^{13}C$ is strongly temperature-dependent. However, it should be noted that one common equilibrium process — that of dissolution and reprecipitation — does not substantially fractionate carbon isotopes.

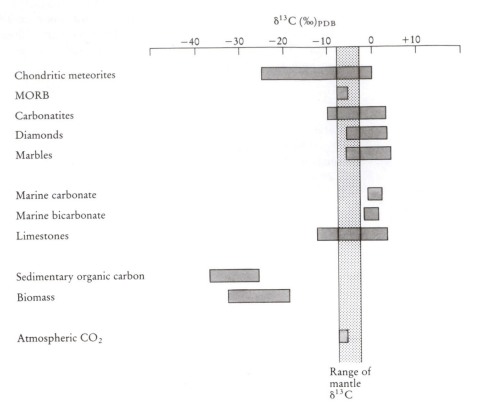

Figure 7.17 Natural $\delta^{13}C$ reservoirs. The ranges of $\delta^{13}C$ values in natural, carbon-bearing samples. Data from Kerridge (1985), Exley *et al.* (1986), Field and Fifarek (1986), Hoefs (1987) and Schidlowski (1987).

Kinetic fractionation is important in biological systems such as the fixing of CO_2 as organic carbon and the evolution of methane from anaerobic fermentation of organic matter during diagenesis. In these cases the fractionation is controlled by reaction rate and the greater readiness of the lighter isotope to react. Where the reaction involves both carbon and oxygen, the kinetic effect will influence the isotopes of both elements in a similar way and a correlation between $\delta^{18}O$ and $\delta^{13}C$ is expected.

7.4.2 Combined oxygen and carbon isotope studies of carbonates — $\delta^{18}O$ vs $\delta^{13}C$ plots

A combined study of carbon and oxygen isotopes in carbonates is a powerful means of distinguishing between carbonates of different origins. Figure 7.19 shows the compositions of terrestrial and meteoritic carbonates plotted on a $\delta^{18}O$–$\delta^{13}C$ diagram. Compositions are plotted relative to the PDB standard, although the SMOW standard is increasingly used for $\delta^{18}O$ measurements in carbonates. The conversion from PDB to SMOW is given in Eqns [7.9] and [7.10] but for the sake of clarity it is worth restating that

$\delta^{18}O$ in *marine carbonate* on the PDB scale is zero, and
$\delta^{18}O$ in *seawater* on the SMOW scale is also zero.

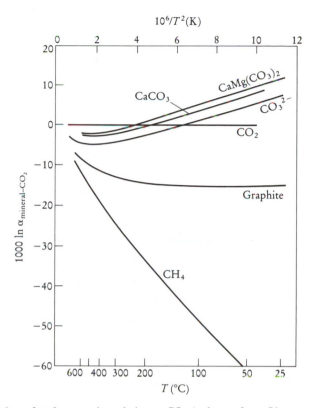

Figure 7.18   Fractionation of carbon species relative to CO_2 (redrawn from Ohmoto and Rye, 1979).

The two are related by the equation of Epstein *et al.* (1953), given as Eqn [7.12] and the two are in equilibrium at approximately 16.9 °C.

Limestone   The $\delta^{18}O$–$\delta^{13}C$ diagram is particularly useful in understanding the processes
diagenesis   involved in the formation of sedimentary carbonates and especially the process of
limestone diagenesis. A study of the carbon isotopes allows the origin of carbon in
the carbonate to be determined and can distinguish between marine, organic and
methane-related carbon (Coleman and Raiswell, 1981). A study of oxygen isotopes
in sedimentary carbonates can be used to determine the origin of fluids in
equilibrium with the carbonates, and provide an estimate of the temperature of
carbonate formation using the Epstein *et al.* (1953) thermometer (Section 7.2.2).
Temperature determinations may give the original seawater temperature or the
temperature of diagenesis, although care should be taken to establish chemical
equilibrium (McConnaughey, 1989). Fluid studies allow the composition of pore
waters in equilibrium with calcite cements to be calculated, provided the
temperature of formation is known.

Hudson (1977), in an excellent review of the application of carbon and oxygen
isotopes to the process of limestone lithification, points out that the different
crystalline forms of carbonate have grown at different times. Hence, the most
fruitful approach to understanding limestone lithification is through the isotopic

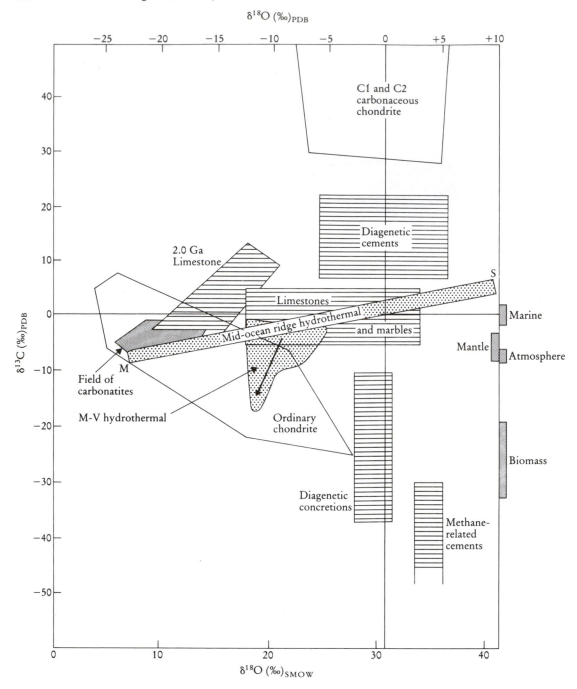

Figure 7.19 $\delta^{18}O$ vs $\delta^{13}C$ plot showing the composition of carbonates from a variety of environments. Note that $\delta^{18}O$ is plotted relative to both the SMOW and PDB scales. The isotopic composition of a number of different carbon reservoirs is plotted along the right-hand side of the diagram. The values for sedimentary carbonates (horizontal rule) are from Hudson (1977) and Baker and Fallick (1989); hydrothermal calcites (stippled ornament) from the mid-ocean ridges show mixing between mantle-derived carbon (M) and seawater carbon (S) (Stakes and O'Neil, 1982); in the field of hydrothermal calcites from Mississippi Valley-type deposits (M-V hydrothermal) the arrow shows the direction of younging (Richardson *et al.*, 1988). Chondrite compositions (unornamented) are from Wright *et al.* (1988). The field of carbonatites is from Deines and Gold (1973).

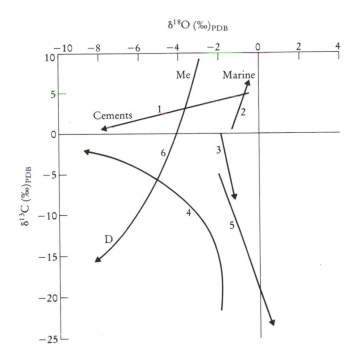

Figure 7.20 $\delta^{18}O$ vs $\delta^{13}C$ plot showing the evolution of limestones and calcareous mudstones during lithification. The trends are as follows: 1, lithification of limestone; compositions lie on a mixing line between original seawater compositions (marine) and carbonate cements (cements); 2, lithification of a modern marine limestones showing enrichment in $\delta^{13}C$ accompanied by a slight decrease in $\delta^{18}O$ — the enrichment in isotopically heavy carbon may be due to a carbon residue remaining after photosynthesis; 3, interaction with isotopically light carbon of organic origin, most probably through interaction with soil-derived CO_2; 4, compositional change in calcareous concretions in mudstone due to methane formation and interaction with pore waters; 5, carbonate cements in clastic sediments containing carbon of methanogenic origin; 6, compositional change in late diagenetic cements in mudstone in the methanogenesis (Me) and decarboxylation (D) zones of progressive anoxic diagenesis. (Data from Hudson, 1977 and Scotchman, 1989.)

analysis of the different generations of carbonate. This allows the construction of an evolutionary pathway on a $\delta^{18}O$–$\delta^{13}C$ diagram and shows the $\delta^{18}O$–$\delta^{13}C$ history of the rock. Traditionally, the separation of finely crystalline intergrowths has been made using a dentist's or jeweller's drill. The more recent application of the laser microprobe, however, offers even greater spatial resolution (Smalley *et al.*, 1989). A number of 'typical' limestone isotopic histories are illustrated in Figure 7.20.

Allan and Matthews (1982) show that the interaction between limestone and meteoric water can be recognized on a $\delta^{18}O$–$\delta^{13}C$ plot. A characteristic signature of such a process is the wide spread of $\delta^{13}C$ values combined with a narrow range of $\delta^{18}O$ values. This pattern arises because the $\delta^{13}C$ value of recrystallized carbonate comes from a combination of two carbon reservoirs (soil-derived CO_2 and the original carbonate) whereas oxygen is dominated by a single reservoir, meteoric water.

Mixing-zone carbonates, formed in the groundwater zone where seawater and meteoric water mix, typically show a marked positive covariation between $\delta^{18}O$ and

δ^{13}C for a single generation of carbonate cement (Allan and Matthews, 1982). The covariation results from the mixing of waters with different compositions — meteoric water and seawater. Meteoric water is depleted in ^{18}O relative to its seawater source, and once it has passed through soil is also enriched in ^{12}C. Searl (1988, 1989) utilized this relationship to identify mixing-zone dolomites in the lower Carboniferous limestones of south Wales.

Hydrothermal Hydrothermal calcites formed by water–rock interaction at a mid-ocean ridge show
calcite a wide range of compositions on a δ^{18}O–δ^{13}C plot (Figure 7.19) chiefly as a function of differing water/rock ratios (Stakes and O'Neil, 1982). At one extreme calcite in a greenstone breccia has mantle-like δ^{13}C values and formed in a rock-dominated environment (low water/rock ratio) at high temperature (145–170 °C). Low-temperature vein calcites have seawater δ^{13}C values and represent a large volume of seawater interacting with the host basalts.

Calcites associated with Mississippi Valley-type lead–zinc mineralization in Carboniferous limestone in Illinois show a marked decrease in δ^{13}C during their growth coupled with only a small decrease in δ^{18}O (Figure 7.19). These changes indicate that early calcites were very similar in composition to carbonates in the limestone, whilst late in the evolution of the hydrothermal system, fluids associated with the degradation of organic carbon become important (Richardson *et al.*, 1988).

7.4.3   The δ^{13}C composition of seawater

Seawater has, by definition, a δ^{13}C value close to 0 $^{0}/_{00}$ and marine carbonate has values between −1 and +2 $^{0}/_{00}$. Ancient seawater, however, has not had such a constant composition and excursions from present-day values have been noted in the Phanerozoic record by Veizer *et al.* (1986) and Hoffman *et al.* (1991). The largest are of the order of 6 $^{0}/_{00}$ (PDB) in the Permian. A smaller excursion (4.8 $^{0}/_{00}$) noted in the Cretaceous (Hart and Leary, 1989) is used as stratigraphic marker. There is a very precise record for the Tertiary based upon bulk carbonate values from the ocean floors which is also used for stratigraphic correlation (Williams *et al.* 1988).

Some anomalous δ^{13}C values for seawater represent local conditions, whereas others are thought to represent global anoxic events in which there is increased deposition of light organic carbon into black shales and a corresponding increase in δ^{13}C in seawater bicarbonate. Baker and Fallick (1989) recently described a massive positive excursion of δ^{13}C (13 $^{0}/_{00}$ PDB) during the Proterozoic (*ca* 2.0 Ga) which appears to be of world-wide significance (Figure 7.19). This event may document an increase in the population of photosynthesizing bacteria.

7.4.4   Biogeochemical evolution

Schidlowski (1987, 1988) has shown how the study of carbon isotopes in the two principal forms of sedimentary carbon — carbonate and organic carbon — can be used to trace ancient biological activity through the geological record. His argument is based upon the observation that isotopically light, organic carbon and isotopically heavy carbonates are complementary reservoirs which have originated from the biological fractionation of mantle carbon. Both reservoirs are therefore responsive to changes in the level of biological activity. On a graph showing the ranges and mean values of $\delta^{13}C$ in carbonates and organic carbon through time, Schidlowski shows that, once the effects of diagenesis and metamorphism are accounted for, there is very little change between the present-day and the earliest part of the geological record at 3.8 Ga. This he argues is evidence for the constancy of biological activity since 3.8 Ga and presence of CO_2-fixing organisms since that time.

7.4.5   Carbon isotopes in CO_2

Since CO_2 is an important component of igneous, metamorphic and mineralizing fluids, it is instructive to try and use the isotopic composition of the carbon to discover the original source of the CO_2. Studies of this type have concentrated on CO_2 released from fluid inclusions and on carbonate minerals formed in equilibrium with CO_2-rich fluids. From Figure 7.18 we might attempt to differentiate between CO_2 derived from organic carbon, from the mantle and from marine carbonate on the basis of the $\delta^{13}C$ values. Unfortunately fractionation processes confuse this simplistic subdivision and rarely can a unique solution to the origin of CO_2 be obtained, as the examples below will show.

CO$_2$ dissolved in igneous melts   It has recently been realized that CO_2 exsolved from igneous melts is an important route for accessing mantle CO_2. Step-heating experiments on oceanic basalts produce two fractions of CO_2, one below 600 °C with $\delta^{13}C$ values around $-25\,^0/_{00}$, the other at above 600 °C with $\delta^{13}C$ values between -5 to $-8\,^0/_{00}$ (Javoy *et al.*, 1986). Experimental studies by Javoy *et al.* (1978) showed that there is a fractionation of about $4\,^0/_{00}$ at about 1200 °C between carbon dissolved in a basaltic melt and CO_2. There are two possible interpretations of the low-temperature light-isotopic carbon fraction: (1) it is organic contamination; or (2) it is a residue remaining after CO_2 fractionation and outgassing.

CO$_2$ in metamor-phic fluids   The decarbonation of a marine limestone during metamorphism leads to lower (more negative) $\delta^{13}C$ values in calcite (Nabelek *et al.*, 1984) and CO_2 which is correspondingly enriched in $\delta^{13}C$. The metamorphism of biogenic, non-carbonate carbon, on the other hand, leads to a loss of methane and higher (less negative) $\delta^{13}C$ values in the residual carbon or graphite.

(a) Granulites   A popular model for the origin of granulites is the pervasive influx of CO_2 at depth in the Earth's crust. This model is based upon the observation that many granulites contain abundant CO_2-rich fluid inclusions. Jackson *et al.* (1988) analysed CO_2-rich fluid inclusions in granulites from south India and found that the fluid had $\delta^{13}C$ composition in the range -5 to $-8\,^0/_{00}$ (PDB). At first sight this result suggests a mantle origin for the CO_2 although similar $\delta^{13}C$ values ($+2$ to $-10\,^0/_{00}$) can be produced by the decarbonation of marine carbonates (initial $\delta^{13}C$ $\sim 0\,^0/_{00}$) during lower crustal metamorphism (Wickham, 1988).

(b) The origin of metamorphic graphite   Metamorphic graphite is produced by the infiltration of, reaction with and mixing of carbon-bearing fluids. For example, Rumble and Hoering (1986) found that hydrothermal vein graphites in high-grade

gneisses in New Hampshire formed by the mixing of two fluids, one carrying organically derived carbon whilst the other contained carbonate carbon. Similarly, Baker (1988) found that Ivrea zone pelites generally contained graphite from an organic source, although higher $\delta^{13}C$ values in the vicinity of marble bodies suggested that some of the graphite came from marble-derived fluids.

CO$_2$ in gold-mineralizing fluids

Carbonate minerals precipitated in association with Archean lode gold deposits are thought to result from the CO_2-rich nature of the auriferous fluids and as such are an indicator of the source of the gold-bearing solutions. There are two schools of thought, one favouring a metamorphic origin for the CO_2-rich fluids whilst the other favours a mantle origin. Carbon isotopes have been used in an attempt to resolve this controversy. Burrows *et al.* (1986) found that the mean $\delta^{13}C$ value of calcite carbon was between -3 and $-4\,^0/_{00}$ (PDB). Since the only local source of carbon was of seawater derivation ($0\,^0/_{00}$) these authors concluded that the CO_2-rich fluids were externally derived and of mantle origin. However, the more recent work of Groves *et al.* (1988) reveals some of the subtleties involved in using $\delta^{13}C$ as a tracer. These authors report a second carbon reservoir in the area of the mineralization with $\delta^{13}C$ values close to those recorded for the mineralization (median value $-4.8\,^0/_{00}$). They conclude that CO_2 in the gold-bearing fluids is metamorphic in origin, and that although originally mantle-derived and deposited in fault zone, it was subsequently reworked during metamorphism.

CO$_2$ fluid–rock interaction

Fluid–rock interaction has been fully discussed in the section on water–rock interaction (Section 7.3) and the same principles apply when considering CO_2-rich fluids. When the fluid/rock ratio is small the $\delta^{13}C$ in the rock dominates the system and it is the fluid composition which is changed, whereas when the fluid/rock ratio is large and the $\delta^{13}C$ of the CO_2 dominates, the $\delta^{13}C$ value of the rock is modified.

   The calculation, based upon mass balance constraints, requires knowledge of the initial $\delta^{13}C$ value for the carbonate and for the fluid, the proportions of the carbon-bearing species and the fractionation factors for ^{13}C between CO_2 and the carbon-bearing species (Figure 7.18). The shift in $\delta^{13}C$ from the original value to that measured in calcite and/or graphite is used to calculate the extent of fluid–rock interaction. Examples are given by Baker (1988) and Kreulen (1988). Baker found very low fluid/rock ratios in the amphibolite and granulite facies rocks of the Ivrea zone whereas Kreulen found fluid/rock ratios of up to 2.0 in schists from Naxos.

7.4.6   Carbon isotope thermometry

Inspection of Figure 7.18 and Box 7.4 show that the fractionation of ^{13}C between species of carbon is in many cases strongly temperature-dependent. Two of these fractionations have been used as thermometers in metamorphic rocks.

The calcite-graphite $\delta^{13}C$ thermometer

Graphite coexists with calcite in a wide variety of metamorphic rocks and is potentially a useful thermometer at temperatures above 600 °C. There are three calibrations currently in use. The calibration of Bottinga (1969) is based upon theoretical calculations and has the widest temperature range. Valley and O'Neil (1981) produced an empirical calibration based upon temperatures calculated from two-feldspar and ilmenite–magnetite thermometry which is valid in the range

Box 7.4

Equations governing the fractionation of ^{13}C (temperature in kelvin units unless otherwise stated)

CO_2–calcite (Bottinga, 1969) (0–700 °C)

$$1000 \ln \alpha = -2.988(10^6/T^2) + 7.6663(10^3/T) - 2.4612$$

Dolomite–calcite (Sheppard and Schwartz, 1970) (100–650 °C)

$$1000 \ln \alpha = 0.18(10^6/T^2) + 0.17$$

Calcite–graphite (Valley and O'Neil, 1981) (610–760 °C)

$$\Delta_{calcite-graphite} = 0.00748T + 8.68 \ (T \text{ in } °C)$$

Calcite–graphite (Wada and Suzuki, 1983) (400–680 °C)

$$\Delta_{calcite-graphite} = 5.6(10^6/T^2) - 2.4$$

Dolomite–graphite (Wada and Suzuki, 1983) (400–680 °C)

$$\Delta_{dolomite-graphite} = 5.9(10^6/T^2) - 1.9$$

600–800 °C. This curve is about 2 ‰ lower than the calculated curve of Bottinga (1969). Wada and Suzuki (1983) also calibrated the calcite–graphite and the dolomite–graphite thermometers empirically using temperatures obtained from dolomite–calcite solvus thermometry. Their calibration is valid in the range 400–680 °C and is close to the Valley and O'Neil (1981) curve at high temperatures. Valley and O'Neil (1981) suggest that equilibrium is not reached between calcite and graphite at temperatures below 600 °C. However, at temperatures above 600 °C organic carbon loses its distinctively negative $\delta^{13}C$ signature in the presence of calcite.

The CO_2–graphite thermometer   The carbon isotope composition of CO_2 in fluid inclusions and that of coexisting graphite can also be used as a thermometer. The exchange was calibrated by Bottinga (1969); see Figure 7.18. The method was used by Jackson *et al.* (1988) who obtained equilibration temperatures close to the peak of metamorphism from CO_2-rich inclusions in quartz and graphite in granulite facies gneisses from south India.

7.5   Using sulphur isotopes

There are four stable isotopes of sulphur which have the following abundances:

$^{32}S = 95.02\,\%$
$^{33}S = 0.75\,\%$
$^{34}S = 4.21\,\%$
$^{36}S = 0.02\,\%$

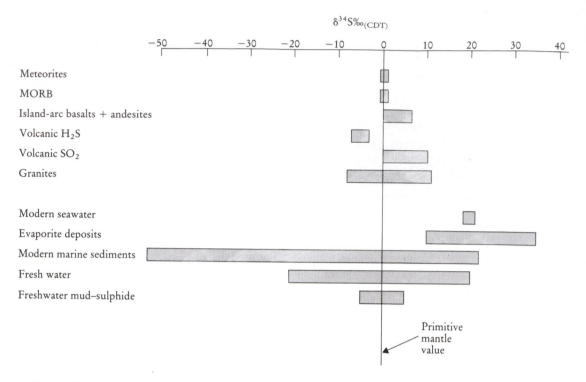

Figure 7.21   Natural sulphur isotope reservoirs. Data from: Sakai *et al.*, 1982; 1984; Ueda and Sakai, 1984; Claypool *et al.*, 1980; Kerridge *et al.*, 1983; Chambers, 1982; Coleman, 1977; Chaussidon *et al.*, 1989. A similar diagram for hydrothermal sulphur-bearing minerals is given in Figure 7.28.

The ratio between the two most abundant isotopes, $^{34}S/^{32}S$, is used in geochemistry and expressed in parts per thousand relative to the reference standard of troilite (FeS) from the Canon Diablo iron meteorite (CDT) in the form

$$\delta^{34}S \,^{0}/_{00} = \left[\frac{^{34}S/^{32}S \,(\text{sample}) - \,^{34}S/^{32}S \,(\text{standard})}{^{34}S/^{32}S \,(\text{standard})} \right] \times 1000 \quad [7.17]$$

Sulphur isotope ratios are measured on SO_2 gas and precision during mass spectrometry is *ca* 0.02 $^{0}/_{00}$ and accuracy about 0.10 $^{0}/_{00}$. The *in situ* analysis of fine-grained sulphide intergrowths is now feasible using the ion microprobe with a precision of between 1.5 and 3.0 $^{0}/_{00}$. (Eldridge *et al.*, 1988) and the laser microprobe with a precision of $\pm$ 1 $^{0}/_{00}$ for $\delta^{34}S$ (Kelley and Fallick, 1990). Sulphur isotope data are most commonly presented in the form of frequency histograms. Bivariate correlation diagrams are rarely used although sometimes a plot of $\delta^{34}S$ vs total sulphur is instructive.

7.5.1   The distribution of sulphur isotopes in nature

Naturally occurring sulphur-bearing species include native sulphur, the sulphate and sulphide minerals, gaseous H_2S and SO_2 and a range of oxidized and reduced suphur ions in solution. A summary of the isotopic compositions of some major rock types is given in Figure 7.21.

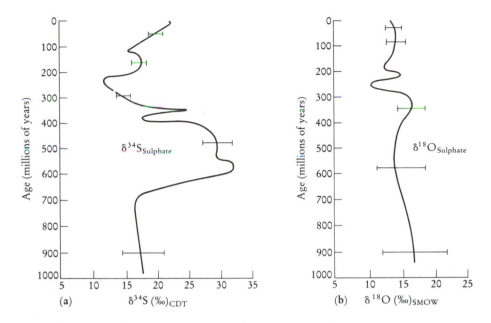

Figure 7.22   Age curves for $\delta^{34}S$ and $\delta^{18}O$ in sulphate in equilibrium with seawater. Note the change in scale; $\delta^{34}S$ is relative to CDT and $\delta^{18}O$ relative to SMOW. The curves are based upon the isotopic composition of sulphate in evaporite deposits and the error bars show the uncertainty in the curves at different time intervals. The oxygen curve is less well constrained than the sulphur curve. The fractionation of $\delta^{34}S$ between dissolved sulphate in seawater and evaporite sulphate is $+1.65\,^o/_{oo}$ and therefore, strictly, this value should be subtracted from the values given. However, the relatively small difference is within the errors on the plotted curve and is usually ignored. On the other hand, the fractionation of $\delta^{18}O$ between seawater and evaporite sulphate is $+3.5\,^o/_{oo}$ and so to obtain the isotopic composition of $\delta^{18}O$ in sulphate in seawater at a given time, $3.5\,^o/_{oo}$ should be subtracted from the plotted values (after Claypool *et al.*, 1980).

There are three isotopically distinct reservoirs of $\delta^{34}S$: (1) mantle-derived sulphur with $\delta^{34}S$ values in the range $0 \pm 3\,^o/_{oo}$ (Chaussidon and Lorand, 1990); (2) seawater sulphur with $\delta^{34}S$ today about $+20\,^o/_{oo}$, although this value has varied in the past, and (3) strongly reduced (sedimentary) sulphur with large negative $\delta^{34}S$ values. The best estimate of the $\delta^{34}S$ composition of the primitive mantle relative to CDT is $+0.5\,^o/_{oo}$ (Chaussidon *et al.*, 1989), slightly but significantly different from that of chondritic meteorites ($0.2 \pm 0.2\,^c/_{oo}$). MORB values, indicative of depleted mantle, are in the narrow range $\delta^{34}S = +0.3 \pm 0.5^o/_{oo}$ (Sakai *et al.*, 1984). Island-arc volcanic rocks have a wider range of $\delta^{34}S$ (-0.2 to $+20.7\,^o/_{oo}$). Granitic rocks are also very variable in composition (-10 to $+15\,^o/_{oo}$; Coleman, 1977) and show a wide range compared to the average value for the continental crust ($\delta^{34}S = +7.0\,^o/_{oo}$) proposed by Chaussidon *et al.* (1989).

The $\delta^{34}S$ value for modern seawater varies between about $18.5\,^o/_{oo}$ and $21.0\,^o/_{oo}$ (Kerridge *et al.*, 1983; Chambers, 1982). Present-day sulphate evaporites are enriched in $\delta^{34}S$ by between 1 and $2\,^o/_{oo}$ relative to seawater, a relationship exploited by Claypool *et al.* (1980) to determine the $\delta^{34}S$ value of ancient seawater. Their 'seawater' curve shows marked excursions from the present value (Figure 7.22) with a particular low ($+10.5\,^o/_{oo}$) in the Permian and a marked high at the base of the Cambrian ($+31.0\,^o/_{oo}$). There is a superficial similarity between the

$\delta^{34}S$-sulphate seawater curve and the $\delta^{18}O$-sulphate seawater curve (Figure 7.22) although the two isotopic systems are not closely coupled. $\delta^{34}S$ in modern marine sediments has an extensive range from values around $+20\,^{o}/_{oo}$, reflecting the composition of seawater, down to $-56\,^{o}/_{oo}$, the product of bacterial sulphate reduction (Section 7.5.2). Some metasediments have values as high as $+40\,^{o}/_{oo}$ CDT.

7.5.2   Controls on the fractionation of sulphur isotopes

Inspection of Figure 7.21 shows that many crustal rocks have wide ranges of $\delta^{34}S$ values, indicative of extensive fractionation. These fractionation mechanisms are outlined below.

Sulphur isotope fractionation in igneous rocks

Sulphur isotope fractionations in an igneous melt are small and take place either between crystals and melt or through solid–gas fractionation. ^{34}S fractionation between primary sulphide minerals and magma is of the order of $1–3\,^{o}/_{oo}$ (Ohmoto and Rye, 1979; Chaussidon *et al.*, 1989). Differentiated felsic and mafic sequences record a maximum decrease in $\delta^{34}S$ of $1\,^{o}/_{oo}$ (Ueda and Sakai, 1984).

An alternative means of sulphur isotope fractionation in igneous rocks is by the degassing of SO_2 from sub-aerial or shallow submarine lavas. This fractionation is controlled by the sulphate/sulphide ratio of the melt, which in turn is controlled by temperature, pressure, water content and oxygen activity. SO_2 outgassed by basic lavas is enriched in $\delta^{34}S$ relative to the melt because the dominant sulphur species in the melt is sulphide and SO_2 is enriched in $\delta^{34}S$ relative to sulphide (see Figure 7.26). The process is documented from Kilauea volcano, Hawaii, by Sakai *et al.* (1982) and quantified by Zheng (1990).

Sulphur isotope fractionation in sedimentary rocks

The sedimentary sulphur cycle and the associated fractionation of $\delta^{34}S$ are summarized in Figure 7.23. There are four possible processes.

(a) The bacterial reduction of sulphate to sulphide   The principal low-temperature control (i.e. below 50 °C) on sulphur isotope fractionation is the reduction of seawater sulphate to sedimentary sulphide, although there is evidence to suggest that this process did not take place during the Archean (Hattori *et al.*, 1983). The reaction is kinetically controlled, for it has been shown that the rate of reaction of the lighter ion, $^{32}SO_4^{2-}$ is 2.2 % greater than that of $^{34}SO_4^{2-}$ (Harrison and Thode, 1957). Hence a reaction of this type taken to completion will produce sulphide depleted by $22\,^{o}/_{oo}$ relative to the seawater source.

In addition to the kinetic control of sedimentary sulphide fractionation, it is also important to consider whether the system is open or closed. In an open system such as a large body of stagnant seawater there is an infinite reservoir of seawater sulphate. Sulphate-reducing bacteria operating in stagnant bottom waters will produce H_2S extremely depleted in ^{34}S whilst $\delta^{34}S$ in the seawater reservoir will be effectively unchanged. Chambers (1982) records fractionations as large as $48\,^{o}/_{oo}$ in modern intertidal sediments. In a closed system in which sulphide is removed by either the loss of H_2S gas or the precipitation of sulphide minerals the isotopic fractionation of sulphur is controlled by Rayleigh fractionation — the equations are given in Ohmoto and Rye (1979). In this case the isotopic composition of the

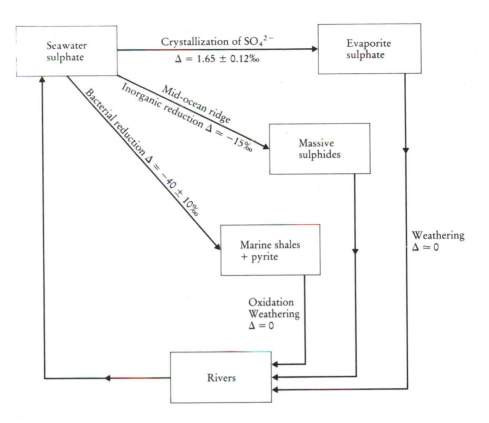

Figure 7.23   The sedimentary sulphur cycle showing the pattern of sulphur-isotope fractionation in sedimentary and hydrothermal processes.

sulphate also changes with decreasing sulphate concentration and the residual sulphate may have very high $\delta^{34}S$ values (see for example Richardson *et al.*, 1988).

(b) The bacterial oxidation of sulphide to sulphate   In contrast to the bacterial reduction of seawater sulphate, the oxidation of sulphide by bacteria produces very little fractionation and is unimportant, except perhaps in the Archaean (Hattori, 1989).

(c) The crystallization of sedimentary sulphate from seawater — evaporite formation   The crystallization of sedimentary sulphate during evaporite formation produces a relatively small $\delta^{34}S$ enrichment of $1.65 \pm 0.12\,^o/_{oo}$ (Thode and Monster, 1965).

(d) The non-bacterial reduction of sulphate to sulphide   Some mineral deposits show clear evidence for the reduction of sulphate-bearing solutions at temperatures above those favourable to sulphate-reducing bacteria, indicating the inorganic reduction of sulphate (Trudinger *et al.*, 1985). There are two possible processes. In the temperature ranges 75–175 °C, inorganic sulphate reduction may take place in the

presence of hydrocarbons. A variant of this may be the breakdown of sulphur-bearing organic compounds in the formation of sulphides associated with coal (Whelan *et al.*, 1988). At temperatures above 250 °C, sulphate reduction can be achieved by reduction with ferrous iron. This process is particularly important in seawater interaction with MORB at a mid-ocean ridge (Section 7.5.4).

Sulphur isotope fractionation in hydrothermal systems

At very high temperatures ($T > 400$ °C) the dominant sulphur species in hydrothermal systems are H_2S and SO_2 and the isotopic composition of the fluid is approximated by

$$\delta^{34}S_{fluid} = \delta^{34}S_{H_2S} \, X_{H_2S} + \delta^{34}S_{SO_2} \, X_{SO_2}$$ [7.18]

where X_{H_2S} etc. is the mole fraction of H_2S relative to total sulphur in the fluid (Ohmoto and Rye, 1979). At these elevated temperatures H_2S and SO_2 are assumed to behave as an ideal gas mixture. The fractionation factor is relatively large and is given in Table 7.4; the fractionation curve is illustrated in Figure 7.24.

At lower temperatures ($T < 350$ °C) the dominant sulphur species in a hydrothermal system are sulphate and H_2S (Ohmoto and Rye, 1979). All sulphate species are assumed to have the same fractionation factors and these are given in Table 7.4. A number of very important studies summarized by Ohmoto and Rye (1979) show that the fractionation of ^{34}S between the different sulphur-bearing species in hydrothermal fluids is not simply controlled by temperature. Rather it is a function of the physico-chemical conditions of the fluid: these include oxygen activity, sulphur activity, pH and the activity of cations associated with sulphate. The effect is illustrated on a log (oxygen activity) vs pH diagram (Figure 7.25) showing the stability relationships for a number of relevant mineral species. The deviation in $\delta^{34}S$ of a mineral from that in the equilibrium hydrothermal fluid is shown as a function of oxygen activity and pH. The diagram predicts that sulphide minerals forming in equilibrium with magnetite will have a $\delta^{34}S$ much lower than the fluid, whereas at low oxygen activities and low pH the fluid and minerals will have approximately the same $\delta^{34}S$ values.

The importance of these studies is firstly that the $\delta^{34}S$ of a hydrothermal fluid cannot be directly estimated from the $\delta^{34}S$ value of sulphide minerals unless variables such as oxygen activity and pH are also known. Secondly, the $\delta^{34}S$ of a sulphide mineral cannot be used directly to determine the source of the sulphur since it is a function of many more variables than simply temperature.

Sulphur isotope fractionation between sulphide and sulphate phases — sulphur isotope thermometry

There are a number of theoretical and experimental determinations of the fractionation of $\delta^{34}S$ between coexisting sulphide phases as a function of temperature (Table 7.4). Sulphide-pair thermometers derived from these results are given in Box 7.5 and in Figure 7.26. However, the partitioning of sulphur isotopes between sulphides is not a particularly sensitive thermometer and requires precise isotopic determinations. More extensive $\delta^{34}S$ fractionation is between sulphide and sulphate phases.

Sulphide mineral-pairs and sulphide–sulphate mineral-pairs are not always in equilibrium. This can arise when (1) the mineral-pair formed at low temperatures ($T < 200$ °C); (2) the isotopic composition of a mineralizing fluid is variable; (3) there was continued isotopic exchange following the formation of the mineral

Table 7.4 Fractionation for the distribution of ^{34}S between H_2S and sulphur compounds. The governing equation is $1000 \ln \alpha_{mineral-H_2S} = A(10^6/T^2) + B$ (temperature in Kelvin units)

Mineral	A	B	Temperature range (°C)	Reference
Anhydrite/gypsum/	6.463	0.56 ± 0.5	200–400	Ohmoto and Lasaga (1982)
Baryte	6.5 ± 0.3		200–400	Miyoshi *et al.* (1984)
Molybdenite	0.45 ± 0.10		Uncertain	Ohmoto and Rye (1979)
Pyrite	0.40 ± 0.08		200–700	Ohmoto and Rye (1979)
Sphalerite	0.10 ± 0.05		50–705	Ohmoto and Rye (1979)
Pyrrhotite	0.10 ± 0.05		50–705	Ohmoto and Rye (1979)
Chalcopyrite	-0.05 ± 0.08		200–600	Ohmoto and Rye (1979)
Bismuthinite	-0.67 ± 0.07		250–600	Bente and Nielsen (1982)
Galena	-0.63 ± 0.05		50–700	Ohmoto and Rye (1979)
SO_2	4.70	-0.5 ± 0.5	350–1050	Ohmoto and Rye (1979)

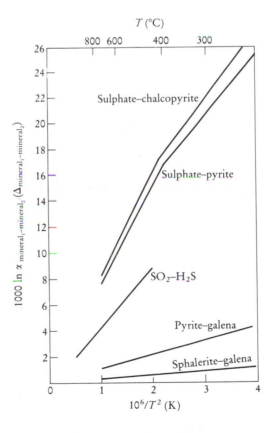

Figure 7.24 Partitioning of ^{34}S between mineral pairs and H_2S–SO_2 as a function of temperature. These fractionation curves can be used in sulphur isotope thermometry and indicate that the greatest fractionations are between sulphides and sulphate minerals.

Box 7.5

Calibrations for sulphur isotope thermometers (temperature in kelvin units), mainly based upon the fractionation factors given in Table 7.4

Pyrite–galena

$1000 \ln \alpha = 1.03(10^6/T^2)$ Ohmoto and Rye (1979)
$1000 \ln \alpha = 1.08(10^6/T^2)$ (150–600 °C)   Clayton (1981)

Pyrite–sphalerite (pyrite–pyrrhotite)

$1000 \ln \alpha = 0.30(10^6/T^2)$ Ohmoto and Rye (1979)

Pyrite–chalcopyrite

$1000 \ln \alpha = 0.45(10^6/T^2)$ Ohmoto and Rye (1979)

Sphalerite–galena (pyrrhotite–galena)

$1000 \ln \alpha = 0.73(10^6/T^2)$ Ohmoto and Rye (1979)
$1000 \ln \alpha = 0.76(10^6/T^2)$ (100–600 °C)   Clayton (1981)

Sulphate–pyrite

$1000 \ln \alpha = 6.063(10^6/T^2) + 0.56$ Ohmoto and Lasaga (1982)

Sulphate–chalcopyrite

$1000 \ln \alpha = 6.513(10^6/T^2) + 0.56$ Ohmoto and Lasaga (1982)

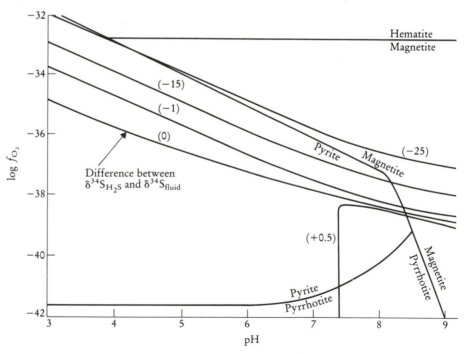

Figure 7.25   Log $f\mathrm{O}_2$ vs pH diagram for sulphur species showing the stability fields of pyrite, pyrrhotite, magnetite and hematite. The boundaries are for $\mu = 1.0$, and molalities for total sulphur = 0.01, $K^+ = 0.1$, $Na^+ = 0.9$, $Ca^{2+} = 0.01$. The contours show deviations of $\delta^{34}S_{H_2S}$ from $\delta^{34}S_{fluid}$ at 250 °C under equilibrium conditions (after Ohmoto and Rye, 1979).

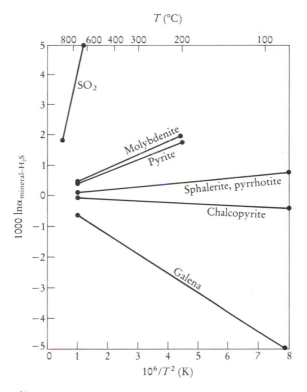

Figure 7.26   Partitioning of ^{34}S between minerals and H_2S as a function of temperature. These fractionation curves are used as a basis for the sulphur isotope thermometers given in Box 7.5. Mineral-pairs showing the greatest separation on this diagram will be the most sensitive thermometers (mainly after Ohmoto and Rye, 1979).

phases. The attainment of isotopic equilibrium is best demonstrated by the determination of temperature estimates between three coexisting minerals. Agreement between the two estimated temperatures may be taken as evidence of equilibrium. If this approach is not possible, there should be clear textural evidence of equilibrium.

7.5.3   Using sulphur isotopes in igneous petrogenesis

In this section three examples are given which illustrate different ways in which sulphur isotope variations may be explained in igneous rocks.

(a) Outgassing of SO_2   The difference in sulphur isotope composition of sub-aerial and submarine lavas from Kilauea volcano was demonstrated by Sakai *et al.* (1982) and attributed to SO_2 outgassing. Submarine basalts have $\delta^{34}S = +0.7$ $^o/_{oo}$ and a high sulphate–sulphide ratio whereas sub-aerial basalts have lower total sulphur, a reduced $\delta^{34}S$ (-0.8 $^o/_{oo}$) and a lower sulphate/sulphide ratio, all features consistent with the rapid outgassing of SO_2. This mechanism was also invoked by Faure *et al.* (1984) and Mensing *et al.* (1984), who found a wide range of $\delta^{34}S$ values (-1.45 to 11.73 $^o/_{oo}$) in the Jurassic Kirkpatrick basalt of Antarctica. In these rocks $\delta^{34}S$ correlates with magnetite concentration, a feature which is indicative of high

activities of oxygen. Faure *et al.* (1984) therefore proposed that the high oxygen activity led to the stabilization of SO_2 as the dominant sulphur species which was degassed, giving rise to the wide range in $\delta^{34}S$ in the basalts.

(b) Contamination The basalt–andesite–dacite suite of the Mariana island arc has $\delta^{34}S$ values in the range +2 to +20.7 ⁰/₀₀. This huge range relative to mantle values is thought by Woodhead *et al.* (1987) to reflect contamination with seawater sulphate ($\delta^{34}S = +20$ ⁰/₀₀).

(c) Crystal fractionation Layered gabbros of the Huntley–Knock intrusion of north–east Scotland contain sulphide horizons of magmatic origin. $\delta^{34}S$ values for sulphides in the cumulate rocks have a mean value of 0.5 ⁰/₀₀, close to the mantle value (Fletcher *et al.*, 1989). $\delta^{34}S$ values within the intrusion, however, vary from the contact zone (mean value of −1.2 ⁰/₀₀) to cumulate and granular rocks (mean +2.4 ⁰/₀₀). This variation is attributed to magmatic isotopic fractionation under conditions of variable oxygen activity.

7.5.4 Using sulphur isotopes to understand the genesis of hydrothermal ore deposits

Sulphur isotope studies of hydrothermal ore deposits offer the opportunity to: (1) determine the origin of the sulphur present in the orebody as sulphides and sulphates; (2) determine the temperature of formation of the sulphides and of the ore-forming fluids; (3) determine the water/rock ratio effective during the mineralization; (4) determine the degree of equilibrium attained; and (5) thus constrain the mechanism of ore deposition.

Modern hydro-thermal mineral-ization at mid-ocean ridges   High-temperature hydrothermal vents currently active at mid-ocean ridges offer a unique opportunity to study a hydrothermal mineral deposit in the process of formation. The current working model assumes that cold seawater sulphate is drawn down into sea-floor basalts, where it is heated in the vicinity of a magma chamber. Some sulphate is precipitated as anhydrite whilst the remainder is reduced to sulphide by reaction with the basalt. The fluid is vented back onto the seafloor at about 350 °C laden with sulphides. On mixing with seawater these are precipitated onto the sea floor as a fine sulphide sediment whilst at the vent site itself the sulphides are built into a 'chimney' a metre or so in height.

Recent studies on the East Pacific Rise (Woodruff and Shanks, 1988; Bluth and Ohmoto, 1988) have determined the sulphur isotope composition of the different components of this system (Figure 7.27). These data allow the model outlined above to be tested and a number of observations to be made about deposits of this type:

(1)   The sulphide phases are not in isotopic equilibrium with each other, for the results of sulphur isotopic thermometry do not agree with the known temperature of the vent fluid.

(2)   The $\delta^{34}S$ of the sulphide phases varies from the inner to outer wall of a chimney.

(3)   The vent fluid is apparently out of isotopic equilibrium with the minerals deposited.

(4)   The $\delta^{34}S$ composition of the vent fluid is produced by the mixing of basalt

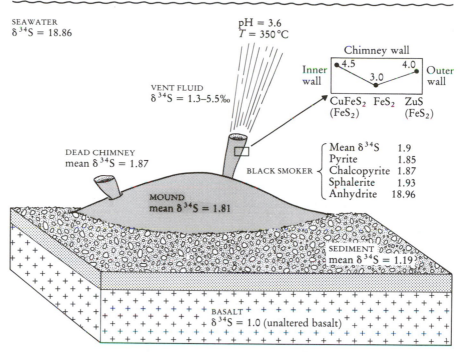

Figure 7.27 Schematic representation of $\delta^{34}S$ values in a modern mid-ocean ridge hydrothermal vent system. Most values are taken from Kerridge *et al.* (1983) but with additions from Bluth and Ohmoto (1988) and Woodruff and Shanks (1988). The enlargement of the chimney wall shows changing $\delta^{34}S$ values and sulphide composition from inner to outer wall.

     sulphur and reduced seawater sulphate. The precise value is a function of the water/rock ratio (Skirrow and Coleman, 1982).

(5) The composition of the hydrothermal fluid is variable.

(6) Cogenetic sulphate minerals (anhydrite) are in equilibrium with seawater sulphate.

These features are explained by Bluth and Ohmoto (1988) in a model in which the $\delta^{34}S$ value of the fluid changes with time during the life of a chimney, as a result of changes in the plumbing and water/rock ratios in the basalts beneath the vent system. In addition, the hydrothermal fluid reacts with the already precipitated sulphides in the chimney walls, resulting in their changing composition during the growth of a chimney.

Ancient hydrothermal mineralization The sulphur isotope composition of hydrothermal ore deposits is reviewed by Ohmoto (1986). The $\delta^{34}S$ values for a selection of hydrothermal ore deposits are summarized in Figure 7.28, from which it can be seen that whilst some types of deposit have a narrow compositional range and therefore a fairly specific origin, others have a wide compositional range and probably multiple origins. An example of the latter are deposits of the Mississippi Valley-type; sulphides from Mississippi Valley-type ores extend across the entire compositional range of Figure 7.28. Individual deposits, however, have a relatively restricted range of $\delta^{34}S$ values,

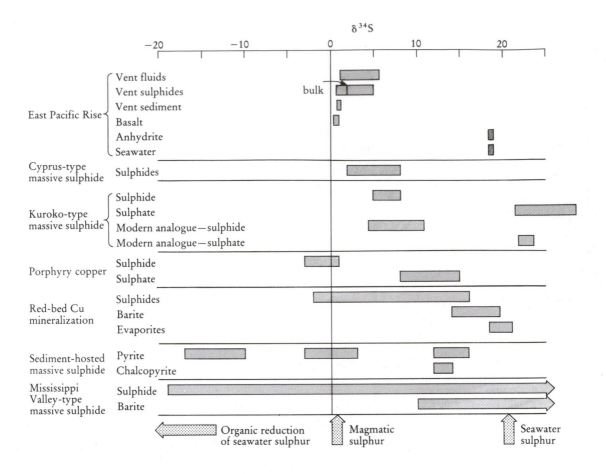

Figure 7.28   The $\delta^{34}S$ values for sulphur-bearing minerals in hydrothermal deposits. Data from Kerridge *et al.* (1988), Ohmoto and Rye (1979), Halbach *et al.* (1989), Naylor *et al.* (1989) and Eldridge *et al.* (1988).

supporting the view that in each case the source of the sulphur and mechanism of H_2S production are different.

Below, a number of different types of hydrothermal deposit are briefly reviewed and classified according to their probable source of sulphur.

(a) High-temperature inorganic reduction of seawater sulphate   This mechanism has been discussed in some detail above in the description mid-ocean ridge massive sulphide deposits. It is equally applicable to massive Pb–Zn sulphides of the Kuroko type hosted in felsic calc–alkaline lavas. Most Kuroko ores have sulphide $\delta^{34}S$ values in the range +5 to +8 ‰ and sulphate values in the range 21.5 to 28.5 ‰. The close relationship between $\delta^{34}S$ in Kuroko sulphates and the value for contemporary seawater (Sangster, 1968) indicates that inorganic sulphate reduction is a plausible mechanism of sulphide formation.

(b) Low-temperature inorganic reduction of sulphate   Red-bed copper mineralization in the Cheshire basin of north-west England is associated with evaporite deposits.

Sulphides show a wide range of $\delta^{34}S$ values (-1.8 to $+16.2\ ^0/_{00}$) and associated barite is in the range 13.8 to 19.3 $^0/_{00}$. These latter values are close to $\delta^{34}S$ for evaporite sulphate (18.4 to 20.8), present in the overlying succession. Naylor *et al.* (1989) suggest that both the barite and the sulphides formed from sulphur-bearing solutions derived from this source. The temperature of mineralization is not well known but is thought to have exceeded the temperatures at which sulphate-reducing bacteria can exist. For this reason the authors suggest the sulphate reduction took place in the presence of gaseous hydrocarbons and the range of sulphide values requires that the reduction took place in a closed system and was incomplete.

(c) Low-temperature bacteriological reduction of sulphate   The Rammelsberg orebody, a sediment-hosted massive sulphide in Germany, is made up primarily of clasts containing varying proportions of pyrite, chalcopyrite and other sulphides together with silicates. $\delta^{34}S$ values in pyrite clasts fall into three groups ($+12$ to $+16\ ^0/_{00}$, $+3$ to $-3\ ^0/_{00}$ and -10 to $-17\ ^0/_{00}$) and on a frequency histogram the values show a skewed distribution, a feature not seen in volcanogenic massive sulphide ores and thought to be characteristic of sediment-hosted massive sulphides (Eldridge *et al.*, 1988). The lower groups of values are explained most easily by bacteriogenic reduction of seawater sulphate (the fractionation factor is around $-45 \pm 20\ ^0/_{00}$) in a partially open system. The higher $\delta^{34}S$ values are probably of hydrothermal origin and the two groups of clasts were mixed during sedimentation.

(d) Sulphur of magmatic origin   Porphyry copper deposits are the most likely candidate for a magmatic, igneous source of sulphur. $\delta^{34}S$ values for sulphides fall in the narrow range -3 to $+1\ ^0/_{00}$ close to the accepted mantle range. The sulphate and sulphide phases are in equilibrium and yield temperatures of 450–650 °C (Ohmoto and Rye, 1979). In addition, oxygen and hydrogen isotope data indicate that the calculated fluid compositions are close to those for magmatic fluids.

References

Abbey S., 1989, The evaluation of reference materials for rock analysis. In: Ahmedali S.T. (ed.), *X-ray fluorescence analysis in the geological sciences: Advances in methodology.* Geol. Assn. Canada: Short course 7, 1–38.

Adam J., 1988, Dry, hydrous and CO_2-bearing liquidus phase relationships in the CMAS system at 28 kb, and their bearing on the origin of alkali basalts. *J. Geol.*, **96**, 709–720.

Addy S.K. and Garlick G.D., 1974, Oxygen isotope fractionation between rutile and water. *Contrib. Mineral. Petrol.*, **45**, 119–121.

Ahmedali S.T. (ed.), 1989, *X-ray fluorescence analysis in the geological sciences: Advances in methodology.* Geol. Assn. Canada: Short course 7.

Aitchison J., 1981, A new approach to null correlations of proportions. *J. Math. Geol.*, **13**, 175–189.

Aitchison J., 1982, The statistical analysis of compositional data (with discussion). *J. Roy. Stat. Soc.*, **44**, 139–177.

Aitchison J., 1984, The statistical analysis of geochemical compositions. *J. Math. Geol.*, **16**, 531–564.

Aitchison J., 1986, *The statistical analysis of compositional data.* Methuen, New York.

Aitchison, J., 1989, Measures of location of compositional data sets. *J. Math. Geol.*, **21**, 787–790.

Allan J.R. and Matthews R.K., 1982, Isotope signatures associated with early meteoric diagenesis. *Sedimentology*, **29**, 797–817.

Allegre C.J., 1987, Isotope geodynamics. *Earth Planet. Sci. Lett.*, **86**, 175–203.

Allegre C.J. and Ben Othman D., 1980, Nd–Sr isotopic relationship in granitoid rocks and continental crust development: a chemical approach to orogenesis. *Nature*, **286**, 335–342.

Allegre C.J., Hart, S.R. and Minster J.-F., 1983a, Chemical structure and evolution of the mantle and continents determined by inversion of Nd and Sr isotopic data, I. Theoretical models. *Earth Planet. Sci. Lett.*, **66**, 177–190.

Allegre C.J., Hart, S.R. and Minster J.-F., 1983b, Chemical structure and evolution of the mantle and continents determined by inversion of Nd and Sr isotopic data, II. Numerical experiments and discussion. *Earth Planet. Sci. Lett.*, **66**, 191–213.

Allegre C.J., Lewin, E. and Dupre B., 1988, A coherent crust–mantle model for the uranium–thorium–lead isotopic system. *Chem. Geol.*, **70**, 211–234.

Allegre C.J. and Minster J.-F., 1978, Quantitative models of trace element behaviour in magnetic processes. *Earth Planet. Sci. Lett.*, **38**, 1–25.

Allegre C.J. and Rousseau D., 1984, The growth of the continents through geological time studied by the Nd isotopic analysis of shales. *Earth Planet. Sci. Lett.*, **67**, 19–34.

Allegre C.J., Treuil M., Minster J.-F., Minster B. and Albarede F., 1977, Systematic use of trace elements in igneous processes. Part I: Fractional crystallisation processes in volcanic suites. *Contrib. Mineral. Petrol.*, **60**, 57–75.

Allen P., Condie K.C. and Narayana B.L., 1985, The geochemistry of prograde and retrograde charnockite–gneiss reactions in southern India. *Geochim. Cosmochim. Acta*, **49**, 323–336.

Anders E. and Ebihara M., 1982, Solar system abundances of the elements. *Geochim. Cosmochim. Acta*, **46**, 2363–2380.

Anderson J.L. and Cullers R.L., 1987, Crust-enriched, mantle-derived tonalites in the early Proterozoic, Penokean orogen of Wisconsin. *J. Geology*, **95**, 139–154.

Apted M.J. and Roy S.D., 1981, Corrections to the trace element fractionation equations of Hertogen and Gijbels (1976). *Geochim. Cosmochim. Acta*, **45**, 777–778.

Arculus R.J., 1987, The significance of source versus process in the tectonic controls of magma genesis. *J. Volc. Geothermal Res.*, **32**, 1–12.

Argast S. and **Donnelly T.W.**, 1987, The chemical discrimination of clastic sedimentary components. *J. Sed. Petrol.*, **57**, 813–823.

Arndt N.T., 1983, Element mobility during komatiite alteration. *Eos*, **64**, 331.

Arndt N.T. and **Goldstein S.L.**, 1987, Use and abuse of crust-formation ages. *Geology*, **15**, 893–895.

Arndt N.T. and **Jenner G.A.**, 1986, Crustally contaminated komatiites and basalts from Kambalda, western Australia. *Chem. Geol.*, **56**, 229–255.

Arth J.G., 1976, Behaviour of trace elements during magmatic processes — a summary of theoretical models and their applications. *J. Res. U.S. Geol. Surv.*, **4**, 41–47.

Arth J.G., 1981, Rare-earth element geochemistry of the island-arc volcanic rocks of Rabaul and Talasea, New Britain. *Bull. Geol. Soc. Amer.*, **92**, 858–863.

Ayalon A. and **Longstaff F.J.**, 1988, Oxygen isotope studies of diagenesis and pore water evolution in the western Canada sedimentary basin: evidence from the Upper Cretaceous basal Belly River sandstone, Alberta. *J. Sed. Petrol*, **58**, 489–505.

Bailey J.C., 1981, Geochemical criteria for a refined tectonic discrimination of orogenic andesites. *Chem. Geol.*, **32**, 139–154.

Baker A.J., 1988, Stable isotope evidence for limited fluid infiltration of deep crustal rocks from the Ivrea Zone, Italy. *Geology*, **16**, 492–495.

Baker A.J. and **Fallick A.E.**, 1989, Evidence from Lewisian limestones for isotopically heavy carbon in two-thousand-million-year-old sea water. *Nature*, **337**, 352–354.

Baker D.R. and **Eggler D.H.**, 1983, Fractionation paths of Atka (Aleutians) high-alumina basalts: constraints and phase relations. *J. Volc. Geotherm. Res.*, **18**, 387–404.

Baker D.R. and **Eggler D.H.**, 1987, Compositions of anhydrous and hydrous melts coexisting with plagioclase, augite, and olivine or low-Ca pyroxene from 1 atm. to 8 kbars: application to the Aleutian volcanic center of Atka. *Amer. Mineral.*, **72**, 12–28.

Barker D.S., 1978, Magmatic trends on alkali–iron–magnesium diagrams. *Amer. Mineral.*, **63**, 531–534.

Barker F., 1979, Trondhjemite: Definition, environment and hypotheses of origin. In: Barker F. (ed.), *Trondhjemites, dacites and related rocks*. Elsevier, Amsterdam, pp. 1–12.

Barnes S-J., 1988, Automated plotting of geochemical data using the Lotus symphony package. *Comput. Geosci.*, **14**, 409–411.

Barnes S-J. and **Naldrett A.J.**, 1986, Variations in platinum group element concentrations in the Alexo mine komatiite, Abitibi greenstone belt, northern Ontario. *Geol. Mag.*, **123**, 515–524.

Barnes S-J., **Naldrett A.J.** and **Gorton M.P.**, 1985, The origin of platinum-group elements in terrestrial magmas. *Chem. Geol.*, **53**, 303–323.

Barth T.W., 1952, *Theoretical petrology: a textbook on the origin and evolution of rocks*. Wiley, New York.

Basu A.R., **Sharma M.** and **DeCelles P.G.**, 1990, Nd, Sr-isotopic provenance and trace element geochemistry of Amazonian foreland basin fluvial sands, Bolivia and Peru: implications for ensialic Andean orogeny. *Earth Planet. Sci. Lett.*, **100**, 1–17.

Beach A. and **Tarney J.**, 1978, Major and trace element patterns established during retrogressive metamorphism of granulite facies gneisses, NW Scotland. *Precambrian Res.*, **7**, 325–348.

Bell K., **Blenkinsop J.**, **Cole T.J.S.** and **Menagh D.P.**, 1982, Evidence from Sr isotopes for long lived heterogeneities in the upper mantle. *Nature*, **298**, 251–253.

Bender J.F., **Langmuir C.H.** and **Hanson G.N.**, 1984, Petrogenesis of basalt glasses from the Tamayo region, East Pacific Rise. *J. Petrol.*, **25**, 213–254.

Ben Othman D., **Polve M.** and **Allegre C.J.**, 1984, Nd–Sr isotopic composition of granulites and constraints on the evolution of the lower continental crust. *Nature*, **307**, 510–515.

Bente K. and **Nielsen H.**, 1982, Experimental S isotope fractionation studies between co-existing bismuthinite (Bi_2S_3) and sulphur (S_o). *Earth Planet. Sci. Lett.*, **59**, 18–20.

Bevins R.E., **Kokelaar B.P.** and **Dunkley P.N.**, 1984, Petrology and geochemistry of lower to middle Ordovician igneous rocks in Wales: a volcanic arc to marginal basin transition. *Proc. Geol. Ass.*, **95**, 337–347.

Bhatia M.R., 1983, Plate tectonics and geochemical composition of sandstones. *J. Geol.*, **91**, 611–627.

Bhatia M.R. and **Crook K.A.W.**, 1986, Trace element characteristics of graywackes and tectonic discrimination of sedimentary basins. *Contrib. Mineral. Petrol.*, **92**, 181–193.

Bickle M.J., Bettenay L.F., Chapman H.J., Groves D.I., McNaughton N.J., Campbell I.H. and **de Laeter J.R.**, 1989, The age and origin of younger granitic plutons of the Shaw Batholith in the Archean Pilbara block, western Australia. *Contrib. Mineral. Petrol.*, **101**, 361–376.

Bjorlykke K., 1974, Geochemical and mineralogical influence of Ordovician island arcs on epicontinental clastic sedimentation. A study of lower Palaeozoic sedimentation in the Oslo region, Norway. *Sedimentology*, **21**, 251–272.

Blanckenburg F.v., Villa I.M., Baur H., Morteani G. and **Steiger R.H.**, 1989, Time calibration of a *P–T*-path from the Tauern window Eastern Alps: the problem of closure temperatures. *Contrib. Mineral. Petrol.*, **101**, 1–11.

Blatt H., Middleton G. and **Murray R.**, 1972, *Origin of sedimentary rocks*. Prentice Hall, New Jersey.

Bloomer S. and **Stern R.**, 1990, Tectonic origins. *Nature*, **346**, 518.

Bloxham W. and **Lewis A.D.**, 1972, Ti, Zr and Cr in some British pillow lavas and their petrogenetic affinities. *Nature (Phys. Science)*, **237**, 134–136.

Bluth G.L. and **Ohmoto H.**, 1988, Sulfide–sulfate chimneys on the East Pacific Rise, 11° and 13°N latitudes. Part II: Sulfur isotopes. *Canad. Mineral.*, **26**, 505–515.

Boles J.R. and **Franks S.G.**, 1979, Clay diagenesis in Wilcox sandstones of southwest Texas: implications of smectite diagenesis on sandstone cementation. *J. Sed. Petrol.*, **49**, 55–70.

Boryta M. and **Condie K.C.**, 1990, Geochemistry and origin of the Archaean Beit Bridge Complex, Limpopo Belt, South Africa. *J. Geol. Soc. Lond.*, **147**, 229–239.

Bottinga Y., 1969, Calculated fractionation factors between carbon and hydrogen isotope exchange in the system calcite–carbon dioxide–graphite–methane–hydrogen–water vapour. *Geochim. Cosmochim. Acta*, **33**, 49–64.

Bottinga Y. and **Javoy M.**, 1973, Comments on oxygen isotope geothermometry. *Earth Planet. Sci. Lett.*, **20**, 250–265.

Bottinga Y. and **Javoy M.**, 1975, Oxygen isotope partitioning among the minerals in igneous and metamorphic rocks. *Rev. Space Phys. Geophys.*, **13**, 401–418.

Bottinga Y., Weill D.F. and **Richet P.**, 1981, Thermodynamic modelling of silicate melts. In: Newton R.C., Navrotsky A. and Wood B.J. (eds.), *Thermodynamics of minerals and melts*. Advances in physical geochemistry series, Springer, Berlin, pp. 207–245.

Bottrell S.H., Greenwood P.B., Yardley B.W.D., Sheppard T.J. and **Spiro B.**, 1990, Metamorphic and post-metamorphic fluid flow in the low-grade rocks of the Harlech dome, north Wales. *J. Metamorph. Geol.*, **8**, 131–143.

Bougault H., Joron J.L. and **Treuil M.**, 1980, The primordial chondritic nature and large-scale heterogeneities in the mantle: evidence from high and low partition coefficient elements in oceanic basalts. *Phil. Trans. R. Soc. Lond.*, **A 297**, 203–213.

Bowen N.L., 1928, *The evolution of the igneous rocks*. Princeton Univ. Press.

Boynton W.V., 1984, Geochemistry of the rare earth elements: meteorite studies. In: Henderson P. (ed.), *Rare earth element geochemistry*. Elsevier, pp. 63–114.

Brevart O., Dupre B. and **Allegre C.J.**, 1986, Lead–lead age of komatiite lavas and limitations on the structure and evolution of the Precambrian mantle. *Earth Planet. Sci. Lett.*, **77**, 293–302.

Brewer T.S. and **Atkin B.P.**, 1989, Element mobilities produced by low grade metamorphic events. A case study from the Proterozoic of southern Norway. *Precambrian Res.*, **45**, 143–158.

Brookins D.G., 1989, Aqueous geochemistry of rare earth elements. In: Lipin B.R. and McKay G.A. (eds.), *Geochemistry and mineralogy of rare earth elements*, Rev. Mineral., **21**, 201–225.

Brooks C., Hart S.R. and **Wendt T.**, 1972, Realistic use of two-error regression treatments as applied to rubidium–strontium data. *Rev. Geophys. Space Phys.*, **10**, 551–577.

Brooks C.K., Henderson P. and **Ronsbo J.G.**, 1981, Rare earth element partitioning between allanite and glass in the obsidian of Sandy Braes, northern Ireland. *Mineral. Mag.*, **44**, 157–160.

Brown G.C., Thorpe R.S. and Webb P.C., 1984, The geochemical characteristics of granitoids in contrasting arcs and comments on magma sources. *J. Geol. Soc. Lond.*, **141**, 411–426.

Brugmann G.E., Arndt N.T., Hoffmann A.W. and Tobschall H.J., 1987, Nobel metal abundances in komatiite suites from Alexo, Ontario and Gorgona Island, Colombia. *Geochim. Cosmochim. Acta*, **51**, 2159–2169.

Burcher-Nurminen K., 1981, The formation of metasomatic reaction veins in dolomitic marble roof pendants in the Bergell intrusion (Province Sondrio, northern Italy). *Amer. J. Sci.*, **281**, 1197–1222.

Burrows D.R., Wood P.C. and Spooner E.T.C., 1986, Carbon isotope evidence a magmatic origin for Archean gold–quartz vein ore deposits. *Nature*, **321**, 851–854.

Butler J.C., 1979, Trends in ternary petrological variation diagrams — fact or fantasy? *Amer. Mineral.*, **64**, 1115–1121.

Butler J.C., 1981, Effect of various transformations on the analysis of percentage data. *J. Math. Geol.*, **13**, 53–68.

Butler J.C., 1982, Artificial isochrons. *Lithos*, **15**, 207–214.

Butler J.C., 1986, The role of spurious correlation in the development of a komatiite alteration model. *J. Geophys. Res.*, **91**, E275–E280.

Butler J.C. and Woronow A., 1986, Discrimination among tectonic settings using trace element abundances of basalts. *J. Geophys. Res.*, **91**, B10289–B10300.

Cabanis B. and Lecolle M., 1989, Le diagramme La/10–Y/15–Nb/8: un outil pour la discrimination des series volcaniques et la mise en evidence des processus de melange et/ou de contamination crustale. *C.R. Acad. Sci. Ser. II*, **309**, 2023–2029.

Cameron M. and Papike J.J., 1981, Structural and chemical variations in pyroxenes. *Amer. Mineral.*, **66**, 1–50.

Campbell A.C. and 10 authors, 1988, Chemistry of hot springs on the Mid-Atlantic Ridge. *Nature*, **335**, 514–519.

Campbell I.H., Naldrett A.J. and Barnes S.J., 1983, A model for the origin of the platinum-rich sulfide horizons in the Bushveld and Stillwater complexes. *J. Petrol.*, **24**, 133–165.

Carmichael I.S.E., 1964, The petrology of Thingmuli, a Tertiary volcano in eastern Iceland. *J. Petrol.*, **5**, 435–460.

Carr P.F., 1985, Geochemistry of late Permian shoshonitic lavas from the southern Sydney Basin. In: Sutherland F.L., Franklin B.J. and Waltho A.E. (eds.), Volcanism in Eastern Australia. *Geol. Soc. Aust., N.S.W. Div. Publ. No. 1*, pp. 165–183.

Chaffey D.J., Cliff R.A. and Wilson B.M., 1989, Characterisation of the St Helena magma source. In: Saunders A.D. and Norry M.J. (eds.), *Magmatism in ocean basins*. Spec. Publ. Geol. Soc. No. 42, pp. 257–276.

Chamberlain C.P. and Rumble D., 1988, Thermal anomalies in a regional metamorphic terrane: an isotopic study of the role of fluids. *J. Petrol.*, **29**, 1215–1232.

Chambers L.A., 1982, Sulfur isotope study of a modern intertidal environment and the interpretation of ancient sulfides. *Geochim. Cosmochim. Acta*, **46**, 721–728.

Chaussidon M., Albarede F. and Sheppard S.M.F., 1989, Sulphur isotope variations in the mantle from ion microprobe analyses of micro-sulphide inclusions. *Earth Planet. Sci. Lett.*, **92**, 144–156.

Chaussidon M. and Lorand J.P., 1990, Sulphur isotope composition of orogenic spinel lherzolite massifs from Ariege (N.E. Pyrenees, France): An ion microprobe study. *Geochim. Cosmochim. Acta*, **54**, 2835–2846.

Chayes F., 1949, On ratio correlation in petrography. *J. Geol.*, **57**, 239–254.

Chayes F., 1960, On correlation between variables of constant sum. *J. Geophys. Res.*, **65**, 4185–4193.

Chayes F., 1971, *Ratio correlation*. Chicago Univ. Press.

Chayes F., 1977, Use of correlation statistics with rubidium–strontium systematics. *Science*, **196**, 1234–1235.

Chayes F. and Velde D., 1965, On distinguishing between basaltic lavas of circumoceanic and ocean-island type by means of discriminant functions. *Amer. J. Sci.*, **263**, 206–222.

Chiba H., Chacko T., Clayton R.N. and Goldsmith J.R., 1989, Oxygen isotope fractionations involving diopside, forsterite, magnetite and calcite: application to geothermometry. *Geochim. Cosmochim. Acta*, 53, 2985–2995.

Chivas A.R., Andrew A.S., Sinha A.K. and O'Neil J.R., 1982, Geochemistry of a Pliocene–Pleistocene oceanic-arc plutonic complex, Guadalcanal. *Nature*, 300, 139–143.

Claisse F., 1989, Automated sample preparation for analysis of geological materials. In: Ahmedali S.T. (ed.), *X-ray fluorescence analysis in the geological sciences: Advances in methodology.* Geol. Assn. Canada: Short course 7, pp. 39–54.

Claoue-Long J.C., Thirlwall M.F. and Nesbitt R., 1984, Sm–Nd systematics of Kambalda greenstones revisited and revised. *Nature*, 307, 697–701.

Claypool G.E., Holser W.T., Kaplan I.R., Sakai H. and Zak I., 1980, The age curves of sulfur and oxygen isotopes in marine sulfate and their mutual interpretation. *Chemical Geol.*, 28, 199–260.

Clayton R.N., 1981, Isotopic thermometry. In: Newton R.C., Navrotsky A. and Wood B.J. (eds.), *Thermodynamics of minerals and melts.* Springer-Verlag, New York, pp. 85–109.

Clayton R.N., 1991, Oxygen isotope thermometer calibrations. In: Taylor H.P., O'Neil J.R., Kaplan I.R. (eds.), *Stable isotope geochemistry: a tribute to Samuel Epstein*, Geochemical Soc. Spec. Publ. No. 3, 3–10.

Clayton R.N., O'Neil J.R. and Mayeda T.K., 1972, Oxygen isotope exchange between quartz and water. *J. Geophys. Res.*, 77, 3057–3067.

Clayton R.N., Goldsmith J.R. and Mayeda T.K., 1989, Oxygen isotope fractionation in quartz, albite, anorthite and calcite. *Geochim. Cosmochim. Acta*, 53, 725–733.

Cliff R.A., 1985, Isotopic dating in metamorphic belts. *J. Geol. Soc. Lond.*, 142, 97–110.

Cliff R.A., Baker P.E. and Mateer N.J., 1991, Geochemistry of inaccessible island volcanics. *Chem. Geol.*, 92, 251–260.

Cocherie A., Auge T. and Meyer G., 1989, Geochemistry of the platinum-group elements in various types of spinels from the Vourinos ophiolitic complex, Greece. *Chem. Geol.*, 77, 27–39.

Cohen A.S., O'Nions R.K., Siegenthaler R. and Griffin W.L., 1988, Chronology of the pressure–temperature history recorded by a granulite terrain. *Contrib. Mineral. Petrol.*, 98, 303–311.

Coleman M.L., 1977, Sulphur isotopes in petrology. *J. Geol. Soc. Lond.*, 133, 593–608.

Coleman M.L. and Raiswell R., 1981, Carbon, oxygen and sulphur isotope variations in concretions from the Upper Lias of N.E. England. *Geochim. Cosmochim. Acta*, 45, 329–340.

Collerson K.D., Campbell L.M., Weaver B.L. and Palacz Z.A., 1991, Evidence for extreme mantle fractionation in early Archaean ultramafic rocks from northern Labrador. *Nature*, 349, 209–214.

Colson R.O., McKay G.A. and Taylor L.A., 1988, Temperature and composition dependencies of trace element partitioning: Olivine/melt and low Ca-pyroxene/melt. *Geochim. Cosmochim. Acta*, 52, 539–553.

Compston W., Williams I.S. and Clement S.W.J., 1982, U–Pb ages with single zircons using a sensitive high mass-resolution ion microprobe. *30th Ann. Conf. Mass Spectrometry*, 593–585.

Compston W., Williams I.S. and Meyer C., 1984, U–Pb geochronology of zircons from lunar breccia 73217 using a sensitive high mass-resolution ion microprobe. *J. Geophys. Res.*, 89 Supplem., B525–B534.

Condie K.C. and Crow C., 1990, Early Precambrian within-plate basalts from the Kaapvaal craton in southern Africa: a case for contaminated komatiites. *J. Geol.*, 98, 100–107.

Condie K.C., Wilks M., Rosen D.M. and Zlobin V.L., 1991, Geochemistry of metasediments from the Precambrian Hapschan series, eastern Anabar Shield, Siberia. *Prec. Res.*, 50, 37–47.

Condie K.C. and Wronkiewicz D.S., 1990, The Ce/Th ratio in Precambrian pelites from the Kaapvaal Craton as an index of cratanic evolution. *Earth Planet. Sci. Lett.*, 97, 256–267.

Connolly C. and Muehlenbachs K., 1988, Contrasting oxygen diffusion in nepheline, diopside and other silicates and their relevance to isotopic systematics in meteorites. *Geochim. Cosmochim. Acta*, **52**, 1585–1591.

Coplen T.B., Kendall C. and Hopple J., 1983, Comparison of stable isotope reference standards. *Nature*, **302**, 236–238.

Coryell C.G., Chase J.W. and Winchester J.W., 1963, A procedure for geochemical interpretation of terrestrial rare-earth abundance patterns. *J. Geophys. Res.*, **68**, 559–566.

Cox K.G., Bell J.D. and Pankhurst R.J., 1979, *The interpretation of igneous rocks*. George, Allen and Unwin, London.

Cox K.G. and Clifford P., 1982, Correlation coefficient patterns and their interpretation in three basaltic suites. *Contrib. Mineral. Petrol.*, **79**, 268–278.

Cross W., Iddings J.P., Pirsson L.V. and Washington H.S., 1903, *Quantitative classification of igneous rocks*. Univ. Chicago Press.

Cullers R.L., 1988, Mineralogical and chemical changes of soil and stream sediment formed by intense weathering of the Danburg granite, Georgia, U.S.A. *Lithos*, **21**, 301–314.

Cullers R.L., Barrett T., Carlson R. and Robinson B., 1987, Rare-earth element and mineralogic changes in Holocene soil and stream sediment: a case study in the Wet Mountains, Colorado, U.S.A. *Chem. Geol.*, **63**, 275–297.

Cullers R.L., Basu A. and Suttner L.J., 1988, Geochemical signature of provenance in sand-mixed material in soils and stream sediments near the Tobacco Root batholith, Montana, U.S.A. *Chem. Geol.*, **70**, 335–348.

Cumming G.L. and Richards J.R., 1975, Ore lead isotope ratios in a continuously changing earth. *Earth Planet. Sci. Lett.*, **28**, 155–171.

Date A.R. and Jarvis K.E., 1989, The applications of ICP–MS in the earth sciences. In: Date A.R. and Gray A.L. (eds.), *The applications of inductively coupled plasma mass spectrometry*. Blackie, Glasgow, pp. 43–70.

Davidson J.P., 1983, Lesser Antilles isotopic evidence of the role of subducted sediment in island arc magma genesis. *Nature*, **306**, 253–256.

Davies G.R., Gledhill A. and Hawkesworth C., 1985, Upper crustal recycling in southern Britain: evidence from Nd and Sr isotopes. *Earth Planet. Sci. Lett.*, **75**, 1–12.

Deines P. and Gold D.P., 1973, The isotopic composition of carbonatite and kimberlite carbonates and their bearing on the isotopic composition of deep seated carbon. *Geochim. Cosmochim. Acta*, **37**, 1709–1733.

De la Roche H. and Leterrier J., 1973, Transposition du tetraedre mineralogique de Yoder et Tilley dans un diagramme chimique de classification des roches basaltique. *C.R. Acad. Sci. Paris, Ser. D.*, **276**, 3115–3118.

De la Roche H., Leterrier J., Grande Claude P. and Marchal M., 1980, A classification of volcanic and plutonic rocks using R1–R2 diagrams and major element analyses — its relationships and current nomenclature. *Chem. Geol.*, **29**, 183–210.

Deloule E., Albarede F. and Sheppard S.M.F., 1991, Hydrogen isotope heterogeneities in the mantle from ion probe analyses of amphiboles from ultramafic rocks. *Earth Planet. Sci. Lett.*, **105**, 543–553.

DePaolo D.J., 1981a, Neodymium isotopes in the Colorado Frant range and crust-mantle evolution in the Proterozoic. *Nature*, **291**, 193–196.

DePaolo D.J., 1981b, Trace element and isotopic effects of combined wallrock assimilation and fractional crystallisation. *Earth Planet. Sci. Lett.*, **53**, 189–202.

DePaolo D.J., 1982, Sm–Nd, Rb–Sr and U–Th–Pb systematics of granulite facies rocks from Fyfe Hills, Enderby Land, Antarctica. *Nature*, **298**, 614–618.

DePaolo D.J., 1988, *Neodymium isotope geochemistry: An introduction*. Springer Verlag, New York.

DePaolo D.J. and Wasserburg G.J., 1976, Nd isotopic variations and petrogenetic models. *Geophys. Res. Lett.*, **3**, 249–252.

DePaolo D.J. and Wasserburg G.J., 1979, Petrogenetic mixing models and Nd–Sr isotopic patterns. *Geochim. Cosmochim. Acta*, **43**, 615–627.

Dickin A.P., 1987, La–Ce dating of lewisian granulites to constrain the ^{138}La β-decay half life. *Nature*, **325**, 337–338.

Dickson J.A.D., 1991, Disequilibrium carbon and oxygen isotope variations in natural calcite. *Nature*, **353**, 842–844.

Dodson M.H., 1973, Closure temperature in cooling geochronological and petrological systems. *Contrib. Mineral. Petrol.*, **40**, 259–274.

Dodson M.H., 1979, Theory of cooling ages. In: Jager E. and Hunziker J.C. (eds.), *Lectures in isotope geology*. Springer-Verlag, New York, pp. 194–202.

Dodson M.H., 1982, On 'spurious' correlations in Rb–Sr isochron diagrams. *Lithos*, **15**, 215–219.

Doe B.R. and **Zartman R.E.**, 1979, Plumbotectonics. In: Barnes H. (ed.), *Geochemistry of hydrothermal ore deposits*. Chapter 2. Wiley, New York.

Dostal J., **Dupuy C.**, **Carron J.P.**, **Le Guen de Kerneizon M.** and **Maury R.C.**, 1983, Partition coefficients of trace elements: application to volcanic rocks of St Vincent, West Indies. *Geochim. Cosmochim. Acta*, **47**, 525–533.

Downes H. and **Dupuy C.**, 1987, Textural, isotopic and REE variations in spinel peridotite xenoliths, Massif central, France. *Earth Planet. Sci. Lett.*, **82**, 121–135.

Downes H. and **Leyreloup A.**, 1986, Granulitic xenoliths from the French Massif Central — petrology, Sr and Nd isotope systematics and model age estimates. In: Dawson J.B., Carswell D.A., Hall J. and Wedepohl K.H. (eds.), *The nature of the lower crust*. Geol. Soc. Lond. Spec. Publ. 24, pp. 319–330.

Drake M.J. and **Holloway J.R.**, 1981, Partitioning of Ni between olivine and silicate melt: the 'Henry's Law problem' reexamined. *Geochim. Cosmochim. Acta*, **45**, 431–437.

Drake M.J. and **Weill D.F.**, 1975, Partition of Sr, Ba, Ca, Y, Eu^{2+}, Eu^{3+} and other REE between plagioclase feldspar and magmatic liquid: an experimental study. *Geochim. Cosmochim. Acta*, **39**, 689–712.

Draper N.R. and **Smith H.**, 1981, *Applied regression analysis*. 2nd edition, Wiley, New York.

Duncan A.R., 1987, The Karoo igneous province — a problem area for inferring tectonic setting from basalt geochemistry. *J. Volc. Geotherm. Res.*, **32**, 13–34.

Dunn T., 1987, Partitioning of Hf, Lu, Ti and Mn between olivine, clinopyroxene and basaltic liquid. *Contrib. Mineral. Petrol.*, **96**, 476–484.

Eissen J-P., **Juteau T.**, **Joron J-L.**, **Dupre B.**, **Humler E.** and **Al'Mukhamedov A.**, 1989, Petrology and geochemistry of basalts from the Red Sea axial rift at 18 deg north. *J. Petrol.*, **30**, 791–839.

Elderfield H., 1986, Strontium isotope stratigraphy. *Palaeogeogr. Palaeoclimatol. Palaeoecol.*, **57**, 71–90.

Elderfield H., 1988, The oceanic chemistry of the rare-earth elements. *Phil. Trans. R. Soc. London*, **A325**, 105–126.

Elderfield H. and **Greaves M.J.**, 1981, Negative cerium anomalies in the rare earth element patterns of oceanic ferromanganese nodules. *Earth Planet. Sci. Lett.*, **55**, 163–170.

Elderfield H. and **Greaves M.J.**, 1982, The rare earth element elements in seawater. *Nature*, **296**, 214–219.

Eldridge C.S., **Compston W.**, **Williams I.S.**, **Both R.A.**, **Walshe J.L.** and **Ohmoto H.**, 1988, Sulfur isotope variability in sediment-hosted massive sulfide deposits as determined using the ion-microprobe, SHRIMP: I. An example from the Rammelsberg orebody. *Econ. Geol.*, **83**, 443–449.

Ellam R.M. and **Hawkesworth C.J.**, 1988, Elemental and isotopic variations in subduction related basalts: evidence for a three component model. *Contrib. Mineral. Petrol.*, **98**, 72–80.

Elthon D., 1983, Isomolar and isostructural pseudo–liquidus phase diagrams for oceanic basalts. *Amer. Mineral.*, **68**, 506–511.

Englund J-O. and **Jorgensen P.**, 1973, A chemical classification system for argillaceous sediments and factors affecting their composition. *Geol. Foren. Stockholm Forh.*, **95**, 87–97.

Epstein S., 1970, Antarctic ice sheet: stable isotope analysis of Byrd Station coves and interhemispheric climatic implications. *Science*, **168**, 570–572.

Epstein S., **Buchsbaum R.**, **Lowenstam H.A.** and **Urey H.C.**, 1953, Revised carbonate–water isotopic temperature scale. *Geol. Soc. Amer. Bull.*, **64**, 1315–1326.

Epstein S., Sharp R.P. and Gow A.J., 1965, Six-year record of hydrogen and oxygen isotope variations in South Pole fur. *J. Geophys. Res.*, **70**, 1809–1814.

Ernst R.E., Fowler A.D. and Pearce T.H., 1988, Modelling of igneous fractionation and other processes using Pearce diagrams. *Contrib. Mineral. Petrol.*, **100**, 12.

Eugster H.P. and Wones D.R., 1962, Stability relations of the ferromagnesian biotite, annite. *J. Petrol.*, 3, 82–125.

Evensen N.M., Hamilton P.J. and O'Nions R.K., 1978, Rare earth abundances in chondritic meteorites. *Geochim. Cosmochim. Acta*, **42**, 1199–1212.

Ewart A., 1982, The mineralogy and petrology of Tertiary–Recent orogenic volcanic rocks with special reference to the andesitic–basaltic composition range. In: Thorpe R.S. (ed.), *Andesites*. Wiley, Chichester, pp. 25–87.

Exley R.A., Mattey D.P., Clague D.A. and Pillinger C.T., 1986, Carbon isotope systematics of a mantle 'hot-spot': a comparison of Loihi seamount and MORB glasses. *Earth Planet. Sci. Lett.*, **78**, 189–199.

Farver J.R., 1989, Oxygen self diffusion in diopside with application to cooling rate determination. *Earth Planet. Sci. Lett.*, **92**, 386–396.

Faure G., 1977, *Principles of isotope geology*. Wiley, New York.

Faure G., 1986, *Principles of isotope geology*. 2nd edition, Wiley, New York.

Faure G., Hoefs J. and Mensing T.M., 1984, Effect of oxygen fugacity on sulphur isotope compositions and magnetite concentrations in the Kirkpatrick basalt, Mount Falla, Queen Alexandra Range, Antarctica. *Isotope Geosci.*, **2**, 301–311.

Fears D., 1985, A corrected CIPW norm program for interactive use. *Comput. Geosci.*, **11**, 787–797.

Ferry J.M., 1983, Mineral reactions and element migration during metamorphism of calcareous sediments from the Vasselboro Formation, south-central Maine. *Amer. Mineral.*, **68**, 334–354.

Field C.W. and Fifarek R.H., 1986, Light stable isotope systematics in the epithermal environment. In: Berger B.R. and Bethke P.M. (eds.), *Geology and geochemistry of epithermal systems. Society of Economic Geologists, Rev. Econ. Geol.*, **2**, 99–128.

Fitton J.G., James D., Kempton P.D., Ormerod D.S. and Leeman W.P., 1988, The role of the lithospheric mantle in the generation of late Cenozoic basic magmas in the western United States. *J. Petrol.*, Special lithosphere issue, 331–349.

Fleet A.J., 1984, Aqueous and sedimentary geochemistry of the rare earth elements. In: Henderson P. (ed.), *Rare earth element geochemistry*. Elsevier, pp. 343–373.

Flegal A.R., Itoh K., Patterson C.C. and Wong C.S., 1986, Vertical profile of lead isotopic compositions in the north-east Pacific. *Nature*, **321**, 689–690.

Fletcher I.R. and Rosman K.J.R., 1982, Precise determination of initial ε_{Nd} from Sm–Nd isochron data. *Geochim. Cosmochim. Acta*, **46**, 1983–1987.

Fletcher T.A., Boyce A.J. and Fallick A.E., 1989, A sulphur isotope study of Ni–Cu mineralisation in the Huntly–Knock Caledonian mafic and ultramafic intrusions of northeast Scotland. *J. Geol. Soc. Lond.*, **146**, 675–684.

Floyd P.A. and Winchester J.A., 1975, Magma-type and tectonic setting discrimination using immobile elements. *Earth Planet. Sci. Lett.*, **27**, 211–218.

Floyd P.A., Winchester J.A. and Park R.G., 1989, Geochemistry and tectonic setting of Lewisian clastic metasediments from the early Proterozoic Loch Maree group of Gairloch, NW Scotland. *Precambrian Res.*, **45**, 203–214.

Francis D., 1985, The pyroxene paradox in MORB glasses — A signature of picritic parental magmas? *Nature*, **319**, 586–589.

Friedman I. and O'Neil J.R., 1977, *Data of geochemistry*. Compilation of stable isotope fractionation factors of geochemical interest. U.S. Geological Survey Professional Paper. 440-KK.

Fujimaki H., 1986, Partition coefficients of Hf, Zr and REE between zircon, apatite and liquid. *Contrib. Mineral. Petrol.*, **94**, 42–45.

Fujimaki H., Tatsumoto M. and **Aoki K.**, 1984, Partition coefficients of Hf, Zr and REE between phenocrysts and groundmasses. Proceedings of the fourteenth lunar and planetary science conference, Part 2. *J. Geophys. Res.*, **89**, Suppl. B662–B672.

Galer S.J.G. and **O'Nions R.K.**, 1985, Residence time of thorium, uranium and lead in the mantle with implications for mantle convection. *Nature*, **316**, 778–782.

Garcia M.O., 1978, Criteria for the identification of ancient volcanic areas. *Earth Sci. Rev.*, **14**, 147–165.

Garcia R.B. and **Frias J.M.**, 1990, BITERCLA: GW-basic program to plot classification diagrams. *Comput. Geosci.*, **16**, 265–271.

Gebauer D. and **Grunenfelder M.**, 1979, U–Th–Pb dating of minerals. In: Jager E. and Hunziker J.C. (eds.), *Lectures in isotope geology*. Springer-Verlag, New York, pp. 105–131.

Gelinas L., Mellinger M. and **Trudel P.**, 1982, Archaean mafic metavolcanics from the Rouyn–Noranda district, Abitibi greenstone belt, Quebec. 1. Mobility of the major elements. *Can. J. Earth Sci.*, **19**, 2258–2275.

Gerlach D.C., Cliff R.A., Davies G.R., Norry M.J. and **Hodgson N.**, 1988, Magma sources of the Cape Verdes archipelago: isotopic and trace element constraints. *Geochim. Cosmochim. Acta*, **52**, 2979–2992.

Ghiorso M.S., 1985, Chemical mass transfer in magmatic processes. I. Thermodynamic relations and numerical algorithms. *Contrib. Mineral. Petrol.*, **90**, 107–120.

Ghiorso M.S. and **Carmichael I.S.E.**, 1985, Chemical mass transfer in magmatic processes. II. Applications in equilibrium crystallisation, fractionation and assimilation. *Contrib. Mineral. Petrol.*, **90**, 121–141.

Giletti B.J., 1986, Diffusion effects on oxygen isotope temperatures of slowly cooled igneous and metamorphic rocks. *Earth Planet. Sci. Lett.*, **77**, 218–228.

Gill J.B., 1981, *Orogenic andesites and plate tectonics*. Springer, Berlin.

Girard J-P., Savin S.M. and **Aronson J.L.**, 1989, Diagenesis of the lower Cretaceous arkoses of the Angola margin: petrologic, K/Ar dating and $^{18}O/^{16}O$ evidence. *J. Sed. Pet.*, **59**, 519–538.

Glazner A.F., 1984, A short CIPW norm. *Comput. Geosci.*, **10**, 449–450.

Goldstein S.L., 1988, Decoupled evolution of Nd and Sr isotopes in the continental crust. *Nature*, **336**, 733–738.

Goldstein S.L., O'Nions R.K. and **Hamilton P.J.**, 1984, A Sm–Nd study of atmospheric dusts and particulates from major river systems. *Earth Planet. Sci. Lett.*, **70**, 221–236.

Gonfiantini R., 1978, Standards for stable isotope measurements in natural compounds. *Nature*, **271**, 534–536.

Govindaraju K., 1984, 1984 compilation of working values and sample description for 170 international reference samples of mainly silicate rocks and minerals. *Geostandards Newsletter*, Special Issue no. 8.

Graham C.M. and **Harmon R.S.**, 1983, Stable isotope evidence on the nature of crust–mantle interactions. In: Hawkesworth C.J. and Norry M.J. (eds.), *Continental basalts and mantle xenoliths*. Shiva, Nantwich, pp. 20–45.

Graham C.M., Harmon R.S. and **Sheppard S.M.F.**, 1984, Experimental hydrogen isotope studies: hydrogen isotope exchange between amphibole and water. *Amer. Mineral.*, **69**, 128–138.

Graham C.M., Sheppard S.M.F. and **Heaton T.H.E.**, 1980, Experimental hydrogen isotope studies I: Systematics of hydrogen isotope fractionation in the systems epidote–H_2O, zoisite–H_2O and AlO(OH)-H_2O. *Geochim. Cosmochim. Acta*, **44**, 353–364.

Gray C.M., Cliff R.A. and **Goode A.D.T.**, 1981, Neodymium–strontium isotopic evidence for extreme contamination in a layered basic intrusion. *Earth Planet. Sci. Lett.*, **56**, 189–198.

Green T.H. and **Pearson N.J.**, 1983, Effect of pressure on rare earth element partition coefficents in common magmas. *Nature*, **305**, 414–416.

Green T.H. and **Pearson N.J.**, 1985a, Experimental determination of REE partition coefficients between amphibole and basaltic liquids at high pressure. *Geochim. Cosmochim. Acta*, **49**, 1465–1468.

Green T.H. and **Pearson N.J.**, 1985b, Rare earth element partitioning between clinopyroxene and silicate liquid at moderate to high pressure. *Contrib. Mineral. Petrol.*, **91**, 24–36.

Green T.H. and **Pearson N.J.**, 1986, Rare-earth element partitioning between sphene and coexisting silicate liquid at high pressure and temperature. *Chem. Geol.*, **55**, 105–119.

Green T.H. and **Pearson N.J.**, 1987, An experimental study of Nb and Ta partitioning between Ti-rich minerals and silicate liquids at high pressure and temperature. *Geochim. Cosmochim. Acta*, **51**, 55–62.

Green T.H., Sie S.H., Ryan C.G. and **Cousens D.R.**, 1989, Proton microprobe-determined partitioning of Nb, Ta, Zr, Sr and Y between garnet, clinopyroxene and basaltic magma at high pressure and temperature. *Chem. Geol.*, **74**, 201–216.

Gregory R.T. and **Taylor H.P.**, 1981, An oxygen isotope profile in a section of Cretaceous oceanic crust, Semail ophiolite Oman: evidence for $\delta^{18}O$ buffering of the oceans by deep (> 5 km) seawater—hydrothermal circulation at mid-ocean ridges. *J. Geophys. Res.*, **86**, 2737–2755.

Gregory R.T., Criss R.E. and **Taylor H.P.**, 1989, Oxygen isotope exchange kinetics of mineral pairs in close and open systems: applications to problems of hydrothermal alteration of igneous rocks and Precambrian iron formations. *Chem. Geol.*, **75**, 1–42.

Gromet L.P., Dymek R.F., Haskin L.A. and **Korotev R.L.**, 1984, The "North American Shale Composite": its compilation, major and trace element characteristics. *Geochim. Cosmochim. Acta*, **48**, 2469–2482.

Grove T.L., Gerlach D.C. and **Sando T.W.**, 1982, Origin of late calc–alkaline series lavas at Medicine Lake Volcano by fractionation, assimilation and mixing. *Contrib. Mineral. Petrol.*, **80**, 160–182.

Grove T.L., Gerlach D.C. and **Sando T.W.**, 1983, Origin of late calc–alkaline series lavas at Medicine Lake Volcano by fractionation, assimilation and mixing: corrections and clarifications. *Contrib. Mineral. Petrol.*, **82**, 407–408.

Groves D.I., Golding S.D., Rock N.M.S., Barley M.E. and **McNaughton N.J.**, 1988, Archean carbon reservoirs and their relevance to the fluid source for gold deposits. *Nature*, **331**, 254–257.

Hagen H. and **Neumann E-R.**, 1990, Modelling of trace element distribution in magma chambers using open system models. *Comput. Geosci.*, **16**, 549–586.

Halbach P. and 17 other authors, 1989, Probable modern analogue of Kuroko-type massive sulphide deposits in the Okinawa Trough back-arc basin. *Nature*, **338**, 496–499.

Hall A., 1987, *Igneous petrology*. Longman, London.

Hall W.E., Friedman I. and **Nash J.T.**, 1974, Fluid inclusion and light stable isotope study of the Climax molybdenum deposits, Colorado. *Econ. Geol.*, **69**, 884–901.

Hamilton D.L. and **MacKenzie W.S.**, 1965, Phase equilibria studies in the system $NaAlSiO_4$ (nepheline)–$KAlSiO_4$(kalsilite)–SiO_2–H_2O. *Mineral. Mag.*, **34**, 214–231.

Hamilton P.J., Evensen N.M., O'Nions R.K. and **Tarney J.**, 1979a, Sm–Nd systematics of Lewisian gneisses implications for the origin of granulites. *Nature*, **277**, 25–28.

Hamilton P.J., Evensen N.M., O'Nions R.K., Smith H.S. and **Erlank A.J.**, 1979b, Sm–Nd dating of Onverwacht group volcanics, southern Africa. *Nature*, **279**, 298–300.

Hanson G.N., 1978, The application of trace elements to the petrogenesis of igneous rocks of granitic composition. *Earth Planet. Sci. Lett.*, **38**, 26–43.

Hanson G.N. and **Langmuir C.H.**, 1978, Modelling of major and trace elements in mantle–melt systems using trace element approaches. *Geochim. Cosmochim. Acta*, **42**, 725–741.

Harker A., 1909, *The natural history of igneous rocks*. Methuen, London.

Harmer R.E. and **Eglington B.M.**, 1987, The mathematics of geochronometry: Equations for use in regression calculations. National Physical Research Laboratory, Geochronology Division C.S.I.R., South Africa.

Harris N.B.W., Pearce J.A. and **Tindle A.G.**, 1986, Geochemical characteristics of collision-zone magmatism. In: Coward M.P. and Reis A.C. (eds.), *Collision tectonics*. Spec. Publ. Geol. Soc., **19**, 67–81.

Harris P.G., 1974, Origin of alkaline magmas as a result of anatexis. In: Sorenson H. (ed.), *The alkaline rocks*, Wiley, London.

Harrison A.G. and **Thode H.G.**, 1957, The kinetic isotope effect in the chemical reduction of sulphate. *Trans. Faraday Soc.*, **53**, 1648–1651.

Harrison T.M. and McDougall I., 1980, Investigation of an intrusive contact in NW Nelson, New Zealand. I. Thermal, chronological and isotopic constraints. *Geochim. Cosmochim. Acta*, **44**, 1985–2003.

Harrison W.J. and Wood B.J., 1980, An experimental investigation of the partitioning between garnet and liquid with reference to the role of defect equilibria. *Contrib. Mineral. Petrol.*, **72**, 145–155.

Hart M.B. and Leary P.N., 1989, The stratigraphic and palaeogeographic setting of the late Cenomanian 'anoxic' event. *J. Geol. Soc. Lond.*, **146**, 305–310.

Hart S.R., 1984, A large-scale isotope anomaly in the southern hemisphere mantle. *Nature*, **309**, 753–757.

Hart S.R. and Davis K.E., 1978, Nickel partitioning between olivine and silicate melt. *Earth Planet. Sci. Lett.*, **40**, 203–219.

Haskin L.A., Haskin M.A., Frey F.A. and Wildman T.R., 1968, Relative and absolute terrestrial abundances of the rare earths. In: Ahrens L.H. (ed.), *Origin and distribution of the elements*, vol. 1. Pergamon, Oxford, pp. 889–911.

Haskin M.A. and Frey F.A., 1966, Dispersed and not-so-rare earths. *Science*, **152**, 299–314.

Haskin M.A. and Haskin L.A., 1966, Rare earths in European shales: a redetermination. *Science*, **154**, 507–509.

Hattori K., 1989, Barite–celestine intergrowths in Archaen plutons: the product of oxidising hydrothermal activity related to alkaline intrusions. *Amer. Mineral.*, **74**, 1270–1277.

Hattori K., Campbell F.A. and Krouse H.R., 1983, Sulphur isotope abundances in Aphebian clastic rocks: implications for the coeval atmosphere. *Nature*, **302**, 323–326.

Haughton P.D.W., 1988, A cryptic Caledonian flysch terrane in Scotland. *J. Geol. Soc. Lond.*, **145**, 685–703.

Hawkesworth C.J., Hammill M., Gledhill A.R., van Calsteren P. and Rogers G., 1982, Isotope and trace element evidence for late-stage intra-crustal melting in the High Andes. *Earth Planet. Sci. Lett.*, **58**, 240–254.

Hawkesworth C.J., Marsh J.S., Duncan A.R., Erlank A.J. and Norry M.J., 1984, The role of continental lithosphere in the generation of the Karoo volcanic rocks: evidence from combined Nd- and Sr-isotope studies. *Spec. Publ. Geol. Soc. S. Africa*, **13**, 341–354.

Hawkesworth C.J. and van Calsteren P.W.C., 1984, Radiogenic isotopes — some geological applications. In: Henderson P. (ed.), *Rare earth element geochemistry*. Elsevier, Amsterdam, pp. 375–421.

Helz R.T., 1987, Diverse olivene types in lava in the 1959 eruption of Kilauea volcano and their bearing on eruption dynamics. In: Decker R.W., Wright T.L. and Stauffer P.H. (eds), *Volcanism in Hawaii*. US Geol. Surv. Prof. Paper 1350, pp. 691–722.

Henderson P., 1982, *Inorganic geochemistry*. Pergamon, Oxford.

Henderson P. and Pankhurst R.J., 1984, Analytical chemistry. In: Henderson P. (ed.), *Rare earth element geochemistry*. Elsevier, Amsterdam, pp. 467–499.

Hergt J.M., Chappell B.W., McCulloch M.T., McDougall I. and Chivas A.R., 1989, Geochemical and isotopic constraints on the origin of Jurassic dolerites of Tasmania. *J. Petrol.*, **30**, 841–883.

Herron M.M., 1988, Geochemical classification of terrigenous sands and shales from core or log data. *J. Sed. Petrol.*, **58**, 820–829.

Herron M.M. and Herron S.L., 1990, Geological applications of geochemical well logging. In: Hurst A., Lovell M.A. and Morton A.C. (eds.), *Geological applications of wireline logs*, Spec. Publ. Geol. Soc. No. 48, pp. 165–175.

Hertogen J. and Gijbels R., 1976, Calculation of trace element fractionation during partial melting. *Geochim. Cosmochim. Acta*, **40**, 313–322.

Herzberg C.T., 1992, Depth and degree of melting of komatiites. *J. Geophys. Res.*, **97**, 4521–4540.

Hickson C.J. and Juras S.J., 1986, Sample contamination and grinding. *Canadian Mineralogist*, **24**, 585–589.

Hildreth W., 1981, Gradients in silicic magma chambers: implications for lithospheric magmatism. *J. Geophys. Res.*, **86**, B10153–B10192.

Hinton R.W., 1990, Ion microprobe trace-element analysis of silicates: measurement of multi-element glasses. *Chem. Geol.*, **83**, 11–25.

Hirt B., Tilton G.R., Herr W. and Hoffmeister W., 1963, The half-life of ^{187}Re. In: Geiss J. and Goldberg E.D. (eds.), *Earth science and meteorites*. North Holland, Amsterdam, pp. 273–280.

Hoefs J., 1987, *Stable isotope geochemistry*. 3rd edition, Springer-Verlag, Berlin.

Hoernle K., Tilton G. and Schmincke H-U., 1991, Sr–Nd–Pb isotopic evolution of Gran Canaria: evidence for shallow enriched mantle beneath the Canary Islands. *Earth Planet. Sci. Lett.*, **106**, 44–63.

Hoffman A., Gruszczynski M. and Malkowski K., 1991, On the interrelationship between temporal trends in δ^{13}C, δ^{18}O and δ^{34}S in the world ocean. *J. Geol.*, **99**, 355–370.

Hofmann A.W. and Hart S.R., 1978, An assessment of local and regional isotopic equilibrium in the mantle. *Earth Planet. Sci. Lett.*, **38**, 44–62.

Holm P.E., 1982, Non-recognition of continental tholeiites using the Ti–Y–Zr diagram. *Contrib. Mineral. Petrol.*, **79**, 308–310.

Holm P.E., 1990, Complex petrogenetic modelling using spreadsheet software. *Comput. Geosci.*, **16**, 1117–1122.

Hooker P.J., Hamilton P.J. and O'Nions R.K., 1981, An estimate of the Nd isotopic composition of iapetus seawater from *ca* 490 Ma metalliferous sediments. *Earth Planet. Sci. Lett.*, **56**, 180–188.

Hostetler C.J. and Drake M.J., 1980, Predicting major element mineral/melt equilibria: a statistical approach. *J. Geophys. Res.*, **85**, 3789–3796.

Hoyle J., Elderfield H., Gledhill A. and Greaves M., 1984, The behaviour of the rare-earth elements during the mixing of river and sea waters. *Geochim. Cosmochim. Acta*, **48**, 143–149.

Huang W-L. and Wyllie P.J., 1975, Melting relations in the system $NaAlSi_3O_8$–$KAlSi_3O_8$–SiO_2 to 35 kilobars, Dry and excess water. *J. Geol.*, **83**, 737–748.

Hudson J.D., 1977, Stable isotopes and limestone lithification. *J.Geol. Soc. Lond.*, **133**, 637–660.

Huebner M., Kyser T.K. and Nisbet E.G., 1986, Stable-isotope geochemistry of the high-grade metapelites from the Central zone of the Limpopo belt. *Amer. Mineral.*, **71**, 1343–1353.

Humphries S.E., 1984, The mobility of the rare earth elements in the crust. In: Henderson P. (ed.), *Rare earth element geochemistry*. Elsevier, Amsterdam, pp. 315–341.

Innocenti F., Manetti P., Mazzuuoli R., Pasquare G. and Villari, 1982, Anatolia and north-western Iran. In: Thorpe R.S. (ed.), *Andesites*. Wiley, Chichester, pp. 327–349.

Irvine T.N. and Baragar W.R.A., 1971, A guide to the chemical classification of the common volcanic rocks. *Can. J. Earth Sci.*, **8**, 523–548.

Irving A.J., 1978, A review of experimental studies of crystal/liquid trace element partitioning. *Geochim. Cosmochim. Acta*, **42**, 743–770.

Irving A.J. and Frey F.A., 1978, Distribution of trace elements between garnet megacrysts and host volcanic liquids of kimberlitic to rhyolitic composition. *Geochim. Cosmochim. Acta*, **42**, 771–787.

Jackson D.H., Mattey D.P. and Harris N.B.W., 1988, Carbon isotope compositions of fluid inclusions in charnockites from southern India. *Nature*, **333**, 167–170.

Jacobsen S.B. and Wasserburg G.J., 1980, Sm–Nd isotopic evolution of chondrites. *Earth Planet. Sci. Lett.*, **50**, 139–155.

Jager E. and Hunziker J.C. (eds.), 1977, *Lectures in isotope geology*. Springer-Verlag, New York.

Jagoutz E., Palme H., Baddenhausen H., Blum K., Cendales M., Dreibus G., Spottel B., Lorenz V. and Wanke H., 1979, The abundances of major, minor and trace elements in the earth's mantle as derived from primitive ultramafic nodules. *Proc. Lunar and Planet. Sci. Conf.* No. 10, *Geochim. Cosmochim. Acta*, Supplement 11, 2031–2050.

Jahn B.-M., Vidal P. and Tilton G.R., 1980, Archaean mantle heterogeneity: evidence from chemical and isotopic abundances in Archaean igneous rocks. *Phil. Trans. R. Soc. Lond.*, **A297**, 353–364.

James D.E., 1981, The combined use of oxygen and radiogenic isotopes as indicators of crustal contamination. *Ann. Rev. Earth Planet. Sci.*, **9**, 311–344.

James, R.S. and Hamilton D.L., 1969, Phase relations in the system $NaAlSi_3O_8$–

KAlSi$_3$O$_8$–CaAl$_2$Si$_3$O$_8$–SiO$_2$ at 1 kilobar water vapour pressure. *Contrib. Mineral. Petrol.*, **21**, 111–141.

Jarvis K.E. and Williams J.G., 1989, The analysis of geological samples by slurry nebulisation inductively coupled plasma-mass spectrometry (ICP–MS). *Chem. Geol.*, **77**, 53–63.

Javoy M., 1977, Stable isotopes and geothermometry. *J. Geol. Soc. Lond.*, **133**, 609–636.

Javoy M., Fourcade S. and Allegre C.J., 1970, Graphical method for examining ^{18}O/^{16}O fractionation in silicate rocks. *Earth Planet. Sci. Lett.*, **10**, 12–16.

Javoy M., Pineau F. and Delorme H., 1986, Carbon and nitrogen isotopes in the mantle. *Chem. Geol.*, **57**, 41–62.

Javoy M., Pineau F. and Ilyama I., 1978, Experimental determination of the isotopic fractionation between gaseous CO$_2$ and carbon dissolved in tholeiitic magma: a preliminary study. *Contrib. Mineral. Petrol.*, **67**, 35–39.

Jenkin G.R.T., Fallick A.E., Farrow C.M. and Bowes G.E., 1991, COOL: a FORTRAN–77 computer program for modelling stable isotopes in cooling closed systems. *Comput. Geosci.*, **17**, 391–412.

Jenner G.J., Longerich H.P., Jackson S.E. and Fryer B.J., 1990, ICP–MS a powerful tool for high precision trace-element analysis in earth sciences; evidence from analysis of selected U.S.G.S. reference samples. *Chem. Geol.*, **83**, 133–148.

Jensen L.S., 1976, *A new cation plot for classifying subalkalic volcanic rocks.* Ontario Div. Mines. Misc. Pap. 66.

Jensen L.S. and Pyke D.R., 1982, Komatiites in the Ontario portion of the Abitibi belt. In: Arndt N.T. and Nisbet E.G. (eds.), *Komatiites.* George Allen and Unwin, London, pp. 147–157.

Jochum K.P. and Hofmann A.W., 1989, Fingerprinting geological material using SSMS — comment. *Chem. Geol.*, **75**, 249–251.

Jochum K.P., Seufert H.M. and Thirlwall M.F., 1990, High-sensitivity Nb analysis by spark source mass spectrometry (SSMS) and calibration of XRF Nb and Zr. *Chem. Geol.*, **81**, 1–16.

Johannes W., 1980, Metastable melting in the granite system Qz–Or–Ab–An–H$_2$O. *Contrib. Mineral. Petrol.*, **72**, 73–80.

Johannes W., 1983, Metastable melting in granite and related systems. In: Atherton M.P. and Gribble C.D. (eds.), *Migmatites, melting and metamorphism.* Shiva, pp. 27–36.

Johannes W., 1984, Beginning of melting in the granite system Qz–Or–Ab–An–H$_2$O. *Contrib. Mineral. Petrol.*, **86**, 264–273.

Kay R.W. and Hubbard N.J., 1978, Trace elements in ocean ridge basalts. *Earth Planet. Sci. Lett.*, **38**, 95–116.

Kelley S.P. and Fallick A.E., 1990, High precision spatially resolved analysis of δ^{34}S in sulphides using a laser extraction technique. *Geochim. Cosmochim. Acta*, **54**, 883–888.

Kelsey C.H., 1965, Calculation of the CIPW norm. *Mineral. Mag.*, **34**, 276–282.

Kenny B.C., 1982, Beware spurious self-correlations! *Water Resources Res.*, **18**, 1041.

Kermack K.A. and Haldane J.B.S., 1950, Organic correlation in allometry. *Biometrika*, **37**, 30–41.

Kerrich R. and Fryer B.J., 1979, Archaean precious-metal hydrothermal systems, Dome Mine, Abitibi greenstone belt. II, REE and oxygen isotope relations. *Can. J. Earth Sci.*, **16**, 440–458.

Kerridge J.F., 1985, Carbon, hydrogen and nitrogen in carbonaceous chondrites: abundances and isotopic compositions in bulk samples. *Geochim. Cosmochim. Acta*, **49**, 1707–1714.

Kerridge J.F., Haymon R.M. and Kastner M., 1983, Sulfur isotope systematics at the 21°N site, East Pacific Rise. *Earth Planet. Sci. Lett.*, **66**, 91–100.

Kilinc A., Carmichael I.S.E., Rivers M.L. and Sack R.O., 1983, The ferric–ferrous ratio of natural silicate liquids equilibrated in air. *Contrib. Mineral. Petrol.*, **83**, 136–140.

Kramers J.D., Smith C.B., Lock N.P., Harmon R.S. and Boyd F.R., 1981, Can Kimberlites be generated from ordinary mantle? *Nature*, **291**, 53–56.

Kreulen R., 1988, High integrated fluid–rock ratios during metamorphism at Naxos: evidence from carbon isotopes of calcite in schists and fluid inclusions. *Contrib. Mineral. Petrol.*, **98**, 28–32.

Kronberg B.I., Murray F.H., Daddar R. and Brown J.R., 1988, Fingerprinting geological materials using SMSS. *Chem. Geol.*, **68**, 351–359.

Kroner A., Williams I.S., Compston W., Baur N., Vitanage P.W. and Perera L.R.K., 1987, Zircon ion microprobe dating of high-grade rocks in Sri Lanka. *J. Geol.*, **95**, 775–791.

Kuno H., 1966, Lateral variation of basalt magma types across continental margins and island arcs. *Bull. Volcanol.*, **29**, 195–222.

Kuno H., 1968, Differentiation of basalt magmas. In: Hess H.H. and Poldervaart A. (eds.), *Basalts: The Poldervaart treatise on rocks of basaltic composition*, Vol. 2. Interscience, New York, pp. 623–688.

Kyser T.K. and O'Neil J.R., 1984, Hydrogen isotope systematics of submarine basalts. *Geochim. Cosmochim. Acta*, **48**, 2123–2133.

Kyser T.K., O'Neil J.R. and Carmichael I.S.E., 1981, Oxygen isotope thermometry of basic lavas and mantle nodules. *Contrib. Mineral. Petrol.*, **77**, 11–23.

Kyser T.K., O'Neil J.R. and Carmichael I.S.E., 1982, Genetic relations among basic lavas and ultramafic nodules: evidence from oxygen isotope compositions. *Contrib. Mineral. Petrol.*, **81**, 88–102.

Land L.S. and Dutton S.P., 1978, Cementation of a Pennsylvanian deltaic sandstone: isotopic data. *J. Sed. Petrol.*, **48**, 1167–1176.

Langmuir C.H., 1989, Geochemical consequences of *in situ* crystallisation. *Nature*, **340**, 199–205.

Langmuir C.H., Bender J.F., Bence A.E., Hanson G.N. and Taylor S.R., 1977, Petrogenesis of basalts from the FAMOUS area: mid-Atlantic ridge. *Earth Planet. Sci. Lett.*, **36**, 133–156.

Langmuir C.H. and Hanson G.N., 1980, An evaluation of major element heterogeneity in the mantle sources of basalts. *Phil. Trans. R. Soc. Lond.*, **A297**, 383–407.

Langmuir C.H., Vocke R.D., Hanson G.N. and Hart S.R., 1978, A general mixing equation with applications to icelandic basalts. *Earth Planet. Sci. Lett.*, **37**, 380–392.

Leake B.E., Hendry G.L., Kemp A., Plant A.G., Harvey P.K., Wilson J.R., Coats J.S., Aucott J.W., Lunel T. and Howarth R.J., 1969, The chemical analysis of rock powders by automated X-ray fluorescence. *Chem. Geol.*, **5**, 7–86.

Lechler P.J. and Desilets M.O., 1987, A review of the use of loss on ignition as a measurement of total volatiles in whole rock analysis. *Chem. Geol.*, **63**, 341–344.

Leeman W.P. and Lindstrom D.J., 1978, Partitioning of Ni^{2+} between basaltic melt and synthetic melts and olivines — an experimental study. *Geochim. Cosmochim. Acta*, **42**, 801–816.

Le Maitre R.W., 1968, Chemical variation within and between volcanic rock series — a statistical approach. *J. Petrol.*, **9**, 220–252.

Le Maitre R.W., 1976, The chemical variability of some common igneous rocks. *J. Petrol.*, **17**, 589–637.

Le Maitre R.W., 1981, GENMIX — a generalised petrological mixing model program. *Comput. Geosci.*, **7**, 229–247.

Le Maitre R.W., 1982, *Numerical petrology; statistical interpretation of geochemical data*. Elsevier, Amsterdam.

Le Maitre R.W., Bateman P., Dudek A., Keller J., Lameyre Le Bas M.J., Sabine P.A., Schmid R., Sorensen H., Streckeisen A., Woolley A.R. and Zanettin B., 1989, *A classification of igneous rocks and glossary of terms*. Blackwell, Oxford.

Leterrier J., Maury R.C., Thonon P., Girard D. and Marchal M., 1982, Clinopyroxene composition as a method of identification of the magmatic affinities of palaeo-volcanic series. *Earth Planet. Sci. Lett.*, **59**, 139–154.

Liew T.C. and McCulloch M.T., 1985, Genesis of granitoid batholiths of Peninsular Malaysia and implications for models of crustal evolution: Evidence from a Nd–Sr isotopic and U–Pb lead study. *Geochim. Cosmochim. Acta*, **49**, 587–600.

Lindstrom D.J. and Weill D.F., 1978, Partitioning of transition metals between diopside and coexisting silicate liquids. I. Nickel, cobalt and manganese. *Geochim. Cosmochim. Acta*, **42**, 817–831.

Long J.V.P., 1967, Electron probe microanalysis. In: Zussman J. (ed.), *Physical methods in determinative mineralogy*. Academic Press, New York. pp. 215–260.

Long P.E., 1978, Experimental determination of partition coefficients for Rb, Sr and Ba between alkali feldspar and silicate liquid. *Geochim. Cosmochim. Acta*, **42**, 833–846.

Lugmair G.W. and Marti K., 1978, Lunar initial $^{143}Nd/^{144}Nd$: differential evolution of the lunar crust and mantle. *Earth Planet. Sci. Lett.*, **39**, 349–357.

Lugmair G.W., Scheinin N.B. and Marti K., 1975, Search for extinct ^{146}Sm, 1. The isotopic abundance of ^{142}Nd in the Juvinas meteorite. *Earth Planet. Sci. Lett.*, **27**, 79–84.

Luth W.C., 1969, The systems $NaAlSi_3O_8–SiO_2$ and $KAlSi_3O_8–SiO_2$ to 20 kb and the relationship between H_2O content, P_{H_2O} and P_{total} in granitic magmas. *Amer. J. Sci.*, **267A**, 325–341.

Luth W.C., 1976, Granitic rocks. In: Bailey D.K. and MacDonald R. (eds.), *The evolution of the crystalline rocks*. Academic Press, New York, pp. 335–417.

Luth W.C., Jahns R.H. and Tuttle O.F., 1964, The granite system at pressures of 4–10 kbar. *J. Geophys. Res.*, **69**, 759–773.

Maas R. and McCulloch M.T., 1991, The provenance of Archaean clastic metasediments in the Narryer Gneiss Complex, western Australia: trace element geochemistry, Nd isotopes and U–Pb ages for detrital zircons. *Geochim. Cosmochim. Acta*, **55**, 1915–1932.

McConnaughey T., 1989, ^{13}C and ^{18}O isotopic disequilibrium in biological carbonates: I. Patterns. *Geochim. Cosmochim. Acta*, **53**, 151–162.

McCulloch M.T. and Black L.P., 1984, Sm–Nd isotopic systematics of Enderby Land granulites and evidence for the redistribution of Sm and Nd during metamorphism. *Earth Planet. Sci. Lett.*, **71**, 46–58.

McCulloch M.T. and Chappell B.W., 1982, Nd isotopic characteristics of S- and I-type granites. *Earth Planet. Sci. Lett.*, **58**, 51–64.

McCulloch M.T., Jaques A.L., Nelson D.R. and Lewis J.D., 1983, Nd and Sr isotopes in kimberlites and lamproites from western Australia: an enriched mantle origin. *Nature*, **302**, 400–403.

McDermott F. and Hawkesworth C.J., 1991, Th, Pb and Sr isotopic variations in young island arc volcanics and oceanic sediments. *Earth Planet. Sci. Lett.*, **104**, 1–15.

MacDonald G.A., 1968, Composition and origin of Hawaiian lavas. In: Coats R.R., Hay R.L. and Anderson C.A. (eds.), *Studies in volcanology: a memoir in honour of Howel Williams*. Geol. Soc. Amer. Mem. **116**, 477–522.

MacDonald G.A. and Katsura T., 1964, Chemical composition of Hawaiian lavas. *J. Petrol.*, **5**, 83–133.

McDonough W.F., Sun S., Ringwood A.E., Jagoutz E. and Hofmann A.W., 1991, K, Rb and Cs in the earth and moon and the evolution of the earth's mantle. *Geochim. Cosmochim. Acta*, Ross Taylor Symposium volume.

MacGeehan P.J. and MacLean W.H., 1980, An Archaean sub-seafloor geothermal system, 'calc–alkali' trends, and massive sulphide genesis. *Nature*, **286**, 767–771.

McIntire W.L., 1963, Trace element partition coefficients — a review of theory and applications to geology. *Geochim. Cosmochim. Acta*, **27**, 1209–1264.

McKenzie D., 1985, The extraction of magma from the crust and mantle. *Earth Planet. Sci. Lett.*, **74**, 81–91.

McLennan S.M., 1989, Rare earth elements in sedimentary rocks: influence of provenance and sedimentary processes. In: Lipin B.R. and McKay G.A. (eds.), *Geochemistry and mineralogy of rare earth elements. Reviews in Mineralogy*, **21**, pp. 169–200.

McLennan S.M. and Taylor S.R., 1991, Sedimentary rocks and crustal evolution revisited: tectonic setting and secular trends. *J. Geol.*, **99**, 1–21.

McLennan S.M., Taylor S.R., McCulloch M.T. and Maynard J.B., 1990, Geochemical and Nd–Sr isotopic composition of deep sea turbidites: crustal evolution and plate tectonic associations. *Geochim. Cosmochim. Acta*, **54**, 2015–2050.

Magaritz M., Whitford D.J. and James D.E., 1978, Oxygen isotopes and the origin of high $^{87}Sr/^{86}Sr$ andesites. *Earth Planet. Sci. Lett.*, **40**, 220–230.

Mahood G. and Hildreth W., 1983, Large partition coefficients for trace elements in high-silica rhyolites. *Geochim. Cosmochim. Acta*, **47**, 11–30.

Manning D.A.C., 1981, The effect of fluorine on liquidus phase relationships in the system Qz–Ab–Or with excess water at 1 kb. *Contrib. Mineral. Petrol.*, **76**, 206–215.

Marshall B.D. and DePaolo D.J., 1982, Precise age determinations and petrogenetic studies using the K–Ca method. *Geochim. Cosmochim. Acta*, **46**, 2537–2545.

Marshall J.D. and Middleton P.D., 1990, Changes in marine isotopic composition and the late Ordovician glaciation. *J. Geol. Soc. Lond.*, **147**, 1–4.

Marumo K., Nagasawa K. and Kuroda Y., 1980, Mineralogy and hydrogen isotope chemistry of clay minerals in the Ohunuma geothermal area, NE Japan. *Earth Planet. Sci. Lett.*, **47**, 255–262.

Mason B. (ed.), 1971, *Handbook of elemental abundances inmeteorites.* Gordon and Breach, New York.

Masuda A., 1962, Regularities in variation of relative abundances of lanthanide elements and an attempt to analyse separation-index patterns of some minerals. *J. Earth Sci. Nagoya Univ.*, **10**, 173–187.

Masuda A., Nakamura N. and Tanaka T., 1973, Fine structures of mutually normalised rare-earth patterns of chondrites. *Geochim. Cosmochim. Acta*, **37**, 239–248.

Matsuhisa Y., Goldsmith J.R. and Clayton R.N., 1979, Oxygen isotope fractionation in the systems quartz–albite–anorthite–water. *Geochim. Cosmochim. Acta*, **43**, 1131–1140.

Matsui A., Onuma N., Nagasawa H., Higuchi H. and Banno S., 1977, Crystal structure control in trace element partition between crystal and magma. *Bull. Soc. Fr. Mineral. Cristallogr.*, **100**, 315–324.

Matthews A. and Katz A., 1977, Oxygen isotope fractionation during the dolomitisation of calcium carbonate. *Geochim. Cosmochim. Acta*, **41**, 1431–1438.

Matthews A., Goldsmith J.R. and Clayton R.N., 1983, Oxygen isotope fractionations involving pyroxenes: the calibration of mineral-pair thermometers. *Geochim. Cosmochim. Acta*, **47**, 631–644.

Mauche R., Faure G., Jones L.M. and Hoefs J., 1989, Anomalous isotopic compositions of Sr, Ar and O in the mesozoic diabase dykes of Liberia, West Africa. *Contrib. Mineral. Petrol.*, **101**, 12–18.

Mearns E.W., Knarud R., Raestad N., Stanley K.O. and Stockbridge C.P., 1989, Samarium–neodymium isotope stratigraphy of the Lunde and Statfjord formations of Snorre oil field, northern North Sea. *J. Geol. Soc. Lond.*, **146**, 217–228.

Meisch A.T., 1969, The constant sum problem in geochemistry. In: Merriam D.F. (ed.), *Computer applications in the earth sciences.* Plenum Press, New York, pp. 161–176.

Mensing T.M., Faure G., Jones L.M., Bowman J.R. and Hoefs J., 1984, Petrogenesis of the Kirkpatrick basalt, Solo Nunatak, northern Victoria Land Antarctica, based upon isotopic compositions of strontium, oxygen and sulfur. *Contrib. Mineral. Petrol.*, **87**, 101–108.

Menzies M.A., 1989, Cratonic, circumcratonic and oceanic mantle domains beneath the western United States. *J. Geophys. Res.*, **94**, B7899–B7915.

Menzies M.A. and Halliday A., 1988, Lithospheric domains beneath the Archaean and Proterozoic crust of Scotland. *J. Petrol.*, Special Lithosphere Issue, 275–302.

Merlet C. and Bodinier J.-L., 1990, Electron microprobe determination of minor and trace transition metal elements in silicate minerals: a method and its application to mineral zoning in the peridotite nodule PHN 1611. *Chem. Geol.*, **83**, 55–69.

Merrill R.B., Robertson J.K. and Wyllie P.J., 1970, Melting reactions in the system $NaAlSi_3O_8$–$KAlSi_3O_8$–SiO_2–H_2O to 20 kilobars compared with results for other feldspar–quartz–H_2O and rock–H_2O systems. *J. Geol.*, **78**, 558–569.

Meschede M., 1986, A method of discriminating between different types of mid-ocean ridge basalts and continental tholeiites with the Nb–Zr–Y diagram. *Chem. Geol.*, **56**, 207–218.

Michael P.J., 1988, Partition coefficients for rare earth elements in mafic minerals of high silica rhyolites: the importance of accessory mineral inclusions. *Geochim. Cosmochim. Acta*, **52**, 275–282.

Michard A., 1989, Rare earth element systematics in hydrothermal fluids. *Geochim. Cosmochim. Acta*, **53**, 745–750.

Michard A., Gurriet P., Soudant M. and Albarede F., 1985, Nd isotopes in French phanerozoic shales: external vs internal aspects of crustal evolution. *Geochim. Cosmochim. Acta*, **49**, 601–610.

Middlemost E.A.K., 1975, The basalt clan. *Earth Sci. Rev.*, **11**, 337–364.

Middlemost E.A.K., 1985, *Magmas and magmatic rocks.* Longman, London.

Middlemost E.A.K., 1989, Iron oxidation ratios, norms and the classification of volcanic rocks. *Chem. Geol.*, **77**, 19–26.

Middleton G.V., 1960, Chemical composition of sandstones. *Bull. Geol. Soc. Amer.*, **71**, 109–126.

Miller K.G. and Fairbanks R.G., 1985, Cainozoic $\delta^{18}O$ record of climate and sea level. *S. Afr. J. Sci.*, **81**, 248–249.

Miller R.G. and O'Nions R.K., 1985, Source of Precambrian chemical and clastic sediments. *Nature*, **314**, 325–330.

Minster J.F. and Allegre C.J., 1978, Systematic use of trace elements in igneous processes. Part III: Inverse problem of batch partial melting in volcanic suites. *Contrib. Mineral. Petrol.*, **68**, 37–52.

Minster J.F., Minster J.B., Treuil M. and Allegre C.J., 1977, Systematic use of trace elements in igneous processes. Part II. Inverse problem of the fractional crystallisation process in volcanic suites. *Contrib. Mineral. Petrol.*, **61**, 49–77.

Miyoshi T., Sakai H. and Chiba H., 1984, Experimental study of sulphur isotope fractionation factors between sulphate and sulphide in high temperature melts. *Geochem. J.*, **18**, 75–84.

Moorbath S., Powell J.L. and Taylor P.N., 1975, Isotopic evidence for the age and origin of the "grey gneiss" complex of the southern Outer Hebrides, Scotland. *J. Geol. Soc. Lond.*, **131**, 213–222.

Moran P.A.P., 1971, Estimating structural and functional relationships. *J. Multivariate Anal.*, **1**, 232–255.

Morrison G.W., 1980, Characteristics and tectonic setting of the shoshonite rock association. *Lithos*, **13**, 97–108.

Morrison M.A., 1978, The use of "immobile" trace elements to distinguish the palaeotectonic affinities of metabasalts: applications to the Palaeocene basalts of Mull and Skye, northwest Scotland. *Earth Planet. Sci. Lett.*, **39**, 407–416.

Morse S.A., 1969, Syenites. *Carn. Inst. Wash. Ybk*, **67**, 112–120.

Morse S.A., 1970, Alkali feldspars with water at 5 kb pressure. *J. Petrol.*, **11**, 221–253.

Mottl M.J., 1983, Metabasalts, axial hot springs, and the structure of hydrothermal systems at mid-ocean ridges. *Geol. Soc. Amer. Bull.*, **94**, 161–180.

Muecke G.K., 1980, *Neutron activation analysis in the geosciences. Mineral. Assoc. Canada short course handbook 5.*

Muehlenbachs K. and Byerly G., 1982, ^{18}O-enrichment of silicic magmas caused by crystal fractionation at the Galapagos spreading center. *Contrib. Mineral. Petrol.*, **79**, 76–79.

Muehlenbachs K. and Clayton R.N., 1976, Oxygen isotope composition of the oceanic crust and its bearing on seawater. *J. Geophys. Res.*, **81**, 4365–4369.

Muenow D.W., Garcia M.O., Aggrey K.E., Bednarz U. and Schmincke H.U., 1990, Volatiles in submarine glasses as a discriminant of tectonic origin: application to the Troodos ophiolite. *Nature*, **343**, 159–161.

Mukasa S.B. and Henry D.S., 1990, The San Nicolas batholith of coastal Peru: early Palaeozoic continental arc or continental rift magmatism? *J. Geol. Soc. Lond.*, **147**, 27–39.

Mullen E.D., 1983, $MnO/TiO_2/P_2O_5$: a minor element discriminant for basaltic rocks of oceanic environments and its implications for petrogenesis. *Earth Planet. Sci. Lett.*, **62**, 53–62.

Nabelek P.I., Labotka T.C., O'Neil J.R. and Papike J.J., 1984, Contrasting fluid/rock interaction between Notch peak granite intrusion and argillites and limestones in western Utah: evidence from stable isotopes and phase assemblages. *Contrib. Mineral. Petrol.*, **86**, 25–34.

Nakai S., Shimizu H. and Masuda A., 1986, A new geochronometer using lanthanum–138. *Nature*, **320**, 433–435.

Nakamura N., 1974, Determination of REE, Ba, Fe, Mg, Na and K in carbonaceous and ordinary chondrites. *Geochim. Cosmochim. Acta*, **38**, 757–775.

Naldrett A.J. and Duke J.M., 1980, Platinum metals in magmatic sulfide ores. *Science*, **208**, 1417–1424.

Naldrett A.J., Hoffman E.L., Green A.H., Chou C-L., Naldrett S.R. and Alcock R.A., 1979, The composition of Ni-sulfide ores with particular reference to their content of PGE and Au. *Canad. Mineral.*, **17**, 403–415.

Nash W.P. and Crecraft H.R., 1985, Partition coefficients for trace elements in silicic magmas. *Geochim. Cosmochim. Acta*, **49**, 2309–2322.

Nathan H.D. and Van Kirk C.K., 1978, A model of magmatic crystallisation. *J. Petrol.*, **19**, 66–94.

Naylor H., Turner P., Vaughan D.J., Boyce A.J. and Fallick A.E., 1989, Genetic studies of redbed mineralisation in the Triassic of the Cheshire basin, northwest England. *J. Geol. Soc. Lond.*, **146**, 685–699.

Neilsen R.L., 1988, A model for the simulation of combined major and trace element liquid lines of descent. *Geochim. Cosmochim. Acta*, **52**, 27–38.

Nelson B.K. and DePaolo D.J., 1984, 1 700 Myr greenstone volcanic successions in southwestern North America and isotopic evolution of Proterozoic mantle. *Nature*, **312**, 143–146.

Nesbitt H.W., 1979, Mobility and fractionation of rare earth elements during weathering of a granodiorite. *Nature*, **279**, 206–210.

Nesbitt H.W., MacRae N.D. and Kronberg B.I., 1990, Amazon deep-sea fan muds: light REE enriched products of extreme chemical weathering. *Earth Planet. Sci. Lett.*, **100**, 118–123.

Nesbitt H.W. and Young G.M., 1982, Early Proterozoic climates and plate motions inferred from major element chemistry of lutites. *Nature*, **299**, 715–717.

Nesbitt H.W. and Young G.M., 1984, Prediction of some weathering trends of plutonic and volcanic rocks based upon thermodynamic and kinetic considerations. *Geochim. Cosmochim. Acta*, **48**, 1523–1534.

Nesbitt H.W. and Young G.M., 1989, Formation and diagenesis of weathering profiles. *J. Geol.*, **97**, 129–147.

Nicholls J., 1988, The statistics of Pearce element diagrams and the Chayes closure problem. *Contrib. Mineral. Petrol.*, **99**, 11–24.

Nisbet E.G. and 10 others, 1987. Uniquely fresh 2.7 Ga komatiites from the Belingwe greenstone belt, Zimbabwe. *Geology*, **15**, 1147–1150.

Nisbet E.G., Deitrich V.J. and Esenwein A., 1979, Routine trace element determination in silicate minerals and rocks by X-ray fluorescence. *Forschr. Miner.*, **57**, 264–279.

Nisbet E.G. and Pearce J.A., 1977, Clinopyroxene composition in mafic lavas from different tectonic settings. *Contrib. Mineral. Petrol.*, **63**, 149–160.

Norman M.D. and De Deckker P., 1990, Trace metals in lacustrine and marine sediments: a case study from the gulf of Carpentaria, northern Australia. *Chem. Geol.*, **82**, 299–318.

Norman M.D. and Leeman W.P., 1990, Open-system magmatic evolution of andesites and basalts from the Salmon Creek volcanics, southwest Idaho, U.S.A. *Chem. Geol.*, **81**, 167–189.

Norman M.D., Leeman W.P., Blanchard D.P., Fitton J.G. and James D., 1989, Comparison of major and trace element analyses by ICP, XRF, INAA and ID methods. *Geostandards Newsletter*, **13**, 283–290.

Norrish K. and Chappell B.W., 1967, X-ray fluorescence spectrography. In: Zussman J. (ed.), *Physical methods in determinative mineralogy*. Academic Press, New York. pp. 161–214.

Norrish K. and Chappell B.W., 1977, X-ray fluorescence spectrometry. In: Zussman J. (ed.), *Physical methods in determinative mineralogy*, 2nd edition. Academic Press, New York, pp. 201–272.

Norrish K. and Hutton J.T., 1969, An accurate X-ray spectrographic method for the analysis of a wide range of geological samples. *Geochim. Cosmochim. Acta*, **33**, 431–453.

O'Connor J.T., 1965, A classification for quartz-rich igneous rock based on feldspar ratios. *U.S. Geol. Surv. Prof. Paper* **525B**, B79-B84.

O'Hara M.J., 1968, The bearing of phase equilibria studies on the origin and evolution of basic and ultrabasic rocks. *Earth Sci. Rev.*, **4**, 69–133.

O'Hara M.J., 1977, Geochemical evolution during fractional crystallisation of a periodically refilled magma chamber. *Nature*, **266**, 503–507.

O'Hara M.J., 1980, Nonlinear nature of the unavoidable long-lived isotopic, trace and major element contamination of a developing magma chamber. *Phil. Trans. R. Soc.*, **297**, 215–227.

O'Hara M.J. and Matthews R.E., 1981, Geochemical evolution in an advancing, periodically replenished, periodically tapped, continuously fractionated magma chamber. *J. Geol. Soc. Lond.*, **138**, 237–277.

Ohmoto H., 1986, Stable isotope geochemistry of ore deposits. In: Valley J.W., Taylor H.P. and O'Neil J.R. (eds.), *Stable isotopes and high temperature geological processes*. Reviews in Mineralogy 16, Mineral. Soc. Amer., pp. 460–491.

Ohmoto H. and **Lasaga A.C.**, 1982, Kinetics of reactions between aqueous sulphates and sulphides in hydrothermal systems. *Geochim. Cosmochim. Acta*, **46**, 1727–1745.

Ohmoto H. and **Rye R.O.**, 1974, Hydrogen and oxygen isotopic compositions of fluid inclusions in the Kuroko deposits, Japan. *Econ. Geol.*, **69**, 947–953.

Ohmoto H. and **Rye R.O.**, 1979, Isotopes of sulfur and carbon. In: Barnes H.L. (ed.), *Geochemistry of hydrothermal ore deposits*. Wiley, New York, pp. 509–567.

O'Neil J.R., **Clayton R.N.** and **Mayeda T.K.**, 1969, Oxygen isotope fractionation in divalent metal carbonates. *J. Chem. Phys.*, **51**, 5547–5558.

O'Neil J.R. and **Taylor H.P.**, 1967, The oxygen isotope and cation exchange chemistry of feldspars. *Amer. Mineral.*, **52**, 1414–1437.

O'Nions R.K., **Carter S.R.**, **Evensen N.M.** and **Hamilton P.J.**, 1979, Geochemical and cosmochemical applications of Nd isotope analysis. *Ann. Rev. Earth Planet. Sci.*, **7**, 11–38.

O'Nions R.K., **Hamilton P.J.** and **Evensen N.M.**, 1977, Variations in $^{143}Nd/^{144}Nd$ and $^{87}Sr/^{86}Sr$ in oceanic basalts. *Earth Planet. Sci. Lett.*, **34**, 13–22.

O'Nions R.K., **Hamilton P.J.** and **Hooker P.J.**, 1983, A Nd isotope investigation of sediments related to crustal development in the British Isles. *Earth Planet. Sci. Lett.*, **63**, 229–240.

O'Nions R.K. and **McKenzie D.P.**, 1988, Melting and continent generation. *Earth Planet. Sci. Lett.*, **90**, 449–456.

Onuma N., **Clayton R.N.** and **Mayeda T.K.**, 1972, Oxygen isotope cosmothermometer. *Geochim. Cosmochim. Acta*, **36**, 169–188.

Onuma N., **Higuchi H.**, **Wakita H.** and **Nagasawa H.**, 1968, Trace element partitioning between two pyroxenes and the host lava. *Earth Planet. Sci. Lett.*, **5**, 47–51.

Oskarsson N., **Sigvaldson G.E.** and **Steinthorsson S.**, 1982, A dynamic model of rift zone petrogenesis and the regional petrology of Iceland. *J. Petrol.*, **23**, 28–74.

Patchett P.J. and **Tatsumoto M.**, 1980, Lu–Hf total-rock isochron for eucrite meteorites. *Nature*, **288**, 571–574.

Pearce J.A., 1976, Statistical analysis of major element patterns in basalts. *J. Petrol.*, **17**, 15–43.

Pearce J.A., 1980, Geochemical evidence for the genesis and eruptive setting of lavas from Tethyan ophiolites. *Proc. Int. Ophiolite Symp., Cyprus 1979*. Institute of Mining and Metallurgy, pp. 261–272.

Pearce J.A., 1982, Trace element characteristics of lavas from destructive plate boundaries. In: Thorpe R.S. (ed.), *Andesites*. Wiley, Chichester, pp. 525–548.

Pearce J.A., 1983, Role of the sub-continental lithosphere in magma genesis at active continental margins. In: Hawkesworth C.J. and Norry M.J. (eds.), *Continental basalts and mantle xenoliths*. Shiva, Nantwich, pp. 230–249.

Pearce J.A., 1987, An expert system for the tectonic characterisation of ancient volcanic rocks. *J. Volc. Geothermal Res.*, **32**, 51–65.

Pearce J.A., **Alabaster T.**, **Shelton A.W.** and **Searle M.P.**, 1981, The Oman ophiolite as a Cretaceous arc-basin complex: evidence and implications. *Phil. Trans. R. Soc.* **A300**, 299–300.

Pearce J.A. and **Cann J.R.**, 1971, Ophiolite origin investigated by discriminant analysis using Ti, Zr and Y. *Earth Planet. Sci. Lett.*, **12**, 339–349.

Pearce J.A. and **Cann J.R.**, 1973, Tectonic setting of basic volcanic rocks determined using trace element analyses. *Earth Planet. Sci. Lett.*, **19**, 290–300.

Pearce J.A. and **Gale G.H.**, 1977, Identification of ore-deposition environment from trace element geochemistry of associated igneous host rocks. *Geol. Soc. Spec. Publ.*, **7**, 14–24.

Pearce J.A., **Harris N.B.W.** and **Tindle A.G.**, 1984, Trace element discrimination diagrams for the tectonic interpretation of granitic rocks. *J. Petrol.*, **25**, 956–983.

Pearce J.A. and **Norry M.J.**, 1979, Petrogenetic implications of Ti, Zr, Y and Nb variations in volcanic rocks. *Contrib. Mineral. Petrol.*, **69**, 33–47.

Pearce T.H., 1968, A contribution to the theory of variation diagrams. *Contrib. Mineral. Petrol.*, **19**, 142–157.

Pearce T.H., 1970, Chemical variations in the Palisades Sill. *J. Petrol.*, **11**, 15–32.

Pearce T.H., Gorman B.E. and Birkett T.C., 1975, The TiO_2–K_2O–P_2O_5 diagram: a method of discriminating between oceanic and non-oceanic basalts. *Earth Planet. Sci. Lett.*, **24**, 419–426.

Pearce T.H., Gorman B.E. and Birkett T.C., 1977, The relationship between major element chemistry and tectonic environment of basic and intermediate volcanic rocks. *Earth Planet. Sci. Lett.*, **36**, 121–132.

Pearson K., 1896, On a form of spurious self correlation which may arise when indices are used in the measurement of organs. *Proc. R. Soc. London*, **60**, 489–502.

Peccerillo R. and Taylor S.R., 1976, Geochemistry of Eocene calc–alkaline volcanic rocks from the Kastamonu area, northern Turkey. *Contrib. Mineral. Petrol.*, **58**, 63–81.

Perfit M.R., Gust D.A., Bence A.E., Arculus R.J. and Taylor S.R., 1980, Chemical characteristics of island arc basalts: implications for mantle sources. *Chem. Geol.*, **30**, 227–256.

Peterson D.W. and Moore R.B., 1987, Geologic history and evolution of geologic concepts, Island of Hawaii. In: Decker R.W., Wright T.L. and Stauffer P.H. (eds.), *Volcanism in Hawaii*. USGS Prof. Paper 1350, Vol. 1, pp. 149–189.

Pettijohn F.J., Potter P.E. and Siever R., 1972, *Sand and sandstones*. Springer-Verlag, New York.

Peucat J.J., Vidal P., Bernard-Griffiths J. and Condie K.C., 1988, Sr, Nd and Pb isotopic systematics in the Archaean low- to high-grade transition zone of southern India: syn accretion vs. post-accretion granulites. *J. Geol.*, **97**, 537–550.

Philpotts J.A., 1978, The law of constant rejection. *Geochim. Cosmochim. Acta*, **42**, 909–920.

Philpotts J.A. and Schnetzler C.C., 1970, Phenocryst–matrix partition coefficients for K, Rb, Sr and Ba with applications to anorthosite and basalt genesis. *Geochim. Cosmochim. Acta*, **34**, 307–322.

Piccirillo E.M., Civetta L., Petrini R., Longinelli A., Bellieni G., Comin-Chiaramonti P., Marques L.S. and Melfi A.J., 1989, Regional variations within the Parana flood basalt basalts (southern Brazil): evidence for subcontinetal mantle heterogeneity and crustal contamination. *Chem. Geol.*, **75**, 103–122.

Potter P.E., 1978, Petrology and chemistry of modern big river sands. *J. Geol.*, **86**, 423–449.

Potter P.E., Shimp N.F. and Witters J., 1963, Trace elements in marine and fresh-water argillaceous sediments. *Geochim. Cosmochim. Acta*, **27**, 669–694.

Potts P.J., Webb P.C. and Watson J.S., 1990, Exploiting energy dispersive X-ray fluorescence spectrometry for the determination of trace elements in geological samples. *Anal. Proc.*, **27**, 67–70.

Powell R., 1984, Inversion of the assimilation and fractional crystallisation (AFC) equations; characterisation of contaminants from isotope and trace element relationships in volcanic suites. *J. Geol. Soc. Lond.*, **141**, 447–452.

Presnall D.C. and Hoover J.D., 1984, Composition and depth of origin of primary mid-ocean ridge basalts. *Contrib. Mineral. Petrol.*, **87**, 170–178.

Presnall D.C., Dixon J.R., O'Donnell T.H. and Dixon S.A., 1979, Generation of mid-ocean ridge tholeiites. *J. Petrol.*, **20**, 3–35.

Prestvick T., 1982, Basic volcanic rocks and tectonic setting. A discussion of the Zr–Ti–Y discrimination diagram and its suitability for classification purposes. *Lithos*, **15**, 241–247.

Price W.J., 1972, *Analytical atomic absorption spectrometry*. Heyden, London.

Rautenschlein M., Jenner G.A., Hertogen J., Hofmann A.W., Kerrich R., Schmincke H.-U. and White W.M., 1985, Isotopic and trace element composition of volcanic glasses from the Akaki Canyon, Cyprus: implications for the origin of the Troodos ophiolite. *Earth Planet. Sci. Lett.*, **75**, 369–383.

Reed S.J.B., 1989, Ion microprobe analysis — a review of geological applications. *Mineral. Mag.*, **53**, 3–24.

Reuter A. and Dallmeyer R.D., 1989, K–Ar and $^{40}Ar/^{39}Ar$ dating of cleavage formed during very low grade metamorphism: a review. In: Daly J.S., Cliff R.A. and Yardley B.W.D. (eds.), *Evolution of metamorphic belts*, Spec. Publ. Geol. Soc. Lond. 43, pp. 161–172.

Richardson C.K., Rye R.O. and Wasserman M.D., 1988, The chemical and thermal evolution of the fluids in the Cave-in-rock fluorspar district, Illinois: stable isotope systematics at the Deardorff mine. *Econ. Geol.*, **83**, 765–783.

Richardson S.H., Gurney J.J., Erlank A.J. and Harris J.W., 1984, Origin of diamonds in old enriched mantle. *Nature*, **310**, 198–202.

Richter D.H. and Moore J.G., 1966, Petrology of the Kilauea Iki lava lake, Hawaii. *U.S. Geol. Surv. Prof. Paper* 537-B, B1-B26.

Richter F.M., 1986, Simple models for trace element fractionation during melt segregation. *Earth Planet. Sci. Lett.*, **77**, 333–344.

Rickwood P.C., 1989, Boundary lines within petrologic diagrams which use oxides of major and minor elements. *Lithos*, **22**, 247–263.

Robinson P., Higgins N.C. and Jenner J.G., 1986, Determination of rare earth elements, yttrium and scandium in rocks by an ion-exchange X-ray fluorescence technique. *Chem. Geol.*, **55**, 121–137.

Rock N.M.S., 1987a, ROBUST: An interactive Fortran–77 package for exploratory data analysis using parametric robust and nonparametric location and scale estimates, data transformations, normality tests and outlier assessment. *Comput. Geosci.*, **13**, 463–494.

Rock N.M.S., 1987b, The need for standardization of normalised multi-element diagrams in geochemistry: a comment. *Geochem. J.*, **21**, 75–84.

Rock N.M.S., 1988a, *Numerical geology: a source guide, glossary and selective bibliography to geological uses of computers and statistics*. Spinger-Verlag lecture notes in earth sciences, Vol. 18.

Rock N.M.S., 1988b, Summary statistics in geochemistry: a study of the performance of robust estimates. *Math. Geol.*, **20**, 243–275.

Rock N.M.S., 1989, Reply to Aitchison. *Math. Geol.*, **21**, 791–793.

Roeder P.L. and Emslie R.F., 1970, Olivine–liquid equilibrium. *Contrib. Mineral. Petrol.*, **29**, 275–289.

Roelandts I., 1988, Comparison of inductively coupled plasma and neutron activation analysis for precise and accurate determination of nine rare-earth elements in geological materials. *Chem. Geol.*, **67**, 171–180.

Rogers N.W. and Hawkesworth C.J., 1982, Proterozoic age and cumulate origin for granulitic xenoliths, Lesotho. *Nature*, **299**, 409–413.

Rollinson H.R., 1983, The geochemistry of mafic and ultramafic rocks from the Archaean greenstone belts of Sierra Leone. *Mineral. Mag.*, **47**, 267–280.

Rollinson H.R., 1992, Another look at the constant sum problem in geochemistry. *Mineral. Mag.* **56**, 469–475.

Rollinson H.R. and Roberts C.R., 1986, Ratio correlation and major element mobility in altered basalts and komatiites. *Contrib. Mineral. Petrol.*, **93**, 89.

Roser B.P. and Korsch R.J., 1986, Determination of tectonic setting of sandstone–mudstone suites using SiO_2 content and K_2O/Na_2O ratio. *J. Geol.*, **94**, 635–650.

Roser B.P. and Korsch R.J., 1988, Provenance signatures of sandstone–mudstone suites determined using discriminant function analysis of major-element data. *Chem. Geol.*, **67**, 119–139.

Rumble D. III and Hoering T.C., 1986, Carbon isotope geochemistry of graphic vein deposits from New Hampshire USA, *Geochim. Cosmochim. Acta*, **50**, 1239–1247.

Rumble D. III and Spear F.S., 1983, Oxygen isotope equilibration and permeability enhancement during regional metamorphism. *J. Geol. Soc. Lond.*, **140**, 619–628.

Russell J.K., Nicholls J., Stanley C.R. and Pearce T.H., 1990, Pearce element ratios. A paradigm for testing hypotheses. *Eos*, **71**, 234–247.

Rutherford E. and Soddy F., 1903, Radioactive change. *Phil. Mag.* 6, 576–591.

Rye R.O. and Ohmoto H., 1974, Sulfur and carbon isotopes and ore genesis. A review. *Econ. Geol.*, **69**, 826–842.

Rye R.O., Schuiling R.D., Rye D.M. and Jansen J.B.H., 1976, Carbon hydrogen and oxygen isotope studies of the regional metamorphic complex at Naxos, Greece. *Geochim. Cosmochim. Acta*, **40**, 1031–1049.

Ryerson F.J. and Hess P.C., 1978, Implications of liquid–liquid distribution coefficients to mineral–liquid partitioning. *Geochim. Cosmochim. Acta*, **42**, 921–932.

Sachs L., 1984, *Applied statistics: a handbook of techniques*, 2nd edition. Springer-Verlag, New York.

Sack R.O., Carmichael I.S.E., Rivers M. and Ghiorso M.S., 1980, Ferric–ferrous equilibria in natural silicate liquids at 1 bar. *Contrib. Mineral. Petrol.*, **75**, 369–376.

Sakai H., Casadevall T.J. and Moore J.G., 1982, Chemistry and isotope ratios of sulfur in basalts and volcanic gases at Kilauea volcano, Hawaii. *Geochim. Cosmochim. Acta*, **46**, 729–738.

Sakai H., Des Maris D.J., Ueda A. and Moore J.G., 1984, Concentrations and isotope ratios of carbon, nitrogen and sulfur in ocean–floor basalts and volcanic gases at Kilauea volcano, Hawaii. *Geochim. Cosmochim. Acta*, **48**, 2433–2441.

Sangster D.F., 1968, Relative sulfur isotope abundance of ancient seas and stratabound sulphide deposits. *Proc. Geol. Assoc. Canada*, **19**, 79–86.

Saunders A.D., Norry M.J. and Tarney J., 1988, Origin of MORB and chemically depleted mantle reservoirs: trace element constraints. *J. Petrol.*, Special Lithosphere Issue, 415–445.

Saunders A.D. and Tarney J., 1984, Geochemical characteristics of basaltic volcanism within back-arc basins. In: Kokelaar B.P. and Howells M.F. (eds.), *Marginal basin geology*, Spec. Publ. Geol. Soc. London 16, pp. 59–76.

Savin S.M. and Lee M., 1988, Isotopic study of phyllosilicates. In: Bailey S.W. (ed.), *Hydrous phyllosilicates (exclusive of muscovite)*. Min. Soc. Amer. Rev. Mineralogy. 19, pp. 189–223.

Schairer J.F., 1950, The alkali feldspar join in the system $NaAlSiO_4$–$KAlSiO_4$–SiO_2. *J. Geol.*, **58**, 512–517.

Schairer J.F. and Bowen N.L., 1935, Preliminary report on equilibrium relations between feldspathoids, alkali feldspars and silica. *Trans. Amer. Geophys. Union*, 16th Ann. Meeting, pp. 325–328.

Schidlowski M., 1987, Application of stable isotopes to early biochemical evolution on earth. *Ann. Rev. Earth Planet. Sci.*, **15**, 47–72.

Schidlowski M., 1988, A 3 800-million-year isotopic record of life from carbon in sedimentary rocks. *Nature*, **333**, 313–318.

Schiffman P., Smith B.M., Varga R.J. and Moores E.M., 1987, Geometry, conditions and timing of off-axis hydrothermal metamorphism and ore-deposition in the Solea graben. *Nature*, **325**, 423–425.

Schnetzler C.C. and Philpotts J.A., 1970, Partition coefficients of rare earth elements between igneous matrix material and rock-forming mineral phenocrysts — II. *Geochim. Cosmochim. Acta*, **34**, 331–340.

Schock H.H., 1979, Distribution of rare-earth and other trace elements in magnetites. *Chem. Geol.*, **26**, 119–133.

Scotchman I.C., 1989, Diagenesis of the Kimmeridge clay formation, onshore UK. *J. Geol. Soc. Lond.*, **146**, 285–303.

Searl A., 1988, Mixing-zone dolomites in the Gully Oolite, lower Carboniferous, south Wales. *J. Geol. Soc. Lond.*, **145**, 891–899.

Searl A., 1989, Diagenesis of the Gully Oolite (lower Carboniferous), south Wales. *Geol. J.*, **24**, 275–293.

Seewald J.S. and Seyfried W.E., 1990, The effect of temperature on metal mobility in subseafloor hydrothermal systems: constraints from basalt alteration experiments. *Earth Planet. Sci. Lett.*, **101**, 388–403.

Shannon R.D., 1976, Revised effective ionic radii and systematic studies of interatomic distances in halides and chalcogenides. *Acta Crystallogr., Sect. A*, **32**, 751–767.

Sharpe M.R., 1985, Strontium isotope evidence for preserved density stratification in the main zone of the Bushveld Complex, South Africa. *Nature*, **316**, 119–126.

Sheppard S.M.F., 1977, The Cornubian batholith, SW England: D/H and $^{18}O/^{16}O$ studies of kaolinite and other alteration minerals. *J. Geol. Soc.*, **133**, 573–591.

Sheppard S.M.F., 1981, Stable isotope geochemistry of fluids. In: Rickard D.T. and Wickman F.E. (eds.), *Chemistry and geochemistry of solutions at high temperatures and pressures*. Phys. Chem. Earth 13/14, pp. 419–445.

Sheppard S.M.F. and Schwartz H.P., 1970, Fractionation of carbon and oxygen isotopes and magnesium between metamorphic calcite and dolomite. *Contrib. Mineral. Petrol.*, **26**, 161–198.

Shervais J.W., 1982, Ti–V plots and the petrogenesis of modern and ophiolitic lavas. *Earth Planet. Sci. Lett.*, **59**, 101–118.

Shimizu N. and Hart S.R., 1982, Applications of the ion microprobe to geochemistry and cosmochemistry. *Ann. Rev. Earth Planet. Sci.*, **10**, 483–526.

Shirey S.B., Bender J.F. and Langmuir C.H., 1987, Three-component isotopic heterogeneity near the Oceanographer transform, Mid-Atlantic ridge. *Nature*, **325**, 217–223.

Skala W., 1979, Some effects of the constant sum problem in geochemistry. *Chem. Geol.*, **27**, 1–9.

Skirrow R. and Coleman M.L., 1982, Origin of sulphur and geothermometry of hydrothermal sulphides from the Galapagos Rift, 86 0W. *Nature*, **299**, 142–144.

Smalley P.C., Stijfhoorn D.E., Raheim A., Johansen H. and Dickson J.A.D., 1989, The laser microprobe and its application to the study of C and O isotopes in calcite and aragonite. *Sediment. Geol.*, **65**, 211–221.

Smith R.E. and Smith S.E., 1976, Comments on the use of Ti, Zr, Y, Sr, K, P and Nb in classification of basaltic magmas. *Earth Planet. Sci. Lett.*, **32**, 114–120.

Song Y. and Frey F.A., 1989, Geochemistry of peridotite xenoliths in basalts from Hannuoba eastern China: implications for sub-continental mantle heterogeneity. *Geochim. Cosmochim. Acta*, **53**, 97–114.

Spooner E.T.C., Chapman H.J. and Smewing J.D., 1977, Strontium isotopic composition and oxidation during ocean floor hydrothermal metamorphism of the ophiolitic rocks of the Troodos Massif, Cyprus. *Geochim. Cosmochim. Acta*, **41**, 873–890.

Spooner E.T.C. and Gale N.H., 1982, Pb-isotopic composition of ophiolitic volcanogenic sulphide deposits, Troodos Complex, Cyprus. *Nature*, **296**, 239–242.

Stacey J.S. and Kramers J.D., 1975, Approximation of terrestrial lead isotope evolution by a two-stage model. *Earth Planet. Sci. Lett.*, **26**, 207–221.

Stakes D.S. and O'Neil J.R., 1982, Mineralogy and stable isotope geochemistry of hydrothermally altered oceanic rocks. *Earth Planet. Sci. Lett.*, **57**, 285–304.

Staudigel H. and Bryan W.B., 1981, Contrasted glass–whole rock compositions and phenocryst redistribution, IPOD sites 417 and 418. *Contrib. Mineral. Petrol.*, **78**, 255–262.

Steiger R.H. and Jager E., 1977, Subcommission on geochronology: convention of the use of decay constants in geo- and cosmochronology. *Earth Planet. Sci Lett.*, **36**, 359–362.

Steiner J.C., Jahns R.H. and Luth W.C., 1975, Crystallisation of alkali feldspars and quartz in the haplogranite system $NaAlSi_3O_8$–$KAlSi_3O_8$–SiO_2–H_2O at 4 kb. *Bull. Geol. Soc. Amer.*, **86**, 83–98.

Stille P., Unruh D.M. and Tatsumoto M., 1983, Pb, Sr, Nd and Hf isotopic evidence of multiple sources for Oahu, Hawaii basalts. *Nature*, **304**, 25–29.

Storey M., Saunders A.D., Tarney J., Leat P., Thirlwall M.F., Thompson R.N., Menzies M.A. and Marriner G.F., 1988, Geochemical evidence for plume–mantle interactions beneath Kergulen and Herd Islands, Indian Ocean. *Nature*, **336**, 371–374.

Stormer J.C. and Nicholls J., 1978, XLFRAC: A program for the interactive testing of magmatic differentiation models. *Comput. Geosci.*, **4**, 143–159.

Streckeisen A., 1976, To each plutonic rock its proper name. *Earth. Sci. Rev.*, **12**, 1–33.

Streckeisen A. and Le Maitre R.W., 1979, A chemical approximation to the modal QAPF classification of igneous rocks. *Neues Yahrb. Mineral. Abh.*, **136**, 169–206.

Sun S.S., 1980, Lead isotopic study of young volcanic rocks from mid-ocean ridges, ocean islands and island arcs. *Phil. Trans. R. Soc.*, **A297**, 409–445.

Sun S.S., 1982, Chemical composition and origin of the earth's primitive mantle. *Geochim. Cosmochim. Acta*, **46**, 179–192.

Sun S.S. and McDonough W.F., 1989, Chemical and isotopic systematics of oceanic basalts: implications for mantle composition and processes. In: Saunders A.D. and Norry M.J. (eds.), *Magmatism in ocean basins*. Geol. Soc. London. Spec. Pub. 42, pp. 313–345.

Sun S.S. and Nesbitt R.W., 1977, Chemical heterogeneity of the Archaean mantle, composition of the earth and mantle evolution. *Earth Planet. Sci. Lett.*, **35**, 429–448.

Sun S.S., Wallace D.A., Hoatson D.M., Glikson A.Y. and Keays R.R., 1991, Use of geochemistry as a guide to platinum group element potential of mafic–ultramafic rocks: examples from the west Pilbara Block and Halls Creek Mobile Zone, Western Australia. *Prec. Res.*, **50**, 1–35.

Suzouki T. and Epstein S., 1976, Hydrogen isotope fractionation between OH-bearing minerals and water. *Geochim. Cosmochim. Acta*, **40**, 1229–1240.

Tarney J., 1976, Geochemistry of Archaean high-grade gneisses, with implications as to the origin and evolution of the Precambrian crust. In: Windley B.F. (ed.), *The early history of the earth.* Wiley, New York, pp. 405–418.

Tatsumoto M., Knight R.J. and Allegre C.J., 1973, Time difference in the formation of meteorites as determined from the ratio of lead–207 to lead–206. *Science*, **180**, 1279–1283.

Taylor D. and MacKenzie W.S., 1975, A contribution to the pseudoleucite problem. *Contrib. Mineral. Petrol.*, **49**, 321–333.

Taylor H.P., 1974, The application of oxygen and hydrogen isotope studies to problems of hydrothermal alteration and ore deposition. *Econ. Geol.*, 843–883.

Taylor H.P., 1977, Water/rock interactions and the origin of H_2O in granitic batholiths. *J. Geol. Soc. Lond.*, **133**, 509–558.

Taylor H.P., 1978, Oxygen and hydrogen isotope studies of plutonic granitic rocks. *Earth Planet. Sci. Lett.*, **38**, 177–210.

Taylor H.P., 1979, Oxygen and hydrogen isotope relationships in hydrothermal mineral deposits. In: Barnes H.L. (ed.), *Geochemistry of hydrothermal ore deposits*, 2nd edition. Wiley, New York, pp. 236–277.

Taylor H.P., 1980, The effects of assimilation of country rocks by magmas on $^{18}O/^{16}O$ and $^{87}Sr/^{86}Sr$ systematics in igneous rocks. *Earth Planet. Sci. Lett.*, **47**, 243–254.

Taylor H.P. and Epstein S., 1962, Relationships between ^{18}O and ^{16}O ratios in coexisting minerals of igneous and metamorphic rocks. Part 2. Application to petrologic problems. *Geol. Soc. Amer. Bull.*, **73**, 675–694.

Taylor P.N., Jones N.W. and Moorbath S., 1984, Isotopic assessment of relative contributions from crust and mantle sources to magma genesis of Precambrian granitoid rocks. *Phil. Trans. R. Soc. Lond.*, **A310**, 605–625.

Taylor P.N., Moorbath S., Goodwin R. and Petrykowski A.C., 1980, Crustal contamination as an indicator of the extent of early Archaean continental crust: Pb isotopic evidence from the late Archaean gneisses of west Greenland. *Geochim. Cosmochim. Acta*, **44**, 1437–1453.

Taylor R.P. and Fryer B.J., 1980, Multi-stage hydrothermal alteration in porphyry copper systems in northern Turkey: the temporal interplay of potassic, propylitic and phyllic fluids. *Can. J. Earth Sci.*, **17**, 901–926.

Taylor S.R. and Gorton M.P., 1977, Geochemical applications of spark-source mass spectrography, III. Element sensitivity, precision and accuracy. *Geochim. Cosmochim. Acta*, **41**, 1375–1380.

Taylor S.R. and McLennan S.M., 1981, The composition and evolution of the continental crust: rare earth element evidence from sedimentary rocks. *Phil. Trans. R. Soc.*, **A301**, 381–399.

Taylor S.R. and McLennan S.M., 1985, *The continental crust: its composition and evolution.* Blackwell, Oxford.

Tertian R. and Claisse F., 1982, *Principles of quantitative X-ray fluorescence analysis.* Wiley–Heyden, New York.

Thode H.G. and Monster J., 1965, Sulfur isotope geochemistry of petroleum, evaporites and ancient seas. *Amer. Assoc. Pet. Geol. Mem.*, **4**, 367–377.

Thompson M. and Walsh J.N., 1983, *A handbook of inductively coupled plasma spectrometry.* Blackie, Glasgow.

Thompson R.N., 1982, British Tertiary volcanic province. *Scott. J. Geol.*, **18**, 49–107.

Thompson R.N., 1984, Dispatches from the basalt front. 1. Experiments. *Proc. Geol. Ass.*, **95**, 249–262.

Thompson R.N., 1987, Phase equilibria constraints on the genesis and magmatic evolution of oceanic basalts. *Earth Sci. Rev.*, **24**, 161–210.

Thompson R.N., Morrison M.A., Dickin A.P. and Hendry G.L., 1983, Continental flood basalts . . . Arachnids rule OK? In: Hawkesworth C.J. and Norry M.J. (eds.), *Continental basalts and mantle xenoliths.* Shiva, Nantwich, pp. 158–185.

Thompson R.N., Morrison M.A., Hendry G.L. and Parry S.J., 1984, An assessment of the relative roles of crust and mantle in magma genesis: an elemental approach. *Phil. Trans. R. Soc.*, **A310**, 549–590.

Till R., 1974, *Statistical methods for the earth scientist: an introduction.* Macmillan, Oxford.

Till, R., 1977, The HARDROCK package, a series of FORTRAN IV computer programs for performing and plotting petrochemical calculations. *Comput. Geosci.*, **3**, 185–243.

Tilton G.R., 1973, Isotopic lead ages of chondritic meteorites. *Earth Planet. Sci. Lett.*, **19**, 321–329.

Topley C.G. and **Burwell A.D.M.**, 1984, Triplot: an interactive program in basic for plotting triangular diagrams. *Comput. Geosci.*, **10**, 277–309.

Treuil M. and **Varet J.**, 1973, Criteres volcanologiques, petrologiques et geochimique de la genese et de la differenciation des magmas basaltique: exemple de l'Afar. *Bull. Geol. Soc. France, 7th ser.*, **15**, 401–644.

Troutman B.M. and **Williams G.P.**, 1987, Fitting straight lines in the earth sciences. In: Size W.B. (ed.), *Use and abuse of statistics in the earth sciences*, Int. Assn. Math. Geol., Studies in mathematical geology 1. Oxford Univ. Press, Oxford, pp. 107–128.

Trudinger P.A., **Chambers L.A.** and **Smith J.W.**, 1985, Low-temperature sulphate reduction: biological vs abiological. *Can. J. Earth Sci.*, **22**, 1910–1918.

Tuttle O.F. and **Bowen N.L.**, 1958, *The origin of granite in the light of experimental studies in the system $NaAlSi_3O_8$–$KAlSi_3O_8$–SiO_2–H_2O*. Mem. Geol. Soc. Amer., No. 74.

Twist D. and **Harmer R.E.J.**, 1987, Geochemistry of contrasting siliceous magmatic suites in the Bushveld Complex: genetic aspects and implications for tectonic discrimination diagrams. *J. Volc. Geothermal Res.*, **32**, 83–98.

Ueda A. and **Sakai H.**, 1984, Sulfur isotope study of Quaternary volcanic rocks from the Japanese islands arc. *Geochim. Cosmochim. Acta*, **48**, 1837–1848.

Urey H.C., 1947, The thermodynamic properties of isotopic substances. *J. Chem. Soc.*, 562–581.

Valley J.W. and **O'Neil J.R.**, 1981, $^{13}C/^{12}C$ exchange between calcite and graphite: a possible thermometer in Grenville marbles. *Geochim. Cosmochim. Acta*, **45**, 411–419.

Valley J.W. and **O'Neil J.R.**, 1984, Fluid heterogeneity during granulite facies metamorphism in the Adirondacks: stable isotope evidence. *Contrib. Mineral. Petrol.*, **85**, 158–173.

Valley J.W., **Taylor H.P.** and **O'Neil J.R.**, 1986, *Stable isotopes and high temperature geological processes*. Reviews in Mineralogy, Mineral. Soc. Amer. No. 16.

van Breemen O., **Aftalion M.**, **Pankhurst R.J.** and **Richardson S.W.**, 1979, Age of the Glen Dessary syenite, Invernessshire: diachronous Palaeozoic metamorphism across the Great Glen. *Scott. J. Geol.*, **15**, 49–62.

Vance D., **Stone J.O.H.** and **O'Nions R.K.**, 1989, He, Sr and Nd isotopes in xenoliths from Hawaii and other oceanic islands. *Earth Planet. Sci. Lett.*, **96**, 147–60.

Veizer J., 1989, Strontium isotopes in seawater through time. *Ann. Rev. Earth Planet. Sci.*, **17**, 141–167.

Veizer J., **Fritz P.** and **Jones B.**, 1986, Geochemistry of brachiopods: oxygen and carbon isotopic records of Palaeozoic oceans. *Geochim. Cosmochim. Acta*, **50**, 1679–1696.

Verma M.P., 1986, A program package for major element data handling and CIPW norm calculation. *Comput. Geosci.*, **12**, 381–399.

Vernon R.H., **Flood R.H.** and **D'Arcy W.F.**, 1987, Sillimanite and andalusite produced by base-cation leaching and contact metamorphism of felsic igneous rock. *J. Metamorphic Geol.*, **5**, 439–450.

Vidal Ph. and **Clauer N.**, 1981, Pb and Sr isotopic systematics of some basalts and sulfides from the East Pacific Rise at 21°N (project RITA). *Earth Planet. Sci. Lett.*, **55**, 237–246.

Vitrac A.M., **Albarede F.** and **Allegre C.J.**, 1981, Lead isotopic composition of Hercynian granitic K-feldspars constrains continental genesis. *Nature*, **291**, 460–464.

Wada H. and **Suzuki K.**, 1983, Carbon isotope thermometry calibrated by dolomite–calcite solvus temperatures. *Geochim. Cosmochim. Acta*, **47**, 697–706.

Wainerdi R.E. and **Uken E.A.**, 1971, *Modern methods of geochemical analysis*. Plenum Press, New York.

Wakita H., **Rey P.** and **Schmitt R.A.**, 1971, Abundances of the 14 rare-earth elements and 12 other trace elements in Apollo 12 samples: five igneous and one breccia rocks and four soils. *Proc. 2nd Lunar Sci. Conf.* Pergamon Press, Oxford, pp. 1319–1329.

Walker D., **Shibata T.** and **DeLong S.E.**, 1979, Abyssal tholeiites from the Oceanographer Fracture Zone III. Phase equilibria and mixing. *Contrib. Mineral. Petrol.*, **70**, 111–125.

Walsh J.N., Buckley F. and Barker J., 1981, The simultaneous determination of the rare-earth elements in rocks using inductively coupled plasma source spectrometry. *Chem. Geol.*, **33**, 141–153.

Walsh J.N. and Howie R.A., 1980, An evaluation of the performance of an inductively coupled plasma source spectrometer for the determination of major and trace constituents of silicate rocks and minerals. *Mineral. Mag.*, **47**, 967–974.

Wasserburg G.J., Jacobsen S.B., DePaolo D.J., McCulloch M.T. and Wen J., 1981, Precise determinations of Sm/Nd ratios, Sm and Nd isotopic abundances in standard solutions. *Geochim. Cosmochim. Acta*, **45**, 2311–2323.

Watson E.B., 1976, Two-liquid partition coeficients: experimental data and geochemical implications. *Contrib. Mineral. Petrol.*, **56**, 119–134.

Watson E.B. and Harrison M.T., 1983, Zircon saturation revisited: temperature and composition effects in a variety of crustal magma types. *Earth Planet. Sci. Lett.*, **64**, 295–304.

Watson E.B. and Ryerson F.J., 1986, Partitioning of zircon between clinopyroxene and magmatic liquids of intermediate composition. *Geochim. Cosmochim. Acta*, **50**, 2523–2526.

Weaver B.L., 1991, The origin of ocean island basalt end-member compositions: trace element and isotopic constraints. *Earth Planet. Sci. Lett.*, **104**, 3810–397.

Weaver B. and Tarney J., 1984, Empirical approach to estimating the composition of the continental crust. *Nature*, **310**, 575–57.

Weaver B.L., Tarney J. and Windley B., 1981, Geochemistry and petrogenesis of the Fiskenaesset anorthosite complex southern West Greenland: nature of the parent magma. *Geochim. Cosmochim. Acta*, **45**, 711–725.

Weaver S.D., 1987, Introduction. In: Weaver S.D. and Johnstone R.W. (eds.), *Tectonic controls on magmachemistry. J. Volc. Geothermal Res.*, ix–x.

Whelan J.F., Cobb J.C. and Rye R.O., 1988, Stable isotope chemistry of sphalerite and other mineral matter in coal beds of the Illinois and Forest City basins. *Econ. Geol.*, **83**, 990–1007.

White W.M. and Hofmann A.W., 1982, Sr and Nd isotope geochemistry of oceanic basalts and mantle evolution. *Nature*, **296**, 821–825.

White W.M. and Patchett J., 1984, Hf–Nd–Sr isotopes and incompatible element abundances in island arcs: implications for magma origins and crust–mantle evolution. *Earth Planet. Sci. Lett.*, **67**, 167–185.

Whitehouse M.J., 1989a, Sm–Nd evidence for diachronous crustal accretion in the Lewisian complex of northwest Scotland. *Tectonophysics*, **161**, 245–256.

Whitehouse M.J., 1989b, Pb-isotopic evidence for U–Th–Pb behaviour in a prograde amphibolite to granulite facies transition from the Lewisian complex of north-west Scotland: implications for Pb–Pb dating. *Geochim. Cosmochim. Acta*, **53**, 717–724.

Wickham S.M., 1988, Evolution of the lower crust. *Nature*, **333**, 119–120.

Wickham S.M. and Taylor H.P., 1985, Stable isotope evidence for large-scale seawater infiltration in a regional metamorphic terrane: the Trois Seigneurs Massif, Pyrenees, France. *Contrib. Mineral. Petrol.*, **91**, 122–137.

Wilkinson J.F.G., 1982, The genesis of mid-ocean ridge basalt. *Earth Science Rev.*, **18**, 1–57.

Williams D.F., Lerche I. and Full W.E., 1988, *Isotope chronostratigraphy: theory and methods.* Academic Press, San Diego.

Williams G.P., 1983, Improper use of regression equations in earth sciences. *Geology*, **11**, 195–197.

Williams K.L., 1987, *Introduction to X-ray spectrometry.* Allen and Unwin, London.

Wilson M., 1989, *Igneous petrogenesis.* Unwin Hyman, London.

Winchester J.A. and Floyd P.A., 1976, Geochemical magma type discrimination; application to altered and metamorphosed basic igneous rocks. *Earth Planet. Sci. Lett.*, **28**, 459–469.

Winchester J.A. and Floyd P.A., 1977, Geochemical discrimination of different magma series and their differentiation products using immobile elements. *Chem. Geol.*, **20**, 325–343.

Winchester J.A. and Max M.D., 1989, Tectonic setting discrimination in clastic sequences: an example from the late Proterozoic Erris group, NW Ireland. *Precambrian Res.*, **45**, 191–201.

Windrim D.P. and McCulloch M.T., 1986, Nd and Sr isotopic sytematics of central Australian granulites: chronology of crustal development and constraints on the volution of the lower continental crust. *Contrib. Mineral. Petrol.*, **94**, 289–303.

Winkler H.G.F., 1976, *Petrogenesis of metamorphic rocks*, 4th edition. Springer, New York.

Wood D.A., 1980, The application of a Th–Hf–Ta diagram to problems of tectonomagmatic classification and to establishing the nature of crustal contamination of basaltic lavas of the British Tertiary volcanic province. *Earth Planet. Sci. Lett.*, **50**, 11–30.

Wood D.A., Joron J.L., Treuil M., Norry M. and Tarney J., 1979a, Elemental and Sr isotope variations in basic lavas from Iceland and the surrounding ocean floor. *Contrib. Mineral. Petrol.*, **70**, 319–339.

Wood D.A., Tarney J., Varet J., Saunders A.D., Bougault H., Joron J.L., Treuil M. and Cann J.R., 1979b, Geochemistry of basalts drilled in the North Atlantic by IPOD Leg 49: implications for mantle heterogeneity. *Earth Planet. Sci. Lett.*, **42**, 77–97.

Wood D.A., Joron J-L. and Treuil M., 1979c, A re-appraisal of the use of trace elements to classify and discriminate between magma series erupted in different tectonic settings. *Earth Planet. Sci. Lett.*, **45**, 326–336.

Wood D.A., Tarney J. and Weaver B.L., 1981, Trace element variations in Atlantic ocean basalts and Proterozoic dykes from Northwest Scotland: their bearing upon the nature and geochemical evolution of the upper mantle. *Tectonophysics*, **75**, 91–112.

Woodhead J.D., Harmon R.S. and Fraser D.G., 1987, O, S, Sr and Pb isotope variations in volcanic rocks from the Northern Mariana islands: implications for crustal recycling in intra-oceanic arcs. *Earth Planet. Sci. Lett.*, **83**, 39–52.

Woodruff L.G. and Shanks W.C., 1988, Sulfur isotope study of chimney minerals and vent fluid from 21°N, East Pacific Rise: hydrothermal sulfur sources and disequilibrium sulfate reduction. *J. Geophys. Res.*, **93**, 4562–4572.

Woronow A., 1990, Methods for quantifying, statistically testing and graphically displaying shifts in compositional abundances across data suites. *Comput. Geosci.*, **16**, 1209–1233.

Woronow A., 1991, Endmember unmixing of compositional data. *Geochim. Cosmochim Acta*, **55**, 2351–2353.

Woronow A. and Butler J.C., 1986, Complete subcompositional independence testing of closed arrays. *Comput. Geosci.*, **12**, 267–279.

Woronow A. and Love K.M., 1990, Quantifying and testing differences among means of compositional data. *Math. Geol.*, **22**, 837–852.

Wright I.P., Grady M.M. and Pillinger C.T., 1988, Carbon, oxygen and nitrogen isotopic composition of possible martian weathering products in EETA 79001. *Geochim. Cosmochim. Acta*, **52**, 917–924.

Wright T.L., 1974, Presentation and interpretation of chemical data for igneous rocks. *Contrib. Mineral. Petrol.*, **48**, 233–248.

Wright T.L. and Doherty P.C., 1970, A linear programming and least squares computer method for solving petrologic mixing problems. *Geol. Soc. Amer. Bull.*, **81**, 1995–2008.

Wronkiewicz D.J. and Condie K.C., 1987, Geochemistry of Archaean shales from the Witwatersrand Supergroup, South Africa: source-area weathering and provenance. *Geochim. Cosmochim. Acta*, **51**, 2401–2416.

Yoder H.S. and Tilley C.E., 1962, Origin of basalt magmas: an experimental study of natural and synthetic rock systems. *J. Petrol.*, **3**, 342–532.

York D., 1967, The best isochron. *Earth Planet. Sci, Lett.*, **2**, 479–482.

York D., 1969, Least squares fitting of a straight line with correlated errors. *Earth Planet. Sci, Lett.*, **5**, 320–334.

Zachmann D.W., 1988, Matrix effects in the separation of rare earth elements, scandium and yttrium and their determination by inductively coupled optical plasma emission spectrometry. *Anal. Chem.*, **60**, 420–427.

Zartman R.E. and Haines S.M., 1988, The plumbotectonic model for Pb isotopic systematics among major terrestrial reservoirs — a case for bidirectional transport. *Geochim. Cosmochim. Acta*, **52**, 1327–1339.

Zheng Y-F., 1990, The effect of Rayleigh degassing of a magma on sulphur isotope composition: a quantitative evaluation. *Terra Nova*, **2**, 74–78.

Zhou Di, 1987, Robust statistics and geochemical data analysis. *J. Math. Geol.*, **19**, 207–218.

Zindler A. and **Hart S.R.**, 1986, Chemical geodynamics. *Ann. Rev. Earth Planet. Sci.*, **14**, 493–571.

Zindler A., **Jagoutz E.** and **Goldstein S.**, 1982, Nd, Sr and Pb isotopic systematics in a three-component mantle: a new perspective. *Nature*, **298**, 519–523.

Index